Making Effective Business Decisions Using Microsoft Project

Making Effective Business Decisions Using Microsoft Project

Tim Runcie
Mark Dochtermann

Advisicon, Inc.

JOHN WILEY & SONS, INC.

Cover image: © sureyya akin/iStockphoto

Cover design: Michael Rutkowski

This book is printed on acid-free paper.

Published by John Wiley & Sons, Inc., Hoboken, New Jersey
Published simultaneously in Canada

Library of Congress Cataloging-in-Publication Data available on request

ISBN: 978-1-118-09739-7; 978-1-118-33026-5 (ebk.); 978-1-118-33093-7 (ebk.);
978-1-118-33309-9 (ebk.); 978-1-118-48827-0 (ebk.); 978-1-118-48828-7 (ebk.)
978-1-118-48830-0 (ebk.)

Printed in the United States of America

10 9 8 7 6 5 4 3 2 1

Contents

1 **Business Intelligence: Knowledge of Key Success Ingredients for Project Server 2010 1**
In This Chapter 1
Maximizing PPM Ingredients, Culture, and Technology for Business Success 2
What Is the Project Management Lifecycle? 7
Information: What Fuels a PMO's Success? 16
Stakeholders in a Project Management Environment 21
Technology Meets Strategy: Welcome to the Business User Network 29
Important Concepts Covered in This Chapter 32
References 33

2 **Value Proposition by Role of Project Server 2010 35**
In This Chapter 35
Clairvoyance with Project/Server 2010: Forecast Future Results 37
Important Concepts Covered in This Chapter 75
References 77

3 **Meeting CFO Needs with Project/Server 2010 79**
In This Chapter 79
How the CFO Gets the Attention of the PMO 79
What and Why Is Work Management Critical to Organizational Success? 87
Synchronization of Strategic Objectives to Actual Effort 93
Important Concepts Covered in This Chapter 97
Reference 98

4 **The Business Shakes Hands with the Microsoft Project 2010 Platform 99**
In This Chapter 99
Logical Architecture Is More Natural for Business Users 100
Microsoft Project 2010 Platform Is Highly Extensible 122
Important Concepts Covered in This Chapter 130
References 131

5 End Users' Critical Success Factors: Using MS Project 2010 133
In This Chapter 133
Project Management in Small Business and the Enterprise 134
Initiating and Managing Projects Using the Microsoft Project Desktop Client 143
Being an Effective Enterprise Project Manager Using Microsoft Project Server 161
Fluent Project Management Using the Fluent UI: Introducing the Ribbon 171
Important Concepts Covered in This Chapter 181
References 182

6 Thinking Local, Going Social: Project Teams Can Thrive Using Microsoft Project Server 2010 185
In This Chapter 185
Project Management Looking Ahead 185
PPM Lifecycle 188
Important Concepts Covered in This Chapter 202

7 Better Together: Microsoft Project 2010 Worksites Using SharePoint Server 2010 203
In This Chapter 203
Integration of Collaboration, Social Media, and Project-Related Information 203
SharePoint Server 2010 Offers Critical Business Capabilities 208
Being Social in a Project Environment 217
Important Concepts Covered in This Chapter 221

8 Effective Transition of Strategy and Execution: Program Management Using Microsoft Project Server 2010 223
In This Chapter 223
Projects Are the "How," Programs Are the "Why" 224
Important Concepts Covered in This Chapter 249
References 249

9 Intelligent Business Planning and Controlling Using Microsoft Project 2010 251
In This Chapter 251
Understanding Strategic Planning with Project Server 251
Creating and Managing Portfolio Lifecycle for Project Server 256
Understanding and Building Business Drivers 259
Using Project Server to Master Demand Management 268
Building Project Selection Criteria 286

What the Efficient Frontier Is and How to Use It 292
Working with Constraints in Portfolio Planning 296
Creating and Running Multiple Scenarios for Portfolio Planning 303
Applying Strategic Analysis for Corporate to Departmental Needs 306
Committing New Work Portfolios and Measuring for ROI 309
Project Server Optimizing Governance for PMOs 315
Important Concepts Covered in This Chapter 319
References 320

10 Intelligent Business Planning and Reporting Using Microsoft Project 2010 321
In This Chapter 321
What Is Dynamic Reporting . . . 321
Creating Easy-to-Access Reporting in Project Server/SharePoint BI 324
Important Concepts Covered in This Chapter 361

Index 363

CHAPTER 1

BUSINESS INTELLIGENCE: KNOWLEDGE OF KEY SUCCESS INGREDIENTS FOR PROJECT SERVER 2010

IN THIS CHAPTER

This chapter helps set the stage for the deep-dive and thought-provoking tour we will be taking in establishing a good enterprise project portfolio management (PPM) system.

We focus on the importance of leveraging key technology and methodology components to help create a successful foundation for meaningful reporting and maximization of PPM technologies.

We review different types of lifecycles and how to work toward alignment through business and project lifecycles to leverage the power of Project Server's engine to reinforce best practices. We show how to work toward an end game of simple visuals and dashboards that enable business leaders, project managers, and even team members to participate in the success of their projects and what we refer to as "one version of the truth."

What You Will Learn

- Different key focus areas you need to address in establishing a strong PPM system
- The importance of lifecycle, phases, and stages to simplify and automate management, grouping, and reporting
- How to blend technology with methodology
- Understand the difference between Project 2010 and Project Server 2010
- How to scale Project Server 2010 from top down (portfolio planning) to bottom up (detailed and task planning) and how to leverage either or both

MAXIMIZING PPM INGREDIENTS, CULTURE, AND TECHNOLOGY FOR BUSINESS SUCCESS

In the world of business, the drive to getting good business intelligence (BI) has focused predominantly on the tools used to expose and graphically represent that information. Face it; BI dashboards are cool (for the most part), and the end users want to see whiz-bang graphics, nifty graphs, and other stunning visuals.

While these tools to depict information are absolutely critical to enabling more effective data analysis, they are not what BI is all about. BI is about understanding data to help make your business more productive. The end goal of any BI strategy should be to enable better understanding of the data.

Three key elements facilitate better understanding of the data: technology, process and, most important, people. Technology always gets the front row in the discussion, but it is—in our opinion—the least important. It is relatively easy to deploy technology to support business intelligence; hundreds of vendors can help you do this. However, the process and people parts of the equation are much more complex and require systemic organizational realignment and investment.

BI is an enabler that must be deeply interwoven into core business processes. Similarly, the act of transforming data into intelligence must be executed by professionals who are competent in data analysis. Companies that embed BI techniques into their core business processes and develop competency within each business unit are able to exploit the power of business intelligence. The ones that pursue BI through a technology-driven approach get lots of cool graphs, but they don't get information that allows them to make actionable decisions.

Process Side of the Equation

Companies that see BI as strategic to their success embed BI deeply into their core processes. Just take a look at Wal-Mart. BI is pervasive throughout every aspect of its supply chains, from inventory management, to pricing analysis, to store profitability. Information is centralized, real time, and powers the company's core processes.

Wal-Mart would not be as successful as it is without intelligence as its backbone. There are other examples as well: Continental Airlines and customer loyalty, Dell and direct to customer, just to name two. Each of these companies has intertwined BI into its organization to drive actionable decisions. In the case of Continental, identifying its most loyal customer and determining how to provide them with special treatment continues to grow their continued support and expand their customer base through the use of the information analyzed. In the case of Dell, determining promotion and bundle targeting, this namely being use of information to maximize customer purchases based upon their needs for similar features or components to increase the revenue of each purchase. Companies that view BI as an effort driven by information technology (IT) will extract limited value from it.

BI can be embedded into every core process in an organization. Here are some examples:

- **Human resources (HR) intelligence.** This area involves deriving deep understanding of organizational structure by a number of attributes, including size, cost, level, performance, and so on. As a company needs to grow or shrink, the HR function can easily understand and make recommendations based on deep insights into the organizational structure.
- **Finance.** The finance organization can have deep insight into the firm's financial statements by being able to trace from its balance sheet and income statement down to the lowest level of cost detail. Robust BI can also help with robust Sarbanes-Oxley 404 compliance and with understanding product cost structure.
- **Quality.** Better understanding of product quality can be driven through warranty analysis, defect rates, customer feedback, and the like. By having this information at its fingertips, quality organizations can identify specific root causes of quality issues much more easily.
- **Marketing.** Marketing is probably the most prevalent area where BI is critical, but often it is not tightly woven into key processes. Obvious areas of focus include customer loyalty, targeting promotions, call center marketing, sales force effectiveness, and many others.
- **Supply chain and logistics.** This area also is tremendously dependent on sophisticated BI that can power inventory management, supply chain visibility, and better kanban (just-in-time ordering) practices.

People Side of the Equation

While embedding BI techniques into core processes has been challenging for most companies, having individuals on staff who actually can use BI tools, understand data and analytical results, and make decisions based on the data is even more important.

This is a key weakness at many companies, and it often results in suboptimal usage of business intelligence. BI IT professionals are extremely difficult to find; business professionals who have knowledge in data analysis are even harder to find. Part of the problem is that the American educational system (including many graduate schools) does not educate people to analyze and understand data. How many classes in high school require a focus on data analysis? How many classes in college? If America, or any country, wants to continue to be competitive, it must invest more extensively in the analytical competency at an earlier age.

The authors of this book don't just have degrees in business, nor did we take classes in BI from the local technical training company; in many cases, we were forced to build the infrastructure or engineer tools, technologies, and metadata into common workflows to expose and analyze the data necessary to make good strategic and business decisions. In the 1990s, one of the authors had extensive fieldwork as a database administrator. When that experience is applied to getting to BI and integrating that data to end users, he found that there was a significant gap between what tools could produce and what people could easily grasp. This led to some very deep-dive conversations, and in some cases building systems that could transform information collected and gathered to something that end users could, at a glance, understand and know how to act on.

Major corporations also have largely ignored building an analytical competency. Those corporations that wish to seek an advantage should build a strong group of individuals who understand how to analyze data in each business unit. These individuals should have technical or advanced degrees as well as strong business acumen and be comfortable using highly sophisticated tools to analyze data. They should have deep training in the tool set, have an understanding of their process responsibility, and be empowered to make changes based on the results of analysis.

This is the case at very few companies. Most companies roll out a bunch of tools to a user base that does not possess the skills to use them effectively within the business function. Even where user dependence on a particular BI tool set is prevalent, most users don't use the tools for deep analysis; they simply base decisions on the reports they receive.

BI is less about technology and more about people and process. Those companies that get it at a chief executive level are going to have a key strategic advantage. One recent example of this is Hewlett-Packard. HP made a major announcement that it was building a large data warehouse to consolidate all of its customer information. This effort was driven by the chief executive officer (CEO) of the company and was obviously a strategic enabler.

Companies that have CEOs who understand the value of BI and who back their words with actions around process integration and competency will be much more successful, while those that relegate BI to an IT thing or don't leverage or use that information to shape their forward planning and progress, will continue to derive weak benefits from BI.

Business Case

In our work engagements and through many successful implementations of PPM technologies, we discovered that a foundational and essential tool is often overlooked. This tool is the business case. It provides the necessary facts and data for understanding the value, cost, and benefits of implementing a project. It also lists the assumptions used to reach the touted conclusions, the various options considered, and the required cash flow for implementing the project.

One of the keys to making the best decisions is understanding the criteria used to judge and prioritize projects. A company already has projects under way and usually has a list of possible projects to add to that inventory. How do you decide which ones to add, and when to add them? The business case is the fundamental tool for gaining facts and data about each decision criterion to enable apples-to-apples comparisons among projects.

Let us share one invaluable lesson we have learned the hard way: Even "mandatory" projects have options. ("Mandatory" projects are required to be done, perhaps by law or perhaps by your CEO.) Often people will say, "We don't need to do a business case, we have to do this project because . . . " The truth we have unearthed is that there are multiple ways to meet mandatory requirements. For example, if the requirement is to provide an efficient mode of transport, we could meet it with a motorcycle or a sport utility vehicle. But what are the tradeoffs between these two options? Even though we may have to do it, planning and analysis are still needed; these are accomplished effectively by producing a business case. In addition, a business case coupled with project plans enables scenario and option analysis to aid in the decision-making process.

In particular, note the last part of the definition. A true business case looks at more than just the numbers. It includes financial, strategic, commercial, industrial, or professional outcomes of the project under consideration. Ideally, the business case should have more than one option from which to select, including the do-nothing, or business-as-usual, option. The decision about the project needs to be made by those people with responsibility, accountability, and authority for the resources to be allocated (e.g., people, tools, machines, computers, facilities) to achieve the desired outcome. If you can't wait to learn more about business cases and making good project investment decisions, feel free to go straight to Chapters 3, 8, and 10. A key question around business cases is: Are we optimizing our capacity?

This question puts into fancy words a simple concept: Are we using our limited money, time, equipment, material, and skilled people to get the biggest bang for the buck? Capacity optimization can also be called portfolio resource optimization. There are two key principles to understand here:

1. Optimizing resources is about balancing the demand for resources with the supply.
2. The primary aim of resource optimization is to create an open dialogue, based on factual analysis, between the portfolio management office and the business project sponsors (the decision makers).

Resource optimization is achieved through the balanced management of resources. It is about understanding, managing, and balancing the demand side and the supply side of the resource management equation.

Demand-side resource management, which concerns all the things we need in order to accomplish the projects in the portfolio, entails resisting the desire to control the details. In most demand side resource planning the organization reviews the granularity of resource types and capacity from large to small, sometimes referenced as boulders, rocks, pebbles and sand. To ease the planning for the management of portfolio resources, we group resources into three categories:

1. **Skills.** The availability of a sufficient number of people with the right skills and experience.
2. **Technology environment.** The capacity of the computer systems or platforms to cope with the demands of the portfolio.
3. **Facilities.** The physical infrastructure needed (i.e., networks, office space, real estate, etc., needed to deliver projects). This is much of what will be impacted by the output of the project.

We seek to understand three key planning disciplines:

1. Planning for skills
2. Planning for the technology environment
3. Planning for facilities

In effectively implementing PPM, we can engage four levers that help us to manage resource capacity constraints:

1. **Changing time scales.** Shift projects within the portfolio to flatten resource demands.
2. **Decoupling development from roll-out.** Help to flatten technical resource demand.
3. **Descoping.** Help reduce the absolute need for resources.
4. **Removing projects from the portfolio.** If none of the previous options is sufficient in managing resource capacity, projects may have to be canceled.

Supply-side resource management concerns all the things we currently have available to accomplish the projects in the portfolio. It is important to differentiate between the organization's core competencies (those that give a competitive edge) and those competencies that can be commoditized (general skill sets not necessarily unique to the organization). For supply constraints, core competencies are increased by training and recruiting qualified people from the marketplace. Commodity skill sets are increased internally through cross-training and externally by developing and maintaining relationships with partners having different competencies and geographic footprints. There are several ways to deal with supply-side management of the technology environment: by using an application service provider (ASP) model, virtualization, or duplicate environments to better manage constraints.

In handling constraints in the supply-side management of facilities, it is beneficial to consider creative solutions, such as using temporary accommodations, hotels, regional offices, or taking over a new floor in the office building.

WHAT IS THE PROJECT MANAGEMENT LIFECYCLE?

As your organization prepares to spend significant money on new tools to help you better manage projects, how prepared are you to achieve a return on investment (ROI)? ROI related to project and program campaigns increases as the complexity of program demands increase. The complexities that must be managed to successfully execute projects and programs are perhaps the single greatest challenge facing leadership today. Program complexity is the combined nature of multiple, unique information paths all operating at a variety of phases and stages and all requiring different levels of departmental involvement across the company.

Programs have a longer lifespan than typical projects or they are comprised of different projects all driving toward a higher business goal. This growth in complexity means that the lifecycle and information required to deliver requires a better set of standards and metrics to manage the competing business and strategic demands within an organization.

Convergence theories, along with other economic and business system concepts, are pushing companies to embrace a more democratized project and program management system. Cross-sectional/cross-departmental analysis of challenges and requirements determination within an organization often proves to be a serious obstacle. Regional or departmental convergence of the adjusted processes often is not feasible. A good link is http://project-management-knowledge.com/definitions/p/path-convergence.

Project management (PM) and related business systems are modeling emerging economic systems. These economic systems (theoretical economic models) are moving away from the atomistic agent or single decision maker acting in isolation to make or lead key business decisions. These models are reflecting more and more the socialization of leaders with other stakeholders in business the system. These business drivers leading to business value are creating a more tightly coupled need for PM environments to link and showcase the link directly to the requirements and the metrics associated to delivering the business value. These requirements tied to business drivers should showcase that they will capture and deliver more of the stakeholder expectations, both quantitatively and qualitatively.

A key for many business leaders is to be able to model and visualize demand management. Leveraging demand management for project convergence means that there are strengths and weaknesses in the different approaches to PM systems. On one hand, classic PM supports a strong governance model and best practices, and its maximum efficiencies lie at the lowest-level common denominator. However, corporate globalization initiatives and related agile planning leverage a decentralized approach, more of a think global/act local approach.

PPM movements and related technical infrastructures are adopting more of the human input integrated with tools and processes. Large enterprise companies (over 1,000 employees) have struggled with maintaining control and accountability across the portfolio when launching PPM campaigns that cross departments

Table 1.1 Capital versus Social PPM

Capitalistic-based PPM: Revenue Driven	Socialized-based PPM: Human Driven
Desperate resource use	Full resource utilization
Wide skills pool	Baseline skills pool
Business/process cycles	Predictive evolution
High product/output efficiencies	Market disparity
Strong balance sheet efficiencies	Internalized equity
Client-focused measurements	User-focused measurements

and product lines. Small and mid-size organizations (up to 1,000 employees) have found it nearly impossible to wholly adopt user-input product requirements and process capabilities integration into PPM campaigns. The conflict is that as stakeholders are providing the push for use requirements, process cycles and capital capacity provide the controls—sometimes referred to as project bottlenecks. Companies of all sizes would love to provide virtually infinite delivery and quality to stakeholders, but it is just not possible to make everyone happy.

As an example, a U.S. automotive original equipment manufacturer (OEM) had a vehicle line that offered so many options that it was impossible to offer every possible combination to customers. The obvious question is why the OEM was offering options that are not compatible with one another. Were the program requirements different across the various commodity departments? Table 1.1 shows the strengths and weaknesses of the two conflicting theories often found in economic and governmental models regarding PPM. One side depicts the perspective of a monetary-focused system that is more of a survival of the fittest. The other is a perspective of embracing all social elements for a common good.

Many times, projects are decomposed to a point that the goals and objectives stated in the charter have a marginal impact (the project is barely unique). Conversely, many projects are killed or simply fail and surpass any estimated costs and time objectives because they were unique.

It would be a huge benefit for decision makers in any size organization to finally leverage technology that addresses the need to use diverse resources, budgets, and requirements while maximizing the business life cycles. Demand management delivers this and is a part of PPM that is becoming known as unified PM. (See Figure 1.1.)

Aligning Project with Business Life Cycles

"I don't understand, why aren't these projects delivering as they promised?"

This familiar cry has been heard from business leaders and project managers for some time now. Thousands of books and articles offer answers to this

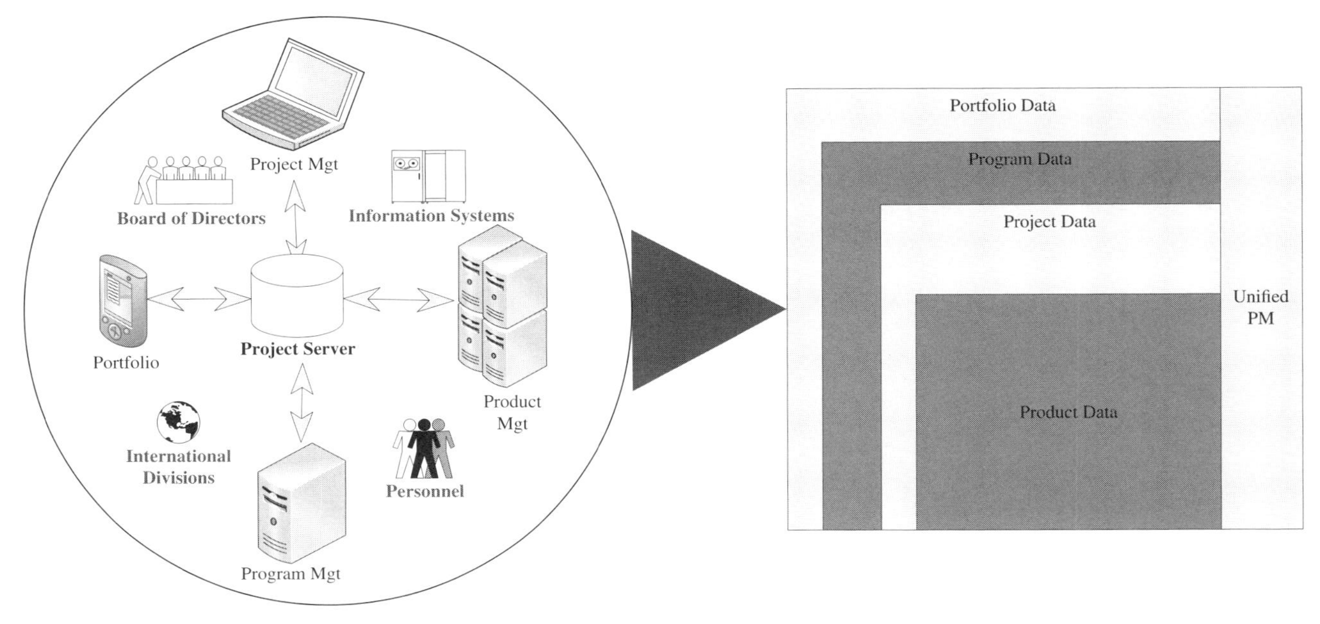

Figure 1.1 Example Unified Project Management and Related Stakeholder Classes Source: Advisicon

question, but the frustration continues. One idea that is gaining ever more traction in answering this question is PPM: the concept of focusing on the selection and management of a set of projects to meet specific business objectives. But when business leaders and project managers review the concept of PPM, their response is often "This portfolio management stuff sounds way too simple. It just can't be the answer!"

However, this response itself begs a question. If PPM is so simple and self-evident, why does it have such limited traction in organizations that are apparently so in need of its help? The logic of simply reviewing all projects under way in an organization and making sure they meet business needs, align with strategy, and provide real value does seem self-evident. Practice and observation tell us that, when properly implemented, PPM does work. Unfortunately, our experience also tells us that a lot of the time, the implementation of PPM leaves much to be desired, and results in responses such as:

- "This process is too complex."
- "We don't have time to go through all this business case stuff; we need to get to work!"
- "This process is really needed for our organization's business projects, but mine are different and don't need to go through all those steps."

Apparently PPM isn't so self-evident after all. So what do we do? Business leaders want the business to be successful. They want sound business processes they can depend on. Project managers want their projects to be successful, so the company will be successful. It sounds like we're all on the same page, right? Wrong. Here's where the age-old dilemma rears its ugly head for business leaders and project managers alike: There are limited resources, lots of ideas and projects, only so much time in a day, and . . . oh yes, things keep changing. This is when it becomes important for us to be able to make tough decisions.

To be successful, which projects do we invest in (and over what time frame)? We need good facts to make the right decisions. We business leaders and project managers need to be able to examine the facts when changes and issues arise that require a decision be made and acted on. And we need to weigh these facts against our gut feel for the situation (sometimes called experience) by. Then we must make a decision.

These needs may seem to be self-evident, but are they really? How do we get the facts and data we need? And how do we know we're making the right decisions? This is where the power of PPM comes into the picture. PPM forces us to think strategically: what we want our organizations to be and what we should be doing to get there. However, it isn't just a simple proposition to turn on, or to just install. When implemented properly, PPM often requires organizational change across the business, and that can be very difficult to carry through, and is a combination of methodology, technology and an applied approach to the analytics metrics gathered.

Successful PPM

PPM invariably changes the culture of the business because it demands that we ask the hard questions. Your ability to answer these questions accurately will determine how well you have implemented PPM in your organization.

Project/Program Phases and Stages

Microsoft's Project Server 2010 has grown to include some new and very powerful elements that help businesses get a handle on demand management, resource capability planning, and strategic impacts of projects.

This functionality out of the box can be integrated into a lifecycle or managed in phases and stages. The exposure of key data elements (fields) and ability to review and give approval by key stakeholders allows for excellent quality control. This is where many project management offices (PMOs) or organizations created to standardize and manage project lifecycles, now with Project Server, have the ability out of the box to leverage a workflow that supports their organization's phase gates, or project stages.

Project Server 2010 includes intuitive demand management capabilities that enable multiple stages of governance workflows, helping to ensure that projects are subject to appropriate controls throughout their lifecycle.

Each workflow may include a series of phases, which in turn includes stages. The phases and workflows establish a blueprint for your organization's governance framework and help ensure that all projects achieve the necessary deliverables and receive managerial sign-off before moving to the next stage. (See Figure 1.2.) This audit functionality keeps stakeholders aware and accountable as projects move from business case creation to consideration to implementation.

Project Server 2010 also provides the flexibility to create custom workflows and templates mapping the organizational governance structure. For additional details, refer to Microsoft Development Network for Project (http://msdn.microsoft.com/project). The following list are some of the key components that allow an organization to align both process with functional elements in Project Server to setup both manual workflows or to automate those workflows.

- **Phase.** Phases represent a collection of stages grouped together to identify a common set of activities in the project lifecycle. Examples of phases are project creation, project selection, and project management. The primary purpose of demand management phases is to provide a smoother user

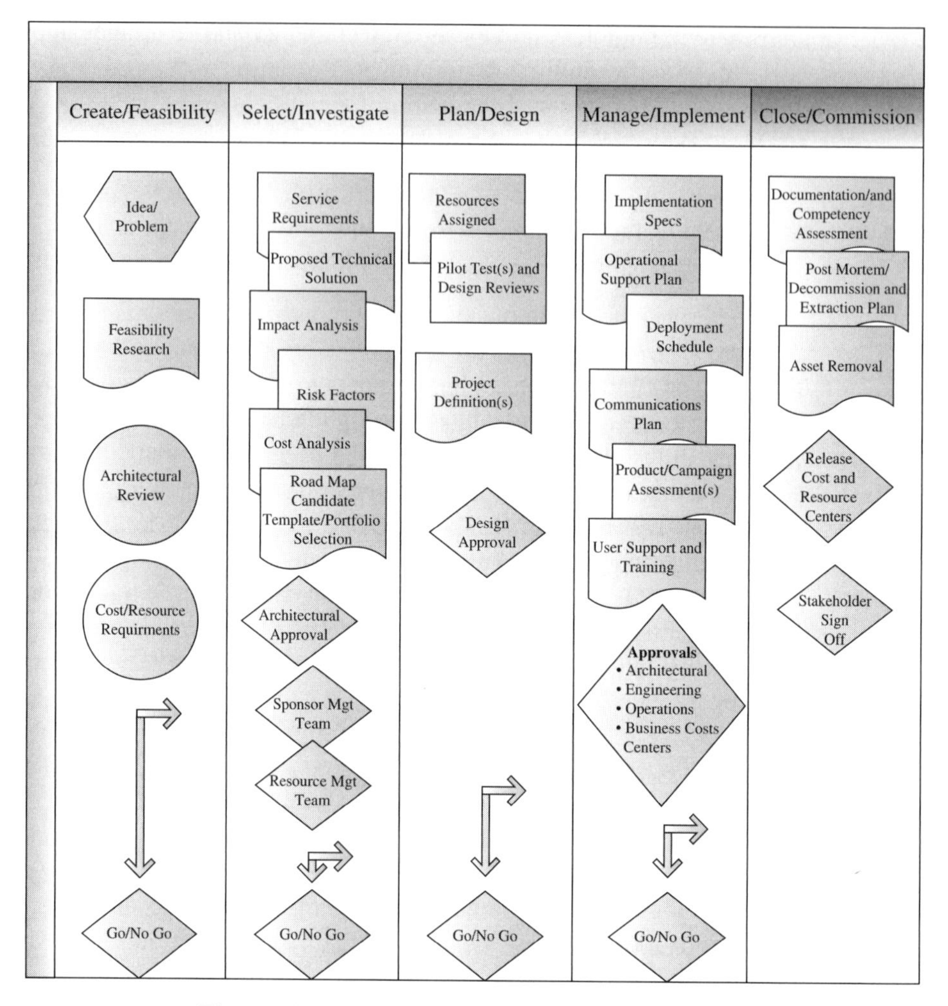

Figure 1.2 Project Lifecycle Flow Source: Advisicon

experience where users have the option of organizing stages into logical groups (e.g., create, select, manage, plan, close).

- **Stage.** Stage represents one step within a project lifecycle. Stages at a user level appear as steps within a project. At each step, data must be entered, modified, reviewed, or processed (e.g., propose idea, initial review).

At a technical level, each stage represents a step where data is entered or calculated or artifacts are approved/rejected before the workflow can move to the next step.

Let us understand each of these phases in detail. The box below helps breakout the key elements and goals that should happen in that stage.

Create
Cost, Benefit, Approach, Resources, Strategic Impact, Risk Assessment

Select
Business Drivers, Strategic Priorities, Scenarios, Impact Standards, Constraints, Analysis

Plan
Phases, Milestones, Dependencies, Resource Management

Manage
Actuals, Change Control, Status Reporting, Forecasts, Issue/Risk Management, Visibility

Closure
Sign-Off, Project Documents, Templates, Lessons Learned, Archive

Planning in a Governed Environment

Project/Program Lifecycles: Demand Management Variations

These demand management or project lifecycle management (PLM, Project/Program Lifecycle Management) variations can be viewed as a hierarchy within a single, unified enterprise context. Specifically, unified context allows for the application of a common semantic foundation, which in turn allows for the coordination of all related data within a single PLM repository. PLM supports active working processes and capabilities already familiar to practitioners of various PLMs.

As the hierarchy (see Figure 1.3) is examined a little more closely, some obvious conclusions arise. Products, systems, or even services can be viewed more or less synonymously as capabilities. A project might consist of one or more capabilities (or capability modules). A portfolio might consist of multiple projects, multiple products/systems, or a mix of both. This hierarchy allows for the ability to group all of these elements together and maintain the relationships between them through reporting chains, requirements, or departmental situations.

Demand derives from written, verbal, or assumed requirements. Requirements represent the information between consumer and producer, between management and developers, and between planning and execution. Visibility emerges by leveraging a lifecycle framework that integrates all of those interests and participants.

PLM enables instant visibility and reconciliation of the many seemingly diverse program elements that exist across a complex enterprise. This visibility usually occurs through visual tracking and automated reports, which illustrate the

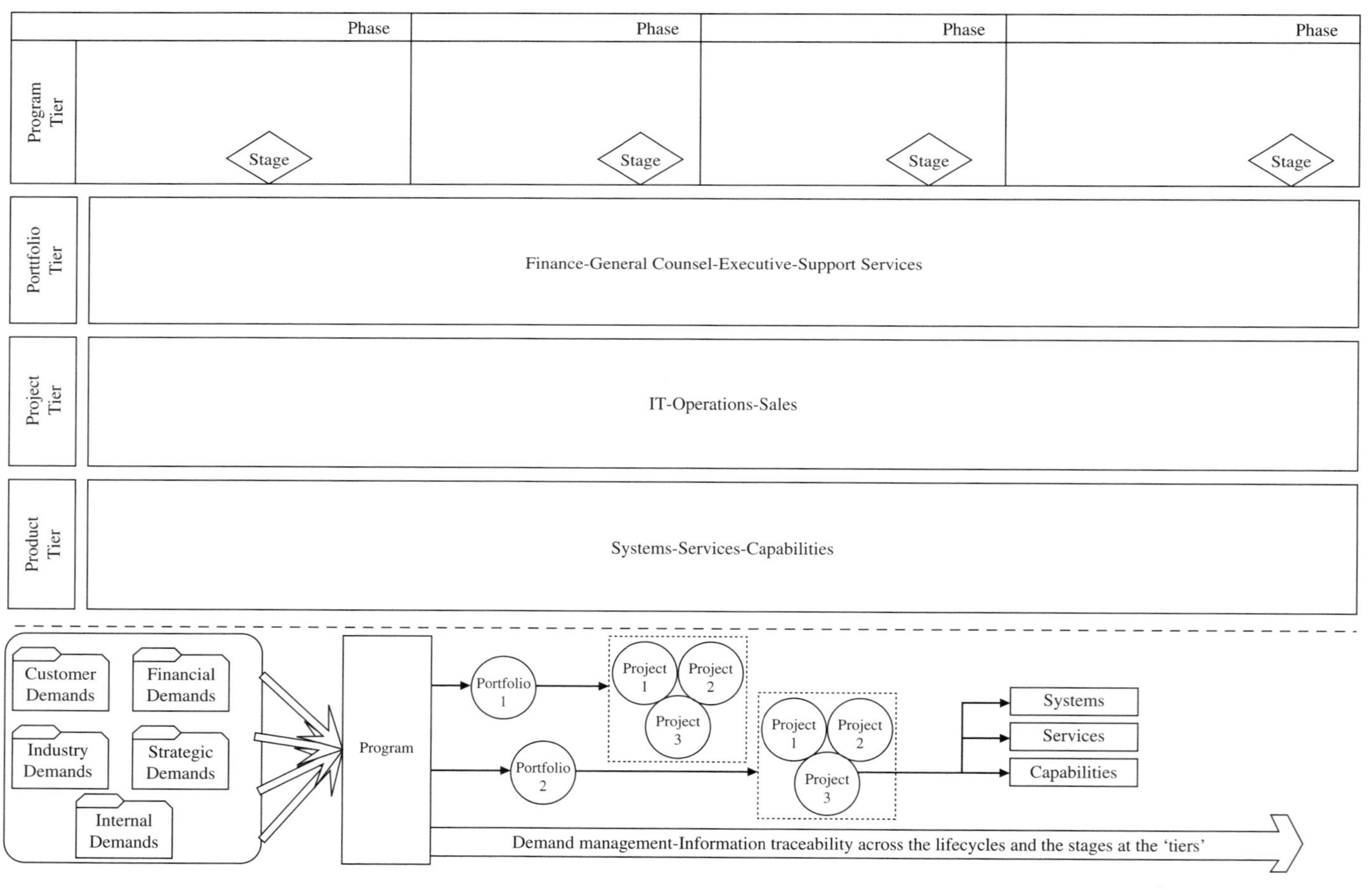

Figure 1.3 Demand Hierarchy Source: Advisicon

potential issues and interdependent relationships between requirements and other program elements. No matter how many systems or component/vendor organizations are involved, if there is a centralized single-instance PLM framework, the various processes and lifecycles associated with an enterprise can be holistically tracked and managed.

Consolidating all of these processes and data centrally eliminates the single most critical problem facing PMOs today: the ability both to see the big picture and to drill down to specific details in an automated fashion. Today's PMOs are essentially integrated on the fly and are top-heavy with manual processes.

The key to PLM is to understand that the PMO runs on information. That information must be easily accessible, transportable, and translatable and available directly to decision makers without going through layers of expert interpretation first. This doesn't mean that other people don't add value to the information; there will always be a need for diverse input, views, and skills in the PMO.

Significance of Portfolio Management and Demand Management in Today's Evolving Market

If you have ever worked in an organization, you understand that time to market or the constant pressure to start new projects and complete existing initiatives creates imbalances of resource capability and availability and high risk. There exists a significant need to bridge the technical or reporting gap that is present between those who are doing the work and what those at the executive level believe is getting done. In defense of senior management, they have heard for years from resource managers and work teams that they are overworked, yet the resources always seem to deliver.

This communication gap is compounded as new work is approved. In many cases, the annual portfolio selection process approves and moves work into the business system without the visibility to see what impact it will have on existing resources or projects.

Project Server 2010 has folded into the same system that manages the existing portfolio of work the ability to review new incoming work and compare it directly with the existing workload, by resources.

This ability is significant. The evaluation and review of demand management of future approved work with existing work enables organizations to see the impact of cost and work on a project, and by extension, the portfolio. As a result, these organizations are able to effectively prioritize the start time of projects within the portfolio and manage the organization's staffing needs both within the context of the current workload and work in the pipeline.

In 2009, we supported a customer in developing a project office and implementation of Microsoft technology to manage, track, and get a handle on more than $300 million worth of projects each year. This customer, due to economic stimulus funds, wanted to increase the number projects to over $600 million a year. The issue was that just because projects were approved and contractors and staff were lined up to do the work, workload and infrastructure was still not getting the work accomplished, causing a backlog of projects that would continue to compound as more work was approved.

In today's evolving market, many companies, agencies, and businesses are driving to get key projects completed in a timely manner, not only to be first to market but also to realize the business value that senior management established. As in the example just mentioned, the need to clearly understand both capacity and demand is critical before projects are started as well as during and after a project has been completed. When a project crosses the finish line, there should be a tie back to the return on investment (ROI) ensuring that it delivered the product as requested, not a project with features or functionality scuttled to make a deadline.

In today's growing business and tightly competitive market, the company or organization that has the ability to manage, view, forecast, and adapt to these types of BI metrics will find a significant competitive advantage.

INFORMATION: WHAT FUELS A PMO'S SUCCESS?

One of the core functional outputs of a PMO is its ability to standardize and measure key metrics across projects. In order to do this, project information must be uniform and measured. This information can cover costs, resources, work, planned, actuals for scope, schedule, and budget.

In Project Server 2010, all of this information can be tied to a project's schedule and its collaboration portal or workspace (a SharePoint site). When combined in a uniform manner in an enterprise-based server system, it enables a PMO not only to measure key information about a single project but to review the entire portfolio of information about all projects and establish BI reporting and trends about key data points or metrics within the project office's managed work portfolio.

Here is what makes the management of project information exciting: Imagine by touching data just once, you've gained the ability to pinpoint resource estimates with their actual work. Or by touching that data once, you are now able to bring in the consolidated time spent by an entire development team (who may be working in an agile system or logging work directly in Team Foundation Server) and can combine the actuals of costs or time spent and tie that information back to a project. Both of these possibilities enable both the project manager and the business managers to forecast the accuracy of the time, costs, and the work required to accomplish the remaining activities in project schedules.

Now imagine taking this information and creating a closed-loop learning process whereby new projects' time, costs, and work estimates can be fine-tuned by existing work portfolio. This closed-loop learning process will essentially combine past, present, and future work estimates to give both project managers and business decision makers the ability to improve their project organization and reduce the effort, cost, and time in delivery of projects.

All of this is possible by touching the data only once in a single system. No wonder project offices are excited about doing more with less. Through Project Server 2010, this is not only possible but is easier than ever before to accomplish, as explained in future chapters, especially chapters 3, 8 and 10.

Overview of Information Acquisition

The pursuit of information is an ancient activity that has always been a part of PM. The acquisition of information covers the entire lifecycle of projects, from estimating, to tracking the progress of work, activities, and deliverables. Even when a project is complete, the information we continue to compile helps us to see if the project delivered the intended results and provides us with the opportunity to learn from the process and improve future efforts.

User Empowerment

It is inaccurate to think that only project managers have PM skills or use PM methodologies. Any person who needs to manage multiple tasks, involving multiple persons, and has a deadline to meet can use PM.

Unfortunately, many organizations restrict the usage of PM tools to project managers and IT departments, as the tools are often viewed as too rigid, and require additional costs for skills and competencies. This practice limits the benefits of PM because other departments are deprived of the efficiency gained by using PM techniques and tools.

Most users are also unwilling to learn an extra tool for managing projects. Projects are subject to spatial dynamics where users or stakeholders end up managing projects on paper, in Outlook or Excel, or even mobile phone calendars. All these methods are inefficient and even inaccurate in some cases.

This means that Microsoft Project is useful not just for project managers but also for:

- Sales and marketing (campaign management, brand management, event management, product lifecycle management, market research, public relations)
- Human resources (training and development scheduling, recruitment planning and execution, appraisal execution, organizational development planning, growth planning)
- Finance (budgeting, variance monitoring, investment planning, mergers and acquisition planning, initial public offering planning and execution, following compliance and disclosure procedures)

- Top management (strategic planning, monitoring initiatives across departments, cash flow monitoring, resource planning, expansion planning, managing value lifecycle, crisis management)
- Manufacturing (process improvement, capacity expansion planning, building new factories, quality management, defect management, kaizen)
- Any role (manage complex interdependent tasks efficiently)
- Home users (planning parties or weddings, building/repairing a house, managing personal finance, tracking investments, buying/selling property, managing cash flow, orchestrating moving from one residence to another)

Ideally, companies are eager to tap the potential user base across all domains. However, as of now, only subsets of project managers use a formal tool, and projects often are not seen as interrelated within the corporate lifecycles. The evolution is now under way as easy-to-use interfaces and features of PM tools are such that users at all levels will find it suitable for its planning, tracking, reporting, and resource management capabilities.

Overview of Knowledge Management

Knowledge management (KM) is a very hot topic these days with the ever-increasing need for connecting and simplifying the steps for project teams and stakeholders to get to information relating to projects, programs, or portfolios. The ability to empower end users to search, find, and quickly respond to information via a pull process versus a push process (i.e., not storing information in e-mails, desktops, hard drives, file drives, etc.) creates a more efficient and more interconnected audience of users around key project information.

While it takes time for a culture to shift from one model of information or KM systems and processes, SharePoint and Project Server 2010 are empowering organizations to expose the widest possible audience to information while maintaining an easy-to-manage, store, and communicate infrastructure that enables IT support groups to manage and do more with less.

End Game: Automation and Getting to Dashboards

KM empowers PM stakeholders to get to information. Executives and business stakeholders have found that the ability to automate work activities and drive information to dashboards help deliver some of the highest value of PM reporting and task/time management.

Project Server 2010's ability to automate workflows (essentially giving programmability to project phases, field setting changes, or key work activities)

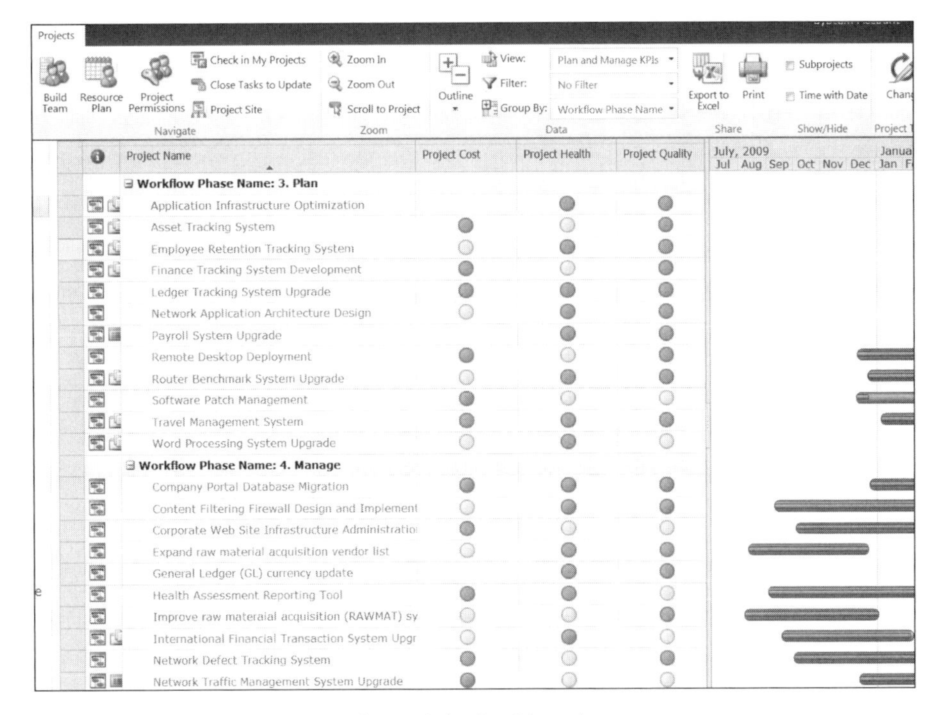

Figure 1.4 Dashboards

enables teams to eliminate many manual activities, focusing time on more key activities around the PM process than just the tool.

Getting to dashboards is always one of the most exciting realizations of organizations that require demand management, resource capacity, and portfolio management. Dashboards (see Figure 1.4) essentially help us focus attention on the correct issues, in a sense, focusing on managing the exceptions, not trying to analyze the entire universe of activities.

Of course, getting to a dashboard requires three key components that are important to take into account with both Project Server and SharePoint 2010:

1. Definition of dashboard thresholds
2. Linkage to key metrics rolled up from the appropriate level of detail up to the highest appropriate level
3. Understanding and creating a process around supporting, caring for, feeding, and addressing the issues that the dashboard metrics present

Many organizations spend significant amounts of time building and deploying systems but have poor results getting the information back out of the system. Project Server and SharePoint offer tools that allow for some quick drilldown and easy-to-build dashboards and reporting. (See Figure 1.5.)

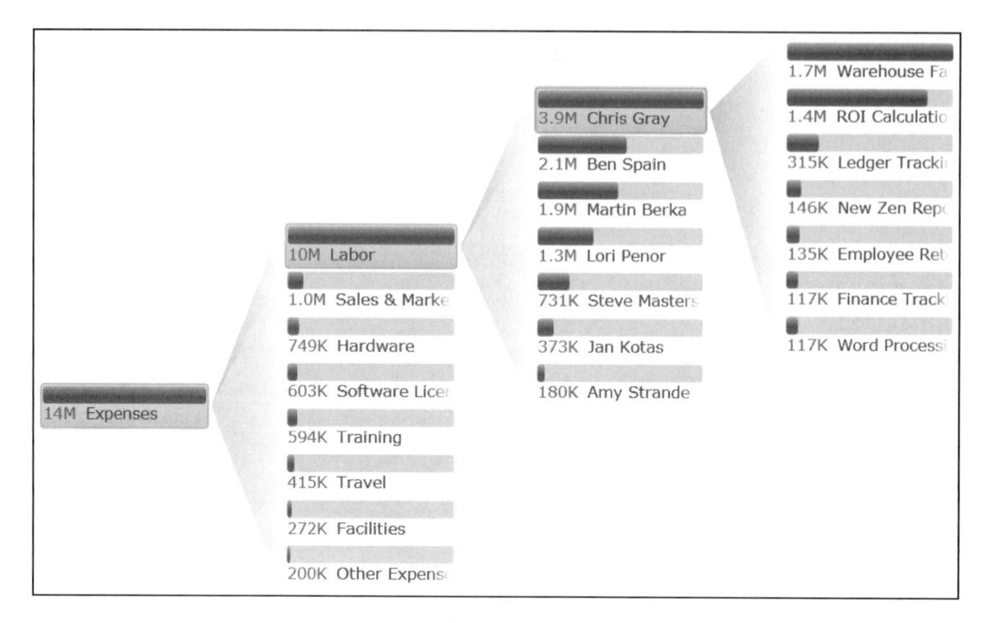

Figure 1.5 Drilldown

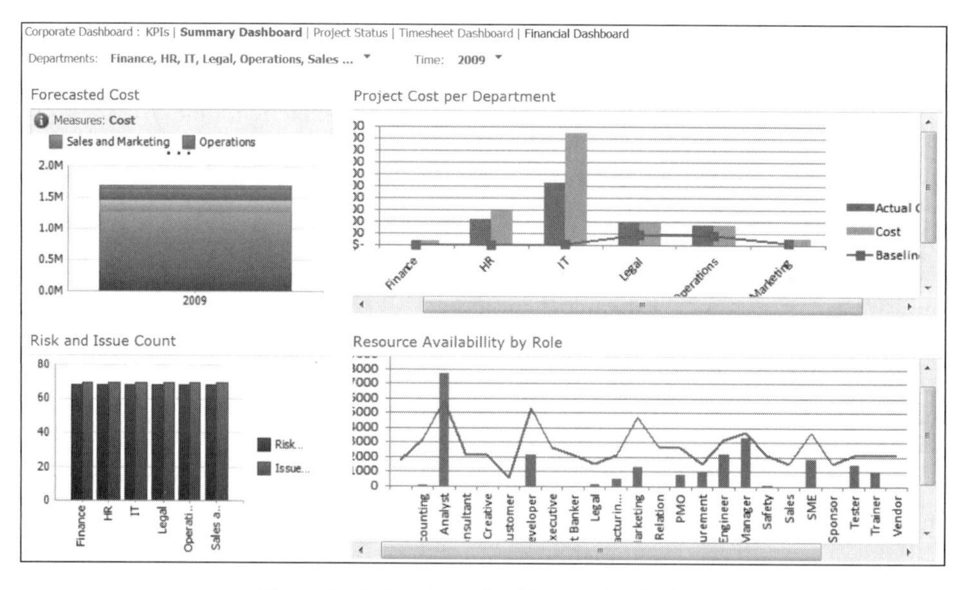

Figure 1.6 Business Intelligence Analysis

Using PowerPivot, Excel Services, Performance Point Server, and other BI tools, you can combine, analyze, and leverage information quickly and easily, leading to more time spent in growing and building a better PMO. (See Figure 1.6.)

STAKEHOLDERS IN A PROJECT MANAGEMENT ENVIRONMENT

Delivering Results, Not Surprises, with Microsoft Project 2010 and Microsoft Project Server 2010

In PM, a key component in any requirements-gathering processes is to ensure that both the stakeholders and the requirements are identified. Even if the scope of the project cannot deliver the requirements, the stakeholders should never be surprised that something will not be delivered.

Project Server 2010 gives end users a fast way to grow and mature an organization with its ability to stage or turn on all or some of the features with an organization's maturity capabilities around project, program, and portfolio management.

Project Server isn't a new tool but in fact is a blend of existing systems integrated to provide a full spectrum of visibility and metrics for making good business decisions and collaborative visibility into existing work.

Microsoft Project Desktop has been in use for almost two full decades, and the collaboration portal (SharePoint) has been leveraging the enterprise (PPM) system since 2002. Project Server features combine and integrate the best of breed of the client version, embedded with SharePoint and its rich feature sets of reporting and document/collaborative Web parts.

The PPM component of Project Server is a powerful blend of features integrated with both the existing enterprise PM system and the built-in metrics of future portfolio work being brought online.

Project Server is integrated with the BI of SharePoint's collaboration portal, business reporting capabilities.

In essence, the ability to see, plan, communicate, report, integrate, connect and review the ROI with other line of business tools (Microsoft Dynamics ERP systems, Visual Studio Team Foundation Server) leverages the richness of the Microsoft stack to enable visibility, not surprises, for project team members to senior stakeholders.

Consolidate Your Project/Program Approach

Large companies benefit from a matrix-style system of project support by scaling projects based on size, revenue, and other strategic factors. Typically, longer and more expensive projects secure executive sponsorship and corporate governance while smaller projects are created in an agile or rolling-wave planning environment.

Small and midsize companies typically view most initiatives as projects because of their limited resource pool and operating capital. Many times, though, things get done through more ad hoc, reactive means. What is missing for both large and small companies is the ability to channel all campaigns through project lifecycles in order to consolidate all costs, resource usage, and requirements. Project Server 2010 supports integrated project and portfolio management capabilities.

Demand management using the tightly integrating project and portfolio management capabilities within the enterprise project management (EPM) solution provides for a consistent user interface, common data storage, and centralized administration. Improving and extending existing capabilities across the solution enables companies to incorporate (just to name a few):

- Project portfolio management
- New product development/project lifecycle management
- Internal workflows/approvals
- Regulatory and compliance management

The features of Project Portfolio Server 2007 are included on a single Project Server 2010 platform. This seamless unification of two products into one offering makes end-to-end project and portfolio management easier than ever.

Using Project Server 2007 with Portfolio Server 2007

A branch banking division of one of the leading financial services organizations in South Africa with 680 branches saw the need to better align projects at its branches with its business strategy and to gain a single view of multiple projects in every branch. Having already implemented the Microsoft EPM solution with Microsoft Project Server, the branch banking was also looking for enhanced portfolio management functions. It chose the Microsoft Portfolio Management solution as the best for future growth, given its acquisition of Microsoft Office Project Server. The 2007 solution gave the branch banking project managers an end-to-end integrated project and portfolio management tool that helped to deliver new business value and ensure excellence in project execution.

They could achieve better collaboration among the project managers and the management team. The solution was also deployed for an IT development division with more than 450 users. Migration to Project Server 2010 will be a much shorter journey because the company will be able to migrate all of the data and processes from the previous version directly to Project Server 2010.

Looking ahead, organizations similar to the branch banking division will be able to leverage Microsoft Project Server 2010—in this case offering international financial organizations unified views of all work in one central location using the Project 2010 demand management capabilities. With Project Server 2010, project and portfolio management departments can:

- Build governance workflows to subject different types of work requests to the appropriate controls throughout the lifecycle of the issue or project.

- Standardize and streamline data collection by using configurable forms and business case templates.
- Capture all requests in a central repository to enhance visibility.

One unified Project Server 2010 system can run through the entire lifecycle of projects, from selection to implementation. A brief example of the 15 steps follows.

1. Create custom proposal templates.
2. Create a Web page to submit proposals.
3. Define the lifecycle for a project.
4. Create the approval process for a proposal.
5. Assess proposals through business strategy alignment.
6. Approve proposals.
7. Create project schedules and assign resources for proposed projects.
8. Define business drivers.
9. Prioritize business drivers.
10. Capture project proposals.
11. Create analyses and prioritize projects.
12. Analyze portfolios based on high-level cost constraints.
13. Analyze portfolios based on high-level resource constraints.
14. Commit selection decisions and communicate to portfolio stakeholders.
15. Create a central repository for project data.

All of this and more can be achieved easily using one centralized, collaborative environment.

Project/Program Governance

In many organizations, there remains a gap in the governing surveillance of project activities. Companies typically operate on the basis of a global or regional matrix of product groups and market territories.

Figure 1.7 illustrates common major business activity via business centers.

Associated with each of these business centers is a corresponding internal department. The key to improving and delivering shareholder value is to understand the requirements, or demand, since everything else flows from this process of demand creation. Years of research and focus on each of these business centers

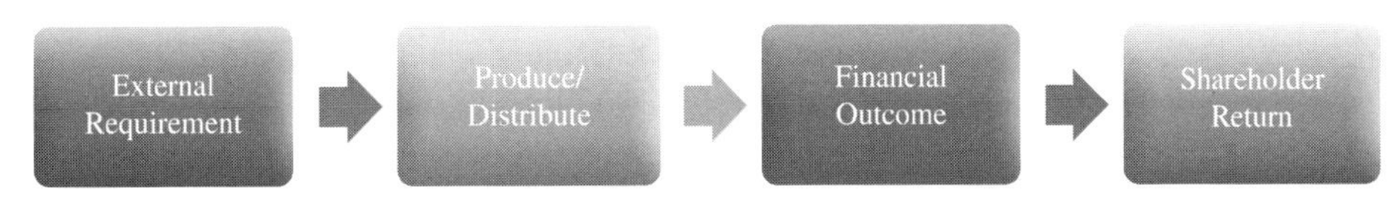

Figure 1.7 Principal Business Activities Source: Advisicon

have provided companies with considerable tools and methodologies to improve the performance of each business center, tools this book will not describe. However, communication and connectivity in relation to each other outside of their respective boundaries has been lacking.

Demand management is really the process of eliciting requirements to determine the goals, objectives, and business drivers that will enable information paths from the beginning. This is essentially doing top-down and then bottom-up planning.

Let's assess a scenario where your organization in the energy industry has to meet a strategic target for operational profitability. The target for operating profit is set as an output from portfolio planning and strategic planning. For our scenario, we will start at shareholder return and set the outcome of an operating profit at a specific target level: 12 percent. Shareholder value related to operating profit is based on economic formulas, such as capital costs associated with staff/resources, tax, interest and expenses, along with formulas such as weighted average cost of capital or similar, which is the capital charge or the amount of money that investors expect as a minimum return from the business (matching the opportunity cost of their capital). Top-down planning using the aggregate of shareholder value/operating profit leads us to financial expense inputs (taxes, invoicing, operating costs, etc.). It also has financial revenue input, such as sales (gross profit, volume, pricing, and market share), distribution (shipments, timing, inventory), and other corporate investment channels.

Now that the top-down planning is complete (essentially a backward pass at a portfolio level), we have some direction to the thresholds and requirements that executives are looking for from projects based on shareholder input. This means that as projects progress through the various lifecycles, stages, phases, and departments, there is a direct link to the strategic objective.

Project Proposals and Strategy

Now that we have done some top-down planning and a backward pass at the portfolio level, it is time to work forward and begin bottom-up planning. In Project Server 2007, project data were dispersed throughout the organization. Even when captured as elements of a project/program, the data resided in a variety of systems, sources, and forms. Targets and requirements, for example, may have been cascaded down to the PMO via e-mail, notes, or data fields in other business systems. New information and change control in Project Server 2007 without a Portfolio Server environment meant using Web forms, Office documents, or other non–project system components to elicit changes/updates to requirements. Proposals and activity Web forms within Project Server 2007 incorporated enterprise-level fields, but the governance and initiation of tasks and projects still resided outside of the Project Server environment (organically). Using our example, as a commercial organization serving the energy industry, we will be required to provide evidence for decisions and status of the projects. Having all the data, workflows, and lifecycle stages attached to each project will be a significant benefit.

Project Server 2010 combines the Proposals feature of Project Server 2007 and the Builder module of Portfolio Server 2007 to give more flexibility and ease of use in one place. This helps program/project managers to strategize, prioritize, and choose the right projects.

A PM need not begin proposal creation right from scratch. Preloaded SharePoint lists or even enterprise project types (EPTs) can be used as a starting point for proposal creation. Furthermore, the proposals/projects can be grouped by department. Thus, a project manager from the HR department now has quick access to proposals belonging only to his or her department.

To bring about more standardization and regulation, templates can be created for project plans and a project workspace site. Hence a proposal can have not only a fixed EPT and workflow, but it can also have a fixed set of predefined activities (which can be further grouped, linked, marked as milestones, etc.) and a workspace for collaboration. An EPT represents a wrapper that encapsulates phases, stages, a single workflow, and the project details pages. Each EPT represents a single project type. Normally, project types are aligned with individual departments—for example, marketing projects, IT projects, HR projects, and so forth. Project types in Project 2010 enables users to categorize projects within the same organization that have a similar project lifecycle.

These proposals can be created and edited directly from the Web. A project can have multiple stakeholders who would like to participate in adding tasks to specific phases of a project. Additionally there might be some changes with more inputs coming from the team members. Web-based editing allows more room for individual stakeholders to contribute their share of information without being required to have Project Professional on their machine.

Once all summary information, deliverables, resource information, and cost estimates are captured for a proposal, it is then submitted for approval through a workflow. Then the approver must access all aspects of a proposal and approve or reject it.

Let us consider a scenario where an HR manager is given a list of project initiatives that have to start according to fiscal year targets:

- HR payroll shared service
- HR employee information
- Training records tracking
- Performance management
- HR management system

It is not an easy job for HR managers to pick up the correct projects for implementation and then implement the projects in the correct order. A number of strategic objectives revolve around selection and prioritization of projects. Factors like business drivers, resource availability, and investment decisions all combine to determine the project kickoff order. (Examples of business drivers include increasing customer satisfaction, increasing employer satisfaction, better team collaboration, and growing revenue.)

Project 2010 capabilities help HR manager to select, prioritize, plan, track, manage, and execute these projects end to end. The tool helps determine each proposal's impact on business drivers and other proposals, in turn generating a priority score. To begin with, each of these proposals can be associated with fixed templates and passed through predefined workflows. The HR manager can have multiple intermediate approvers approving the proposal(s) through multiple stages. Once approved, the proposal will become a potential project and tracked and managed through Microsoft Project Web Application capabilities. This leads to a more scientific method for project selection based on existing data analysis.

Profiles of Business Influencers

Business influencers can be both internal and external to how an organization decides what projects or which project approaches to take. In many companies, customers may drive business standards, processes, and workflows that surround the PM lifecycle.

Business influencers may be market conditions or non–people-oriented, environmental conditions that require certain types of information (e.g., metadata to be tracked and updated within the PM system in order to measure, sort, group, filter, or do detailed analytics). An example of this would be a seasonal or weather conditional factor that might prioritize how projects are to be started or when they are to be completed.

Another type of business influencer is the internal company influencer. In many cases, these are senior stakeholders or departments that have an influence on projects. They may influence which projects are undertaken as well as many of the reporting elements that need to be communicated on. In many cases, these internal influencers are part of the steering committees or business champions who represent the business or organization, and are looking for the value a project can deliver to an organization.

Whether an influencer is internal, external, or environmental, it is important to identify them and ensure that the deployed PPM system can present a clear picture to answer the business influencer's requirements or conditions.

Profiles of Corporate Candidates

As a project stakeholder, you may be asking yourself a few of these questions:

- Does your company need this new system?
- How will your company benefit from Project 2010? Does it really need it?
- What size of organization typically adopts a demand management tool?
- Do demand management tools pertain only to large companies of over 1,000 employees?

The truth is that companies of various sizes have strategic and tactical needs. These needs are directly linked to stakeholder class requirements (e.g., customers, shareholders, and employees). Different requirements, processes, and skill sets often delineate the types of projects companies launch. Typically, the main differences between companies with staffs of 72 versus those with staffs of 52,000 are the number of digits in the balance sheet. Companies of all sizes need to remain competitive, current, solvent, and valuable to their customer base as well as accountable to their shareholders. Additionally, every project failure, regardless of company size or project specifications, has a negative impact on that organization's bottom line, not to mention hits on intangible assets, such as credibility, morale, and customer perception.

Demands on companies today are as complex as ever before. Corporate debt may reach all-time highs, and profit margins continue to be squeezed in a growing global economy. Managing and forecasting resource capacity, ensuring quality delivery to customers, and meeting shareholder expectations means that innovation and planning have become fundamental activities that have to be efficient and effective. Organizations embracing project systems using previous versions of Microsoft Project Server today can select and prioritize projects, obtain better insight into complex, interdependent projects, and reduce project risk. What does Project 2010 provide companies? The next customer example shows how a large organization initiated a PPM solution and how Project 2010 with Demand Management capabilities has and will continue to add value.

Stakeholder Classes

Stakeholder class is a grouping or an identifier that helps quantify what area or functional/nonfunctional grouping a stakeholder represents. For example, accounting and engineering may have some very different reporting, tracking, or prioritization requirements; however, each requirement from each stakeholder group may be mission critical and in some cases diametrically opposed. By organizing your stakeholders into classes, you can help to rate, rank, and prioritize new proposed projects, key project metrics, or project requirements for each

Microsoft Human Resources

Microsoft HR, which supports the 93,000 employees of the global software company, wanted to make better strategic decisions about its portfolio of projects. As HR developed new portfolio management processes, it worked with a prerelease version of Microsoft Project Server 2010, tailoring the software to fit the business processes of its five HR Centers of Excellence (COEs).

In early 2007, HR decided to improve the way the COEs delivered projects. "We had more than 200 projects running," says Bruno Lecoq, director of business process in the Operation Excellence COE at Microsoft HR. "To complete all of them—and deliver them at the right time—was mathematically impossible."

HR generalists sometimes felt overwhelmed by the sheer number of projects, says Joan Wissmann, manager of Compensation/Benefits/Performance Management COE Project Management at Microsoft HR. "One day an HR generalist might get a request for help delivering a training program. The next day, a request to explain a new benefits program. Then the day after that, a request to work on recruiting." If HR generalists felt overwhelmed, their participation could lag, which meant that programs did not perform as well as desired.

The Microsoft HR department has used Project Server 2010 to collect information more easily and to compare and evaluate projects more effectively. The department has pared its project portfolio from 200 to 25 while bringing about greater transparency, accountability, and collaboration. Many users are not even aware that they are using Project Web App because of the ways that HR has adapted Project Server 2010 to the traditional processes and workflows at each COE. These benefits have been noted:

- Better information collection
- Fewer, better-scheduled projects
- Transparency and accountability
- Richer ways to compare projects
- Improved collaboration

group. The goal in this discussion is not to drill into requirements management or best practices in mastering requirements gathering or management but to help readers understand the importance of helping to organize, rate, rank, and in many cases create different portfolio analysis in Project Server, based on stakeholder classes that represent the projects, both current and future, that will be managed or created in the PPM system.

Scalability and Succession Planning with 2010

One of the key success factors with Project 2010 and its related server and collaboration environment is the ability to scale the product based on the maturity needs of the organization. For example, many organizations start from an Excel-based planning model for tracking and reporting on projects. SharePoint combined with Project Professional will allow you to build, manage, and update schedules in SharePoint and sync with Project Professional for rich reporting.

As an organization matures, it can continue migrating information to Project Server 2010, where it can leverage a common resource pool and really begin taking advantage of a common environment for standards and for reporting and managing information. This may come as a surprise, but some organizations don't build schedules; they actually leverage resource plans and track the high-level project information (metadata) associated with each project.

As a project organization continues its journey, it may find itself finally maturing into portfolio management or rating, ranking, and prioritizing new work proposals. Here is the beautiful thing. Many organizations have already started with rating and ranking a portfolio management process but have never achieved a detailed scheduling system or a common resource pool to uniformly track the resources needed for the work to be completed.

Project Server 2010 allows an organization to start at any point in the system and leverage, grow, and mature. Different pieces of the technology can be turned on at different times to move upward or downward to complete a more robust, richer, dynamic scheduling project, program, and portfolio management system and organization.

TECHNOLOGY MEETS STRATEGY: WELCOME TO THE BUSINESS USER NETWORK

Product managers and project managers . . . why separate these roles?

New products are essential for rejuvenating a line of business and are like vitamins for the body. If you provide less than is needed, the organization regresses. If you provide too much, you create waste, and much of this waste will end up being stored in the "great idea that never worked" pile.

Clayton Christensen, Harvard professor and best-selling business author, writes in chapter of *The Innovator's Solution* that more than 60 percent of new product development work is abandoned before market launch. About 40 percent of the introduced products never turn a profit and are pulled back from the market. Therefore, about 25 percent of new product development investments lead to commercial successes, continues Christensen.

His statistics create a clear case for PM. Juggling the triple constraints (time, budget, and scope) minimizes misuse of resources. PM, as a discipline, offers the requisite controls to achieve this goal.

At the same time, creating a new product requires innovation, which is a bit chaotic and often comes with unstable requirements because of changes in the marketplace. Product management enters here, with its tools to build a product that conforms to customers' wants and needs within the shortest possible time frame.

The core issue is control versus speed and innovation. This conflict can be resolved by separating the roles and assigning the responsibility for each area to two different persons.

What Are the Common Traits of the Two Roles?

Individuals who opt for either product management or project management usually share a similar career trajectory. Both functions often are a career choice of technical specialists. After advancing to management level, both leave their functional areas' bastions and face the challenge of working in a cross-functional role, which requires strong political acumen. Thus, the ability to navigate through and get things done in a political minefield is crucial to succeed.

Finally, product managers and project managers are both required to understand the big picture, including the marketplace and their own respective organization's priorities, while keeping an eye on the details.

What Are the Roots of the Differences?

Tracing back the discrepancies between the two roles to their sources, we find that they stem from the key differences between a "product" and a "project."

The Project Management Institute's *Project Management Book of Knowledge* (PMBOK) perfectly describes the discrepancy between the scope of a product and a project. A product's scope is specified through its features and functions. A project's scope is identified by the work itself that needs to be done in order to deliver the product. PMBOK offers further clarification: Product completion is measured against the requirements while project completion is measured against the plan prepared.

A view into their lifecycles reveals another way of seeing the differences. A product's lifecycle is depicted with a chart that has time on its *x*-axis (often with these phases: introduction, growth, maturity and decline) and level of sales on its *y*-axis. The product's lifecycle is most commonly used for determining the appropriate marketing mix (price, distribution channels, product features, and promotion activities). The project lifecycle, in contrast, is illustrated with cost and staff levels on the *y*-axis and time on the *x*-axis. The project's lifecycle is used most commonly for controlling corporate resources.

What Are the Challenges in Working Together?

Recently, while consulting on a technology service development as a project manager, Tim found himself in a constant battle with the product manager.

Let's call this product manager Mike (not his real name). Mike had a tendency to micromanage the cross-functional project team members. The company is a matrix organization, and all the project team members were on loan from functional departments.

The new product development project was stuck in the requirements definition phase for several months. The reason for this was that the Mike-led cross-functional team silently boycotted him as a pushback to his lack of trust in their abilities. They came up with excuses ranging from "I am currently too busy on my other projects" to just plain not showing up to the weekly team meeting. The team slipped into a vicious circle.

After his first serious disagreement with Mike, which was about cosmetic issues relating to my project schedule, Tim realized that we needed to negotiate boundaries that we could both live with.

"Mike," Tim said, "I understand that you have great intentions and would like to see everything go perfectly. However, it is time to leave some tasks for your team members."

"What do you mean?" he replied.

"Well . . . some team members told me that they would be happy if you could focus on product-related issues and let them determine the support requirements and processes in their domain. For example, I would be happy to rearrange the activities on the schedule based on your input; however, you have not provided me a good reason for doing so, and since the schedule is my responsibility, I will not do it."

We never had a chance to find out if Tim was able to convince Mike, because, after a few weeks, Mike was transferred to another product. The new product manager, Chris, was very cooperative. His mantra was "Don't care how you do it, just do it fast."

We completed the requirements in a few weeks and launched the service with a phased approach in three months.

Tim's experience with Mike highlighted a few challenges that are common in project and product management relationships and need careful management:

- **Control versus speed.** In Tim's case with Mike, the situation was a bit backward, because typically project managers are accused of being too controlling. However, the lesson is clear: Control, wherever it comes from, needs to be balanced with flexibility and speed. Too much control oppresses team members' creativity and motivation, resulting in slower problem solving and slipping launch dates.
- **Crossing role boundaries.** Mike meddled with PM and other deliverables, creating animosity within the team, thus damaging teamwork. Project managers who are subject matter experts can fall into the same trap if they don't manage the relationship with the product managers carefully.

What Works, or How to Save This Marriage

Four factors can pave the path for creating a partnership that works for product managers and project managers and can enable a fast delivery of high-quality products.

1. Mutually agreed-on priorities for the new product development project (time, cost, quality) provide the baseline for subsequent decisions.
2. A robust new product development process that defines stakeholders' responsibilities and provides clarity around who owns what deliverables helps in avoiding time-consuming collisions.
3. Open communication builds trust and is the best lubricant for the product development machine.
4. Project managers who view product managers as internal clients and partners in delivering seem to be quicker in building a relationship based on mutual respect, which leads to the ultimate goal: delivering exceptional products to the customer.

Business Users Connected to Business Objectives

Organizations in the twenty-first century, especially in light of the posteconomic environment of the 2008 to 2010 period, are looking at technology platforms that can help them solve issues while being accounted for by the chief executive/financial offices as a solid investment to grow with their business, in full support of the direction the organization wishes to stretch and move. Microsoft Project Server 2010 is meeting expectations and appealing to a wider group of business users to solve planning, forecasting, and financial control needs. This is largely attributed to its ability to slip into the sweet spot of supplying a robust technological platform that can bridge PM methodology and an organization's individual maturity approach for growth. Project Server 2010, built on the business collaboration system SharePoint Server 2010 platform, is delivering enterprise-wide support of aligning work with organizational strategy, strong tactical execution, and meaningful BI to empower organizations to make informed decisions that impact their current situation but also provide a strong foundation for future goal achievement.

Companies are looking to better align strategy with financial planning and look to their corporate diversity, regional presences, and departmental structure for the best way forward. In this book, we address some of the main factors for scalability, best practices, and opportunities for creating wins from the potential challenges companies face when implementing a solution such as Project Server 2010. The aim is to initiate dialogue and thought around the use of and growth with Project 2010 that includes the individual products of Project Client, Project Server, and SharePoint Server.

IMPORTANT CONCEPTS COVERED IN THIS CHAPTER

Key concepts in this chapter will be built on in the rest of this book. Here is a short recap of the core elements covered.

- The importance of leveraging technology and culture for a successful growth of PPM
- The importance of governance, whether process or reinforced with technology and workflows, in ensuring good reporting and visualization of project/program status
- Fueling a PMO's success with good information that is easy to maintain and report on
- Succession and scalability planning with Project Server 2010 (growing the tools with the culture)
- How to leverage business influencers and their key requirements to help ensure a good PPM system

REFERENCES

Christensen, Clayton. 2003. *The Innovator's Solution*. Boston, MA: Harvard Business School Press.

Project Management Institute. 2008. *A Guide to the Project Management Body of Knowledge (PMBOK® Guide)*. Newtown Square, PA: Project Management Institute.

CHAPTER 2

VALUE PROPOSITION BY ROLE OF PROJECT SERVER 2010

IN THIS CHAPTER

This chapter is designed to help you to explore the uses of Project Server based on roles, business perspectives, and different stakeholders. We describe them from six different perspectives. These perspectives represent the majority of end users who work and need technology solutions in the project, program, and portfolio management space.

For each of these perspectives, the importance of being able to view, track, report, see, and predict work and progress has an important relationship to the prospective stakeholders and their working roles and requirements.

What You Will Learn

- How each role plays a part and how that part is connected in Project Server
- How to understand market opportunities and the right timing of leveraging Project Server for organizational success
- How to understand stakeholders' roles and involvement in establishing a successful project portfolio management (PPM) implementation
- Key steps in addressing each stakeholder's needs and avoiding pitfalls while maximizing success factors for a PPM implementation
- Different common challenges faced by different types of organizations that will be implementing PPM and how to overcome these challenges
- Key steps to creating wins for different organizations as well as how to best leverage knowledgeable resources for a successful Project Server 2010 implementation

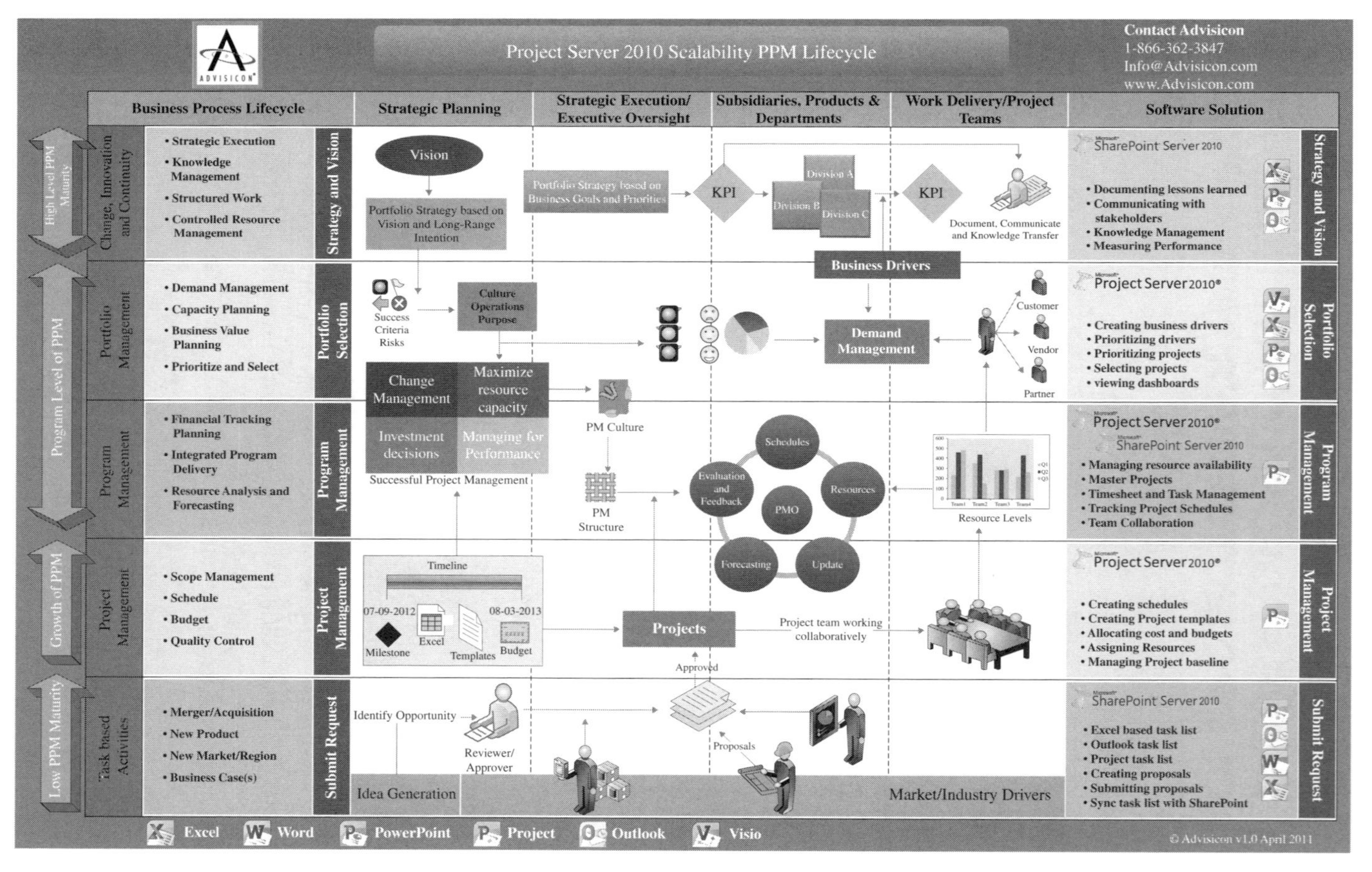

Figure 2.1 Project Scalability Flowchart Source: Advisicon

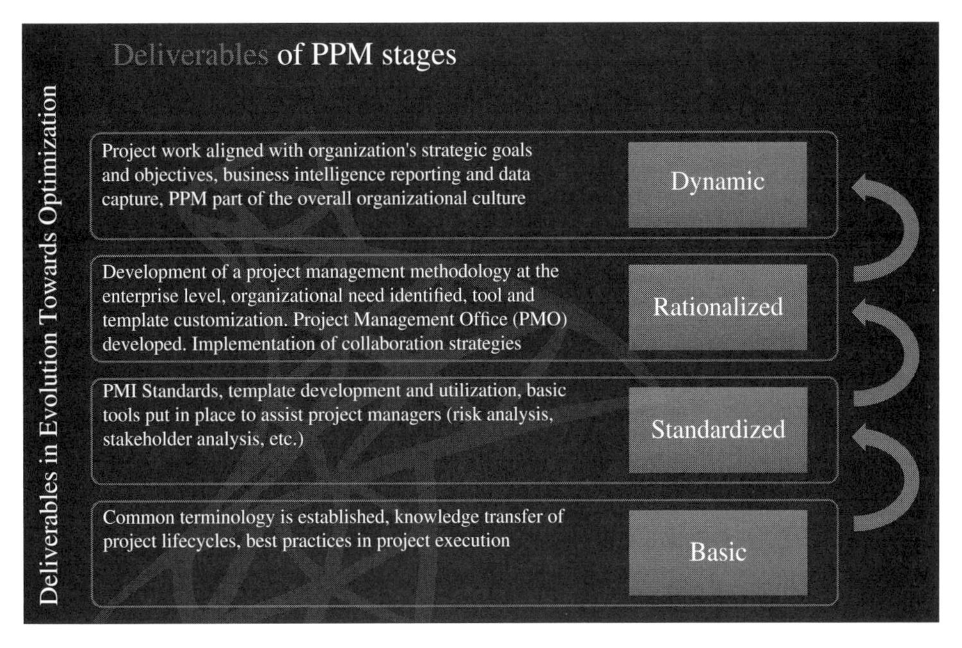

Figure 2.2 Deliverables of PPM Stages Source: Advisicon

CLAIRVOYANCE WITH PROJECT/SERVER 2010: FORECAST FUTURE RESULTS

Project 2010 Scalability Flowchart

Throughout this section, we refer to the Project Scalability flowchart (see Figure 2.1). This chart is designed to clarify the different people and perspectives in dealing with the challenges and the next steps to leveraging Project 2010 and Project Server 2010's capabilities, depending on the users' entrance point for using the application.

Introduction

One size does *not* fit all. Work systems need to be flexible based on the problem you are trying to solve, the culture of the organization, and the time frame in which the solution is required. Project 2010 is an extremely flexible, configurable environment for managing the work, resources, schedules, and reporting and collaboration needs of the enterprise (see Figure 2.2).

Data from business markets* proves that the focus of business in industries today is not about the technology but rather about the processes or the people. It

*Gartner, whose Magic Quadrant report is updated yearly, is an excellent source of reviews on the impact and value of project and portfolio management products. Its analysis of different product suites that support demand, capacity, and resource forecasting and the integration of these data can be a valuable reference point in understanding the best-of-breed tools (including Project Server) that are available to companies today.

is about ensuring that employees have the best possible impact on the business as well as seeing the alignment of their projects to the business and realizing long-term value from leveraging this technology.

This holistic approach is known as project portfolio management (PPM). PPM takes into account project management (PM) methodology (standardized terminology, common project lifecycles, project execution stages), the organization's maturity approach (development of impactful business processes, continuous improvement strategy, management theory), and an enterprise-wide technical platform (Project Server 2010, SharePoint Server 2010) that allows for thorough business use of capturing data, analyzing metrics, and formulating a plan to take appropriate actions.

Although PPM has been most prevalent in select industries, such as construction, aerospace, and some information technology (IT) environments, the demand and benefits have been growing across organizations since the mid-2000s. A few key factors have led to the successful use and adoption of PPM, such as adaptable technology platforms and growing diverse end user profiles (see figure 2.3).

As the market and technical applications have evolved, new challenges and opportunities have emerged for all major industry stakeholders. Stakeholders can be classified into three main groups:

1. Practitioners (customers/consumers)
2. Independent solution partners
3. Services and training providers

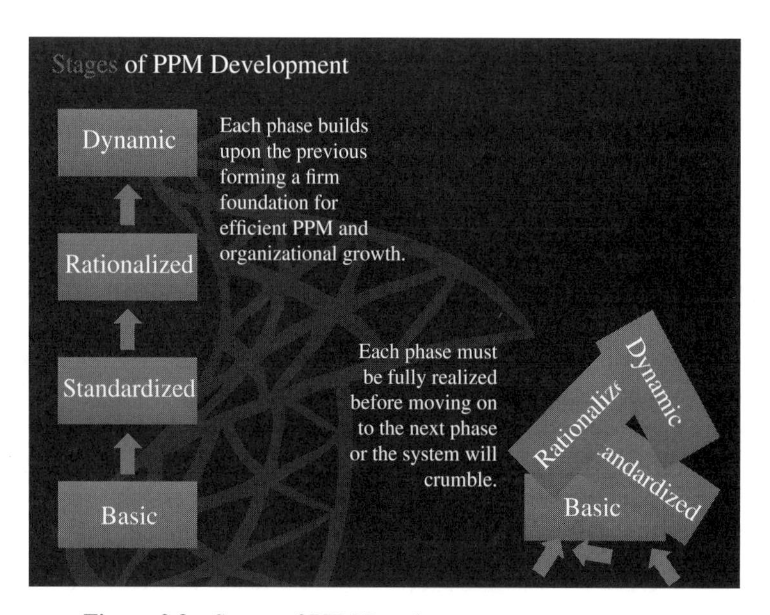

Figure 2.3 Stages of PPM Development Source: Advisicon

As a result of the evolutional change in the PPM environment, all three stakeholder groups have realized growth and demand from previously untapped business segments and corporate entities. Although this is great for business and the PPM competencies, organizations still have to justify spending to acquire technology and mentorship knowledge to fully leverage these capabilities. Organizations are expecting measurable return on investment (ROI) in regard to the capital costs associated with implementing this type of solution. Project 2010 can bring the benefits of PPM to projects in flight and creates long-term returns with alignment to organizational objectives.

We must therefore address PPM from a business perspective and help key users (via their functional role) understand how to leverage Project Server, both now and as they and their organizations mature with the product.

This section covers examples from six different perspectives and highlights scenarios such as initial product entry points, growth options, and assisting diverse stakeholders with challenges that may not be visible to them. This chapter is not designed as a technical how-to or a step-by-step feature review. It is designed to showcase best practices and focus stakeholders on understanding how to get the most business value from Project Server 2010.

Perspectives

This section contains six viewpoints—perspectives that serve as an analysis—related to how and why Microsoft Project 2010 is being selected, implemented and leveraged to meet corporate business needs. These perspectives are developed from information acquired since the 2010 version release in May 2010 and provide valuable insights, whether you are looking to buy, deploy, or serve as the provider of Project 2010 solutions. The six perspectives are:

1. Microsoft Project 2010 Points of Entry and Scalability: Planned and Organic Growth of Technology and Process Systems
2. Know Your PMO: How Stakeholder Classes are Influencing Project 2010 Decisions
3. Roles Played During Project 2010 Acquisition: Views from a Client, Partner and Microsoft
4. Ease of Implementation and Leveragability of Project 2010
5. Decision Threats: What Can Cause Roadblocks or Resistance to Executing a Decision Regarding the Solution?
6. Challenges and Critical Assumptions Related to Project 2010

Each perspective includes a situation, with a set of common challenges and key steps to capture the win in the face of the challenges at the end of its segment.

Perspective 1. Microsoft Project 2010 Points of Entry and Scalability: Planned and Organic Growth of Technology and Process Systems

This section covers points of entry (POEs), the views/roles approaches that a stakeholder will see or use in Project Server 2010, and stakeholder classes. An example is a Team Member will have a different POE than a Project Manager. This highlights the scalability up or down based on the POE for Project 2010. It is divided into two parts:

1. Addressing the needs of customers (decision makers or stakeholder) by role (essentially POE)
2. Business case study examples with Project 2010 customers and scalability challenges, questions, and resolution choices

Recent trends show that organizations are adopting a culture of joint decisions to improve the collaborative effectiveness of business decision making. This means that decision makers from multiple stakeholder classes have visibility into status and actions across parts or all of the organization, viewing metrics from project, program, and even portfolio management. However, we at Advisicon believe that joint decisions are not necessarily a developed target for a PPM culture; rather they are the environment and DNA inherent in every organization.

Within an organization, the cultural attitude can vary from high functioning to mere survival. The state of organizations does not consistently correspond with the business demands. What this means is that a high-functioning team may be part of legacy systems, while those teams trying to survive are doing so with a new product or new direction the organization is taking. An example of this is organizations who need to innovate as technology changes (dial-up Support technology approaches online or broadband technology). Thus, we often find that groups performing very efficiently may not be the first to adopt new tools like Project 2010. So, when working with an organization that is adopting Project 2010 in a department or division, the path of growth may not be as clear as we expect.

Identification of, Responding to, and Driving the Needs of Business Users: Identifying the POE and Path of Growth Often new solutions are introduced due to frustration, incompetence, or outgrowing the capabilities of current platforms. Indeed, many times new technology is implemented as a result of an actual business case and identification of defined needs. This technological shift can provide a POE at a higher level of the organization as executives look for different options to solve the organization's growing pains. Or the POE may come from organizational growth and implementation of a program management office or possible project management office (PMO). These managers may need a more holistic tool that grants them visibility at a different tactical level, requiring different analytics and metrics. This POE is also often supported by a business case.

However, Project 2010 usually is upgraded from a previous version or as a replacement for a previous product (non-Microsoft). This scenario can open the door for introduction of Project 2010 at the information worker level. Information workers reside at the departmental level where a group, team, or class of stakeholders requires the capabilities to map and track work against budgets and time. Project 2010 can appeal to many non-IT departments. Historically, the IT department was tasked with sourcing a technology that would meet business users' needs. Project 2010 exposes business users throughout the organization to the platform's capabilities, and these stakeholders are identifying additional emerging and immediate needs for this solution.

Adhocracies, Islands of Excellence, and First Available More and more organizations are considering and approving capital expenses and process changes based on business cases. This practice is not necessarily new per se but has become more frequent in response to the current external economic forces. That means organizations are initiating changes to counteract these newer, external forces.

Many times this results in the sunsetting of some products and the launch of others. These actions usually create a vortex for "pork" to be added—call them corporate earmarks. Since the organization is launching this new product, new technologies, training, or hiring are often tied to the product's overhead. If a business case is not vetted and validated by multiple stakeholder levels within the organization, quick and ill-informed decisions on lesser solutions may be made to plug the immediate business need or hole.

Without a clear review of the business case, poor decisions may be made based on unseen personal or political partnerships, despite high-functioning or well-performing islands of excellence within organizational departments.

Project 2010 is getting a second look from many stakeholders, even if there isn't yet an official corporate approval. These POEs cannot be ignored, especially as they lead to an identification of the needs of the PPM culture and of potential growth paths for the organization.

Emerging Stakeholders: The PMO's Primary and Secondary Stakeholders Organizations are making significant reinvestments in their internal and external suppliers and in consumer relationships. These investments should also include the reworking of any information systems that are difficult to extract information from, challenging or near impossible to maintain, or just plain archaic.

Organizational remodeling efforts should consider improving the enterprise as a whole rather than focusing on one or two departments. Including the PMO as a stakeholder in these decisions will ensure that the plan of execution for further development will be applied holistically to the PPM environment (Project

Server 2010). This holistic viewpoint will help identify growth needs from the Project 2010 solution and enable an organization to turn on targeted features and functionality as it matures. For example, when a construction or IT organization begins to experience work volumes or increased resource demand outside its normal capacity or physical location, it might consider expanding their use of Project Server from just a desktop scheduling tool to utilizing its more advanced features to leverage a consolidated resource pool and common dashboard roll-up reports that are uniform across all projects.

The economy is creating buckets of market maturities. These market segments are what companies are working hard and fast to capture or to be first to market, delivering goods and services. We're reminded of some of the late-night TV shows that are doing mass ad campaigns for products that address consumer needs. As companies move to take ideas to market, they create or spin-up projects, production manufacturing lines, and distribution channels. All of these (whether stable, unstable) create opportunities to leverage good processes, practices, tools, methodologies, and workflows. The more stable, mature markets can support a number of business efforts but may return smaller results albeit with reduced risk. Conversely, volatile markets pose the highest risks but may deliver substantial results. The importance of identifying a stabilizing market means that there is a need to standardize and measure business or product results and to apply PPM best practices and technologies.

Emerging markets are also somewhat volatile, but they possess some level of stability and history to support a business culture that can benefit from applied PPM methodologies and technologies. Looking across the PPM landscape, Project 2010 can approach corporate penetration by modeling a variety of economic scenarios.

Some organizations are stable; they possess all the fundamental elements to acquire, deploy, and sustain Project 2010. Other organizations are volatile; they are realizing the positive or negative impacts of the economic markets. These volatile organizations may be anything from newer organizations hitting increasing growth curves to stable organizations realizing the effects of significant reduction in sales or market pressures. All of these corporate economic environments present opportunities for Project 2010.

This variety of corporate economic environments is the "pull" or driving force that emerging, volatile, and even stable organizations are looking for to be able to apply a set of standards and measurements to help them navigate future growth or change in their market segments or industries. An example is that stable organizations may react more quickly to the visibility of positive results and metrics from applied PPM. This visibility can help emerging organizations address issues or continue to improve as compared to their rivals and peers, essentially allowing them to focus energy on responding to business metrics delivered through Project/Server 2010 versus spending time trying to unearth that information.

Time to market and visibility early in a project, program, or product lifecycle enables these organizations to be adaptive and responsive, not left behind. Whether large or small, even some of the industry's more stable organizations have opted to make business or product selection choices faster than normal. Chrysler is a good example; it innovated its automotive processes and products to react to the down economy of 2009.

Real-World Example: Large Public Utility Organization in the Energy Industry
We would like to showcase an example of a real-world organization. To protect the innocent, in this example, we call this organization Focused Active Kinetic Energy Corporation (FAKE Corp.). FAKE Corp. is looking to acquire PPM technologies but has diverse PPM requirements for each of its different departments. Each department has its own budget and steering team to select the appropriate technology. Although all of the departments have some level of business interaction, they are each tasked with process improvement and managing independent budgets. At the administration level, FAKE Corp. is also looking to secure a portfolio system to manage costs, improve its use of internal and external resources, and meet requirements set by the government.

FAKE Corp. already practices a number of project and program management processes. It uses a third-party product for cost management, and it intends to continue using this product. It has various timekeeping systems for tracking and reporting time worked by resources and is required to use external vendors to take on some of the workload as per a union contractual agreement.

FAKE Corp. is making these assumptions:

- Adopting a scheduling technology will improve resource capacity planning and forecasting.
- All users in the departments will leverage and find value using the new technology.
- Data in the costing system and other systems will integrate with Project 2010 through simple programming queries.
- One of the main components missing from establishing a robust PPM environment is the technology.

Common Challenges The assumption that data in legacy systems and other planning software are loaded and analyzed as if they were in Project 2010 is wrong. Data outside Project 2010 typically are laid out in a grid format, similar to a Microsoft Excel spreadsheet. But the line items do not have a relationship or any dynamic updating of actuals versus estimates. However, many planning stakeholders use these legacy systems for time-phased purposes (entering forecasts of numbers into the forecast module grid). In doing this, they lack the ability to perform any what-if scenarios of remaining demand and pending capacity. The gap in how the data functions between systems must be considered and addressed prior to implementing the Project 2010 solution; architecting the Project 2010 PPM solution may require some additional efforts in programming that you would not necessarily expect.

Despite your PPM implementation, you can't assume that your end users will stop using their familiar (but inefficient) non-PPM tools and habits. Implementation of any new tool or solution requires organizational development around the change. In our experience, end users usually resist change and rely on conducting business the old way as long as humanly possible. PPM is a relatively new enterprise business solution. It is likely to take some time before adoption begins to grow at a viral rate.

Often organizations assume that project and program data integrated with third-party costing tools and other planning products will automatically roll up nicely to a portfolio view. This is not always the case. How well these data streams and products integrate depends on the line of business (LOB) application, the data architecture, and the manner in which the implementation of the enterprise resource planning (ERP) solution was performed.

Key Steps to Creating the Win Creating the win is not just about the technology, the processes, or the people. Organizational transitions need to look at the people, PPM practices and processes, financial management aspects, technology, and relationships as parts of the organization's gestalt (see Perspective 5 for more on this).

For a successful implementation of Project Server, you will need to determine the key pain points, and be sure to address the enterprise business requirements through the implementation. This means including views, workflows, training, and reporting based on different stakeholder or organization needs.

Look for opportunities that will have the most impact with the least amount of effort (in other words, the low-hanging fruit). Tackling easier issues and seeing some quick results builds confidence in the solution and provides for greater adoption through viral growth (good news gets around). Conversely, be careful to avoid high-cost, low-return initiatives, as these will literally cost you credibility and may cripple the continued efforts of implementing a PPM system. When that one piece of bad press hits, no one will remember all the good things you have done.

Perspective 2. Know Your PMO: How Stakeholder Classes Are Influencing Decisions Regarding Microsoft Project 2010

This section focuses on role-based decisions (the must-haves) and covers what people in functional roles or decision makers need and use to select the right tool. There are two key sections:

1. Internal and external stakeholders to a PMO or a project governance environment
2. Proofs of concept

An organization's culture is more volatile than previously believed. Due to the global economic changes of 2008 to 2010, a number of external factors have impacted the ability of organizations of various sizes to select and adopt updated technology. Large, mature organizations accustomed to functioning in stable markets (and planning their penetration into emerging markets) have found that their stable situation has turned turbulent. Organizational business leaders and PMs are now facing additional challenges regarding their ability to make good, informed decisions while meeting the needs of their key stakeholders. In some cases, the business need of the project changes midstream due to market volatility or industry appetite and demand.

The global economic situation that many organizations are facing showcases short-duration impacts, or dips, in process execution. The impact of these durational dips highlights the need to be adaptable and innovative with projects and project deliverables, even midstream, due to the pace at which solutions, technology, and other rapidly releasing or emerging capabilities or needs arise, both from customer demand or emerging solution capabilities. Organizations of all sizes, locations, and markets will dip into uncertainty throughout their business cycles, but normally they do not stay trapped in the vortex of uncertainty. As organizations climb out of uncertainty to a level of stability, the positive trend is typically the result of a stakeholder class or classes leading the changes.

Often IT departments propose alternative options for managing the metadata and supporting end users, but these departments normally are costs centers within an organization. Thus, it is in fact a specific type of end user that leads the change. Picture IT as the basis for supporting what organizations need. It is the business driver, external stakeholder representatives, or roles within an organization that are the key motivating factors, and represent the face of the change needed.

An example of such a driver within the organization is human resources (HR), which is tasked with recruiting and maintaining specific skill groups and talent levels to meet the needs of department and corporate objectives. Maybe it is an operational division or support organization/department that is constrained by its customer or consumer base but needs to find alternative ways to fulfill work orders (work) with a workforce that is increasingly electronically savvy and used to managing and tracking via electronic formats. This stakeholder group will most likely push the tools and solutions to be more in line with its experience or exposure and consumption of digital information or work processes.

Some stakeholder demands are not determinable based on historic analysis of a PPM environment. The demands for work management and cost controls have been pushed down and disseminated across the entire enterprise. Essentially, more and more divisions within organizations are acquiring access to work, costs, budgets, and results. We are seeing more and more integration with technology and work management than ever before. In the past, in the tool versions of PPM systems, end users (project team members) had little input to their tasks or activities, essentially progressing the work or the remaining effort. Now team members can provide extended input, like issues or risks, additional tasks that were missed and send them with their updates to be reviewed and approved by the team leads, project managers or resource managers. Good task tracking over time is a key ingredient for effective forecasting and timephased reporting (past, present, and future) of task and work status. These metrics (start, finish, estimated remaining work, issues, risks, new tasks, etc.) are key ingredients for dynamic scheduling and forecast reporting of demand, capacity, and earned value analysis for projects. To be effective, a PMO must provide an account of progress and accountability for active projects in its existing work portfolio (through reporting and metrics) for stakeholders who have little understanding of the structures and processes of the PMO or even the details of project work tasks. The technical solution platform must capture stakeholder requirements and factor them into decisions regarding the acquisition of tools that can address their varying needs.

Internal and External Stakeholders of a PMO or Project Governance Environment Project managers and solution process improvement experts have the daunting task of identifying and appealing to many business users. Solution and change initiatives are always more effective when the users respond and adapt to them positively. Now that Project 2010 is a more collaborative business platform, the stakeholder profile and end user classes can help define and tailor their views and reports to their needs with little or no effort. For example, the requirement to meet end user needs may not stop at the direct end user but extend to that end user's network, not just the team members but external stakeholders who may have an interest in the project, its details or status. Thus, the number of affected stakeholders and potential voices influencing decisions is multiplied. Project 2010 captures that networked relationship.

The Project 2010 experience is more than a business user experience. It is a gateway to project metadata access, and it gives end users the ability to interpret and interact dynamically with the data in order to make more informed decisions before taking action. It is this project intelligence that cascades to the secondary level of business users.

As the business world becomes more connected socially, the environment is moving from an it's-who-you-know mind-set to a who-knows-who-you-are-and-what-you-do mind-set. Business initiatives and project approvals are based mainly on visible, quantifiable information and analysis. Often that which benefits one stakeholder group will inevitably benefit other groups as well. Ideally, companies will realize much more value from technology and process investments if they can expose those investments to more parts of the organization.

Proofs of Concept A proof of concept (POC) or pilot project is an opportunity to demonstrate the capabilities of a new approach or solution in a controlled manner in a small organizational area. A POC is an excellent risk mitigation strategy for an agency planning to implement a new system. It can also serve to inform or resolve an analysis of alternatives during the investment planning phase. The pilot helps determine whether the solution is appropriate and how easily it can be configured. It also provides hands-on experience for IT personnel and end users.

For full-scale implementation, a POC should be carefully designed and evaluated. If the scope of the POC is too narrow, the pilot runs the risk of not having a sufficient basis to be useful to end users. If the pilot scope becomes too large or the time scale too long, the decision-making process becomes too drawn out and the business continues to suffer.

POCs are a great way to involve a small cross section of diverse stakeholder class representatives. Organizations can use POCs to their advantage in creating internal champions for the proposed solution; they have touched it and seen its capabilities. But again, careful attention should be paid to the design to ensure using the POC in this way does not backfire and create internal adversaries for the proposed tool.

Activities related to POC projects can be divided into three distinct phases:

1. Preliminary efforts.
    a. Define the purpose, goals, objectives, and scope of the POC pilot demonstration project.
    b. With input from all stakeholders, technical staff, records management staff, and users, establish the success criteria for the POC pilot.
    c. Outline the benefits of conducting a POC pilot as well as the risks of not doing so.
    d. Establish an administrative infrastructure to support and guide the POC pilot project activities.
2. Conduct the POC pilot.
    a. Determine the accuracy of preliminary decisions, assumptions made regarding hardware and software performance, and the service level required by technical staff.
    b. Develop and use tools to facilitate documentation, communication, knowledge transfer, and metadata process utilization.
3. Test and evaluation.
    a. Assess hardware and software, system and database design, and procedures for training, scheduling, system management, and maintenance.

b. Test product(s) in dissimilar locations for functionality, usability, and benefits derived from using PPM.
c. Validate requirements and performance expectations.

Common Challenges If the approach or solution worked in the POC, it should scale to the enterprise. Expanding the POC instance requires diligent evaluation of scalability issues. Remember that the POC is purposely smaller in scale and more tightly controlled than a full-scale enterprise implementation.

Patience is often short when clients are anxious to see the solution in action on a full scale. Often clients say, "The pilot only took a month, why is it taking so long to get everyone enterprise-wide (or department-wide) on to the system?"

Security issues can be prevalent, especially in heavily regulated industries. The need to control who gets access to the data in the new system is a valid concern but can lead to security overkill and champions' disappointment in not seeing the full functionality of the solution. In other words, they ask: "Why did we just spend two years putting this online if we are going to lock up the data?"

An agile approach may work for software development; however, we don't believe that approach would work for managing our enterprise PPM deployment (that methodology also would be problematic in deploying an ERP or customer relationship management system). Using Team Foundation Server or similar products to roll tasks directly up to a more waterfall schedule would be helpful in blending a PPM system. In general, successful integration requires experienced resources and a good understanding of a schedule that can be detailed enough but yet rolled up for higher-level summary reporting. Since PPM is best deployed enterprise wide for the largest ROI, adequate planning is essential. Sprinting is not recommended; measure twice and cut once for best results.

A common attitude that is seen in organizations is that "We can do this deployment ourselves," sometimes referred to the DITS, or "did it themselves." In most cases, this is a very large part of where the project software, commonly called "the bits", may have been installed, but no intelligence exists on how to leverage or use them, causing great frustration for organizations. Our experience with PPM partners has shown PPM systems are different from other more mainstream or more commonly used systems, such as Microsoft Exchange or even SharePoint Server. For even a chance at a successful deployment, Project 2010 requires that business users use, at the very least, cases, configuration, and stakeholder adoption plans Similar to the proven project management approach and methodology, the use of historical data, lessons learned, and requirements planning are tenets of an effective PPM environment.

Key Steps to Creating the Win Utilize a POC approach only if it makes sense for the situation. There is no need to conduct this stage if you already have a well-understood strategy and need to get on with deployment.

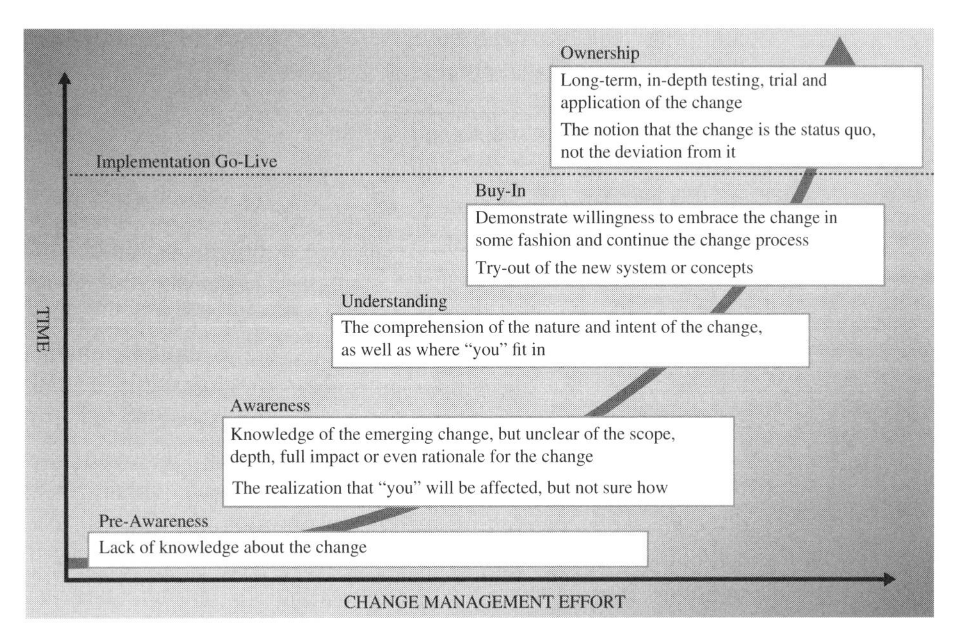

Figure 2.4 Change Management Ensure that the stakeholders have agreed on a common purpose (e.g., strategic alignment [Morgan, Malek, and Levitt, 2007] including long-range Intention, goal set, metrics, and strategies). Source: Advisicon

Ensure that all of the stakeholders are clearly identified, on board, and engaged in the new solution. You don't want to discover months (or years) down the road that you forgot to include the finance folks, or the HR team, or the training department. Stakeholders can be your greatest allies in garnering departmental resources when deploying your solution. Again, don't miss out on the opportunity to create internal champions for the deployment.

Have regular stakeholder communication meetings to keep everyone informed, address objections, and answer any questions. Remember that adults learn through understanding. You need to foster understanding before you can attain buy-in, commitment, or ownership (see figure 2.4).

Change is hard, especially when it is imposed by an external force. Manage the initiative as you would any large project where you expect to get a positive ROI (e.g., excellent sponsorship, great communications, timely risk and issue management, or colleague quality insurance).

Perspective 3. Roles Played During Microsoft Project 2010 Acquisition: Views from a Microsoft Partner and Consumer

This section outlines various perspectives of the PPM acquisition process, from provider to consumer, where end users will be adopting PPM and leveraging technology solutions and processes. The perspectives are based on the three types of primary stakeholders that play a role in the PPM supply chain:

1. Vendor
2. Partner
3. Consumer

Change occurs only when the environment becomes turbulent. Emerging global markets are realizing the growth trends because of an imbalance. This flux pushes people out of their comfort zones. When the environment shifts away from the status quo, innovation, updating, and eventually change occur.

Consumers of Project 2010 ultimately are looking for solutions that are sustainable and that facilitate the blending of corporate political sophistication with the ability to analyze, select, and deliver initiatives against strategic objectives. In the evaluation of PPM solutions, many times organizations or consumers, in the outset of evaluating technical solutions, are hoping to obtain a complete technology package while ensuring the ability for quick adoption. As the global economic market challenges organizations of various sizes and locations, these organizations must be quicker than they have been in the past to review new, alternative technology options to support their PPM and PMO. This situation is great for products just released, such as Microsoft Project 2010. Customer contact with Project 2010 and interest in the new technology is showing a sudden shift upward, but does that indicate that Project 2010 will be selected and deployed?

Customers are looking to acquire Project Server 2010 but in an iterative fashion. In the past, steps for acquiring or upgrading PPM technology began with an initial pilot, then an expanded pilot, followed by an organization-wide roll-out. Often the technology deployment is coordinated with PMO objectives, so an average lifecycle of updating technology and expanding the PMO ranged anywhere from two to five years. The Project 2010 release has increased these steps, adopting an approach normally found in IT and product development environments.

PPM Supply Chain The PPM supply chain is a concept that applies to each of the key stakeholders and describes their native roles to ensure clarity of expectations. Besides describing each PPM supply chain stakeholder, we discuss the challenges and opportunities that they collectivelyare now beginning to address. For example long lead planning or new project evaluation and the impact downstream on production, manufacturing or logistics/operations.

Vendor: Software Products

1. The vendor identifies critical needs and responds to the market by using technology to solve those critical needs.

2. The vendor, in close collaboration with the PPM community, creates the strategic vision and direction for knowledge workers and develops a product roadmap for product development.
3. Then the products (e.g., Office, SharePoint, Project, Project Server, Visio, etc.) are implemented and leveraged.
4. The business marketing organization develops the marketing program for the product suite.
5. The vendor's regional sales teams provide the presales support, help consumers understand the product set, and engage the appropriate level of partner support for the initiative.
6. The vendor manages the interactive relationship among the product, partner, and consumer through the sales cycle.
7. The vendor provides product technical support and validation through its support services.
8. The vendor remains responsible for customer partner experience satisfaction.

Partner: Program, Project, and Portfolio Management Solutions or Consulting Services

1. Vendor-qualified PPM competent partners deliver professional best practices–based program, project, and portfolio management consulting services.
2. Partners have a deep understanding of the holistic approach of project, program, and portfolio management. This understanding includes people, technology, methodologies, and processes that allow implementation of the best solutions available to match consumer needs.
3. The partner works closely with the vendor and the consumer to determine the consumer's business requirements, architects a solution, deploys the technology, trains the end users, and manages the consumer engagement to provide a successful solution that meets the consumer's needs and expectations.

Consumer: Utilizes Business Solutions to Support the Needs of Internal and External Customers

1. The consumer provides the understanding of the organization's long-range business purpose, goal and metric set, strategies, and priorities of the business.
2. The consumer sponsors the PPM initiative at the executive, operational, IT, and business departmental levels.

3. The consumer works closely with the partner to define the specific business requirements, deployment and training schedule, and ongoing maintenance and support of the PPM solution.
4. The consumer assumes responsibility for the ongoing support of the delivered solution. Note: This also might be a contracted component or part of the engagement that includes postimplementation support.

There are critical issues and challenges as well as substantial wins when PPM champions completely understand and leverage the PPM supply chain appropriately. Often there are blurred lines between who does what in the PPM supply chain. One PPM best practice is for business leaders to enforce adherence to each PPM supply chain stakeholder position and responsibility. If each PPM supply chain stakeholder delivers up to the established expectations, the PPM initiative is set for huge success.

Common Challenges Now let's take a look at some significant challenges that might limit our ability to produce quantum leaps in performance and consider some potential opportunities to turn these challenges into success.

Most work in organizations is organized inefficiently around specialists employed in what Hammer and Champy (1993) call "functional silos." This architecture can be very limiting to enterprise-wide collaboration efforts and for visibility into all project work and results.

Many enterprises lack integrated information systems. The 1990s model of acquiring best-of-breed solutions and connecting them together to provide an end-to-end view of the organization's data have largely failed.

Disconnected Excel and PowerPoint files are by and large the most common method of information gathering and reporting in use today. There is no integrated or dynamic reporting mechanism.

All too often, we PMs turn our attention to immediate and ongoing problem solving. This is a distraction that is not only unpleasant but yields little in the way of significant results. The problems never seem to end; they continue to distract the organization from completing or standardizing. This firefighting mode creates internal morale issues and forces organizations to be reactive rather than proactive, the preferred stance.

Organizations in both private and public sectors that recognize the need for a new integrated solution may lack either the internal competence or the resources to make the change

Most reengineering efforts fail to achieve any results primarily because people resist them. Organizations often fail to develop an adequate plan for the full rollout to the enterprise.

There may have been a number of failed attempts to provide both a technical and a methodological solution. Often there is a disconnect between these two pieces, both integral to effective PPM. That is why it is important to seek a qualified partner, who is experienced at delivering an end-to-end solution and ensuring that the technology platform is configured to support the methodology.

Key Steps to Creating the Win A proactive approach to creating the win is that there are no threats, only opportunities. We like to focus on working toward the solution. The one exception is if organizational culture is such that a negative perception is allowed to take over and everything encountered is viewed through that negative lens. Critical thought must come before panic.

This may be a good time to set aside that react mode and put the thinking cap on. Think about the root cause behind all the e-mails. Is your enterprise system failing to channel the communication? Does the organization even have an enterprise system, or just a number of independent, disconnected enterprise systems that need to be replaced? Consider just how many spreadsheets and PowerPoint presentations it takes to run a business of your size. Answering these questions assists in identifying the ROI for using a PPM system as well as the process gaps or integration of different scheduling systems that will produce immediate and long term value for an organization to move to Project Server 2010. These put an organization in a better position to be less reactive and more proactive.

There is a complete body of knowledge on "appreciative inquiry" (Cooperrider and Whitney, 2005). Simply stated, the motto of this field of study is: "Search for the best in people, their organizations, and the relevant world around them." Look at the things you do well, and do more of it!

The dip in the economy is forcing organizations to better understand how to do more with less. What was an important issue at the turn of the century has become a critical business imperative in the second decade of the twenty-first century. Our very jobs and careers are now dependent on identifying and deploying a new means of managing our work, our resources, and our time. Project Server 2010 can assist with building in efficiencies and automations to drive results in less time, using fewer resources. Microsoft realized this need and revamped Project 2010, enabling items such as workflow and business process execution within the basic uses of the tool.

Users are getting better at identifying and using technology to expose relevant information to make decisions. As business trends and needs are forecasted, enterprise PPM solutions will be more widely adopted in organizations by the end of this decade.

Project 2010 introduces a new level of PPM capability to the enterprise to address the critical information needs of the IT and business users. Project 2010 supplies a full enterprise-wide technical platform to support all the aspects of PPM while enabling an organization to address current pain points and build momentum for growth (see figure 2.5).

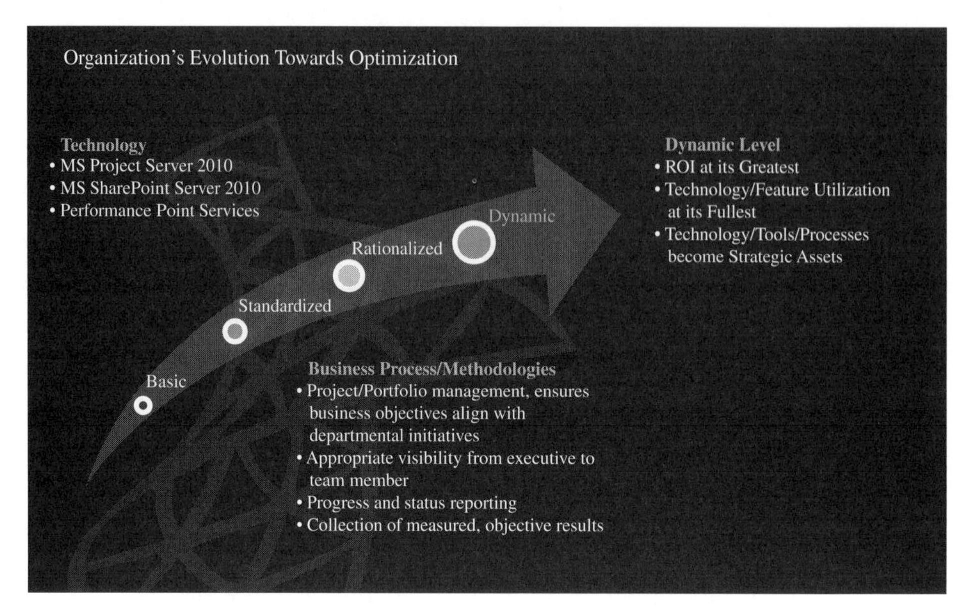

Figure 2.5 Organization's Evolution toward Optimization Source: Advisicon

Perspective 4. Ease of Implementation and Leveragability of Project 2010

This section outlines the ease of implementation and leveragability of Project 2010 from the integration and adoption viewpoint.

Microsoft has taken major steps to improve the implementation process for both Project and SharePoint Server. Fewer implementation steps mean that the organization and its users can leverage the technology more quickly and access more features and capabilities.

Now that Project 2010 is an embedded application within SharePoint Server 2010, PPM metadata natively resides within a collaboration platform. By adding the collaborative mechanisms of SharePoint Server, Project 2010 supports well-rounded, highly functional PPM capabilities within one integrated tool.

Implementing Project 2010 is (and it is not) about the technology. Microsoft's integrated PPM solution leverages the world's fastest-growing collaboration technology infrastructure, SharePoint. Microsoft Project 2010 provides an end-to-end capability that eases implementation hurdles typically associated with other competitive solutions. Project 2010's extensibility with other LOB applications eliminates the need for a lot of third-party applications to fill in gaps, thereby reducing the points of failure that can be present with third-party application POEs.

Project Server 2010 is built on the SharePoint Server 2010 platform and combines powerful business collaboration services with structured execution capabilities to provide flexible work management solutions. (See Figure 2.6.)

Microsoft Project Server 2010 is, at heart, a scheduling engine. It is a great tool for managing projects of all types across an enterprise. It also includes

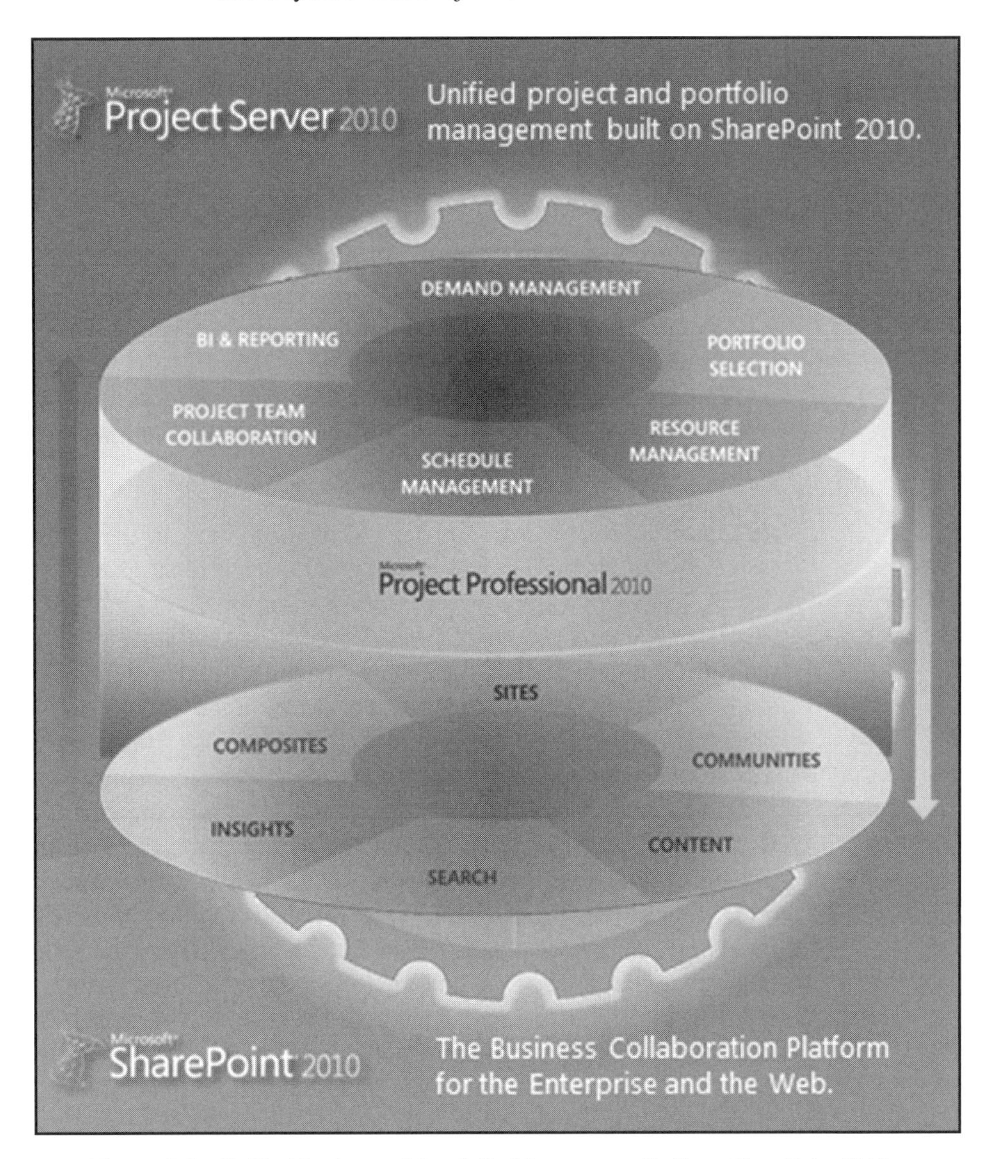

Figure 2.6 Unified Project and Portfolio Management Built on SharePoint 2010

some great portfolio management capabilities and can help you gain ROI from the business need and value assessment stage, all the way through the execution stage.

In addition to the native capabilities of PPM, Project 2010 also provides deep integration with Microsoft Team Foundation Server 2010, an application lifecycle management (ALM) solution for helping manage the software development process. Together they make a great solution for managing your IT investments.

Application Lifecycle Management

Wikipedia defines Application lifecycle management (ALM) as "a continuous process of managing the life of an application through governance, development, and maintenance. ALM is the marriage of business management to software engineering made possible by tools that facilitate and integrate requirements management, architecture, coding, testing, tracking, and release management."

Success in implementing and leveraging PPM technology is all about properly assimilating that technology into the organization. Doing so requires tailoring even feature-rich products like Project Server 2010 to the organization and the stakeholders it will be assisting. The two key areas that need to be taken into consideration are the rate of adoption and the methods that will support the PM processes.

Adoption There is currently a great deal of churn in the solution marketplace, primarily due to the novelty of PPM as an enterprise discipline compared with the disciplines of ERP, human resource information system (HRIS), or customer relationship management (CRM) (whose organizational maturity and adoption tends to be higher). Much of the controversy around maturity models stems from a mismatch between (1) the amount of details and type of processes applied to an organization's basic PM culture and (2) the business market for which those projects are designed to deliver (regardless of market stability).

By helping to build a PM culture and changing the mind-set of the people within that culture, we hope to aid organizations move along the maturity scale to their desired level as quickly as possible. Instead of focusing more time on the tool, mapping an organization's processes and folding those processes into the tool helps reinforce the workflow and the work behavior (since it is already being done by the resources involved). Remember, "Form fits function," sometimes referred to as F3, is a great example of blending processes to technology and also helps rapid adoption for organizations. PPM is not an out-of-the-box solution. It is most successful when deployed in accordance with detailed planning and folding, reinforcing, or enhancing an organization's project or program management processes.

Organizations in stable markets might consider choosing a prepackaged method from a reputable partner. This option offers two benefits.

1. It supplies a fully developed procedural offering that can transition to a higher level of maturity scale. (There are many organizational maturity scales, including Gartner's.)
2. A prepackaged solution can reduce the tendency to go overboard with too many processes. If the organization in a stable industry decides that tools are as important or more important than process, it should consider

purchasing tools that come prepackaged with a methodology. This will aid in establishing consistency and developing a deep understanding of how these processes drive and impact organizational structure, the approval chain, and project team development as well as the business of choosing the most appropriate projects, identifying strategic goals, and selecting tools and technology.

Methodology versus Technical Installation Ease of technical deployment is only one key aspect to consider. Project lifecycles are more easily defined and introduced within organizations when the technology works as intended and is easy to use.

Ease of use leads to viral adoption, which increases the overall potential for establishing an enterprise system for managing an organization's portfolio of projects.

The greater one's familiarity is with project phases and stakeholders, the more easily one can keep the project on track and on budget. For reference, see Figure 2.7.

To make sound PM choices, individuals must understand what a project lifecycle is and what factors can influence it. There are many different "lifeycles" in industry and project/program management (including Project Management Institute [PMI], Prince2, Information Technology Infrastructure Library, etc.).

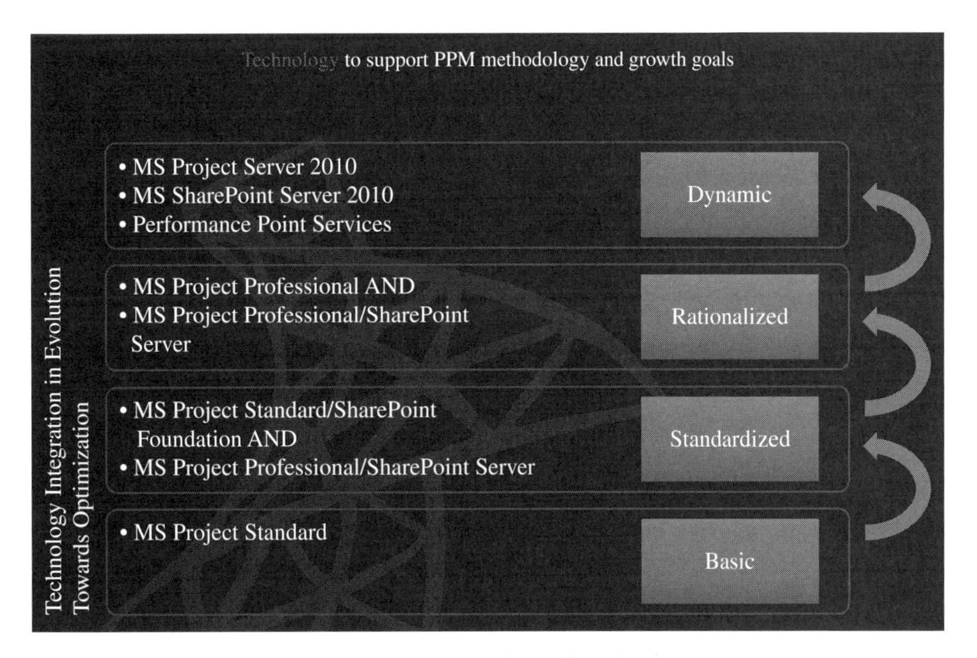

Figure 2.7 Sample Technical Project Lifecycle Source: Advisicon

The lifecycle is a core component to managing projects successfully: identifying project status data across the enterprise for executing well-informed decisions, and building efficiencies in managing multiple projects within an organization.

You the reader need to be able to properly define and manage project phases and recognize the differences between project and product lifecycles.

Breaking out project phases and substages (all part of workflow capabilities in Project Server 2010) enables a project organization using Project Server 2010 to outline and manage the flow of information or data that needs to be filled in, updated based on where a project fits within the project lifecycle.

It is also important to understand how to identify, and factor in, how project stakeholders can affect projects.

According to Wikipedia, when users are presented with a new technology, a number of factors influence their decision about how and when they will use it:

- **Perceived usefulness.** The degree to which a person believes that using a particular system will enhance his or her job performance

Perceived ease of use. The degree to which a person believes that using a particular system will be effortless. Common Challenges An organization adopts the belief that it can deploy a solution itself, either because it feels that it can do it or because it cannot get the financial commitment for partner services. Either way, this approach greatly increases risk to the initiative and to the organization. Because PPM encompasses technology with processes and affects organizational culture, expertise in architecting and configuring the appropriate solution to match the organization's current status while allowing for growth is crucial.

It's not simply about the technology. The organizational change component that accompanies deployment of any PPM solution cannot be overlooked. Even the smallest change is difficult and often resisted.

Users who think "I already have Project on the desktop, how hard can it be to just save my files to the server?" can undermine building a firm foundation to support organization and departmental growth. PPM is bigger than just the project level.

Thoughts of "I wish we had thought of that when we originally deployed the system" (e.g., reference stages of the project, template types, what department or specifically a common set of columns or roll-up values for summary tasks or reporting fields) commonly plague organizations that have rushed to get the product installed and are coming from a localized field approach rather than an enterprise approach to deployment. Project Server partners understand how to solve the current pain points, but also know what to account for in future growth within the organization and the technology.

Another common misconception is assuming that all PPM solutions are similar, and using cost as the only deciding factor to rate or rank projects for selection. This is a dangerous assumption, and unfortunately one that is made all too often. For example, companies have a budget to spend, so they approve new projects each year without rating or ranking the importance of those projects to key business drivers or the impact to the resources already working on a portfolio of projects. Microsoft has made significant progress toward portfolio analysis with the introduction of the new Project 2010 solution. This solution is the only PPM solution on the market today that is fully integrated with Office and other LOB applications, such as Dynamics and Team Foundation Server. Although cost may be higher initially when creating a well-rounded solution, the ROI and adoption is much quicker when the solution is built to integrate with the systems that people are already familiar with and when the solution accounts for all needs, reducing the dependency on multiple third-party add-ons.

Key Steps to Creating the Win Utilize a Microsoft-certified PPM partner for planning, deployment, and training of the PPM solution. Such partners have proven expertise to lead an organization through all aspects of implementing a PPM solution. In the long run, the cost will be lower, end users will be more satisfied with their new system, and the organization will find comfort in astute guidance into unfamiliar territory. In-house deployments are risky; they almost always take longer than planned, and because they do not take full advantage of the tool, they diminish ROI.

It is imperative to ensure that technical support personnel, the PMO team, and end users all receive the necessary training. This training includes the appropriate soft and hard skills for managing critical aspects of projects, programs, and portfolios (e.g., communication training, as well as how to use the tool). The PMI has a number of professional certifications for project and program managers, risk management specialists, and scheduling management specialists. Microsoft also has multiple certification tracks aimed at the individual and organizational level, ensuring that partners have the required skill set to engage in this type of solution delivery.

Identify the key organizational touch points necessary to maintain the enterprise processes required to support a PPM initiative. People do not like to feel left out of the loop or surprised. Appropriate and efficient communication is a must. This means getting the right information to the right people at the right time in the right format.

Make sure that key individuals understand what a project lifecycle is and what factors can influence it. Creating these internal champions will assist in growing adoption across the enterprise. Turn these people that they already know and trust into experts.

Consider the bigger picture from collaboration tools to LOB systems. Enterprise PPM encompasses a broad set of capabilities that affects a number of other areas in the enterprise. When applied correctly, PPM truly reaches across the enterprise, granting visibility into initiatives and ensuring that business objectives are supported.

Perspective 5. Decision Threats and Cultural Change: What Can Cause Roadblocks or Resistance to Deciding on the Solution?

This perspective covers threats to effective decision making, including potential roadblocks or resistance to deciding on the solution. We cover three key aspects:

1. Potential blind spots
2. Social and personal networks
3. Facts and opinions (blogs, research, professional associations)

If organizations attempt to push changes too fast or too far, they often encounter cultural roadblocks and wholesale rejection by all concerned parties. The PPM Maturity Model helps senior management avoid such problems by providing a framework to facilitate communication with executive management by comparing their organization's PPM processes. Examples of this can be found on PMI's and Gartner's Web sites (www.pmi.org; www.gartner.com).

An organization's PPM initiative may stall at Level 1 PPM maturity. Level 2 can be difficult to reach once business units and changes within the organization begin to move from a just-get-it-done environment to a more organized process-driven one. Ultimately, companies looking for ROI from PPM expect that the ad hoc style of PM at Level 1 will give way to a more formal PM discipline. PM initiatives and PPM technology deployments have faced these classic organizational challenges for decades. Typically, some stakeholders are not supportive while others simply do not understand or validate the benefits. The launch of Project 2010 within SharePoint Server 2010 coincides with the evolution of the socialization of business problems and the virtualization of information. Those stakeholders who would not normally be supportive now have wider and deeper access to information that enables them to ask more informed questions and make better decisions. As these normally supportive stakeholders uncover more information, they may have a change of mind. Conversely, the availability of information and the validation of specific types of information by peers may cause the typically resistant stakeholder to become supportive and convinced of the value of PPM. The key is to read the signs as these stakeholder profiles change, as the decision making world we live in has changed.

Microsoft's research and our partners' experiences are listed next.

- **Emerging disciplines.** Organizations are migrating more quickly to a PPM environment in order to realize the ROI of managed work data, including work governance and alignment of projects with business drivers. A growing trend of project and ROI validation is beginning to take root. PPM supports a variety of project and program tracking disciplines, such as Stage Gate (for product development), ALM, and PPM (for professional services). Project 2010 now covers additional competencies and disciplines for a stronger and more richly integrated PPM environment for business and project stakeholders.

- **Initial and annuity costs.** Many of the competing products to Microsoft Project 2010 offer competitive licensing options, such as free access to the data, in an attempt to secure PMO data within that product's environment. However, the long-term costs of these competing products in regard to both acquiring the technology and supporting deployment and adoption by business users is not known up front.

  Microsoft makes a strong case for total cost of ownership (TCO) when taking into account the innate extensibility of the technology (its augmentation and integration with other products) as well as the much larger market of qualified partners and vendors to support deployment. The Microsoft partner program is a proven model for effectively screening people and approaches to determine their ability to support a wide variety of business requirements worldwide. Arguably, competing products carry a higher price tag of consulting and training as well as maintenance fees and support costs when compared to Microsoft. That being said, Microsoft's Project 2010 technology does not support itself. As with all PPM technologies, there is still a requirement for ongoing technology and business user support and maintenance. The question becomes: Which solution provides the most scalability to bring the quickest ROI now and with future growth?
- **Transition from guesstimates to project intelligence.** Organizations worldwide tend to make high-level assumptions or to analyze and make determinations of project status or health as well as look-ahead(s) based on resource planning systems or ad hoc processes. Many organizations compare the efforts of their resources (amount of work specific to a task) and assign a monetary value to the results. As a result, fiscal planning and corporate financial auditing involves both art and science to summarize the organization's financial status.

  There exist several highly complex, comprehensive tools and methodologies to estimate the financial health of organizations; nevertheless, analysis still is involved. Without incorporating the scheduling or time-phased information from a project tracking or PPM system, these financial forecasting modules still just provide best guesses.

  Essentially, if an organization is planning a fixed number of strategic objectives and looking at forecasting requirements such as costs and resources, a time-phased estimate is also required.

The ability to optimize and automate the PPM system, especially around financial or work metrics and processes, really comes down to if-then situations. Decision makers need to identify key events or work packages, assign constraints to those events and packages, then create relationships between them. This ability to integrate strategic plans from a portfolio or ERP system to a PPM system is becoming more essential; it is the inflection point for organizations to transition from coming up with strong guesses and estimates to a more quantitative project-based intelligence that can be updated, tracked, and adjusted throughout project and program lifecycles.

Over time (fiscal year over year), organizations will be able to compare actual historical results or incorporate these data into forecasts.

The analysis, selection, and adoption of PPM technology requires multiple levels of commitment. This is not to suggest that commitment is an all-or nothing situation. Think about any enterprise system or a centralization of doing work a specific way. Classic scenarios show there are those who will embrace change and those who will resist it. Although those archetypes still exist, it is not critical for an organization to alter the hearts and minds of the naysayers or protect the supporters. It is important to understand that a successful implementation of Project 2010 will require training, change management and involving stakeholders, showcasing the strategic benefit to each stakeholder group. Project 2010 and its collaborative capabilities allow for virtually every role-based profile to adopt, understand, and utilize the technology as it was meant to function.

Thus, the threats are less a matter of creating a one-time full adoption of a PPM implementation but more of designing a phased solution that enables organizations to embrace a multilevel environment and scalability that will enable utilization and knowledge transfer while both reinforcing and maintaining PPM governance and structure. Essentially, this approach creates an environment of project intelligence. Regardless of how raw the project metadata are and what the technology or path of adoptability allows, the right type of system will help mature the project metadata to an analytical level.

Potential Blind Spots The effectiveness and use of technology in a business climate is a two-way street: End users have to learn how to the use the tool, and the tool must be configured in a way that users can use it. Microsoft Project 2010 with the SharePoint integration and added social and collaborative elements presents unique solutions where users are looking to shape the technology to more closely meet their needs, essentially leading to an enabling of existing processes or tool approaches, but now centralized in SharePoint. In years past, Microsoft has marketed the notion that its business products enable users to be more productive. However, in other discussions, the term "enabling" takes on negative connotation, indicating that external sources can foster bad habits and encourage poor results by employing kludges to work the technology rather than letting the technology work as it was intended.

Stakeholder profiles are morphing. For years, organizations and external stakeholders, such as consultants, have relied on organizational charts, role definitions, and other tactical data for use in objective planning, forecasting, recovery, and other corporate requirements. Matrix hierarchies have added an additional complexity but still are manageable through the use of visual charts and other tools.

These days, more is accomplished through management by objectives (MBO) approaches. Objectives still may be aligned in relation to a specific role or department, but as organizations are looking to increase efficiencies and output through a reduced workforce, the elements of skill set, positional visibility, and other factors are leveraged in a MBO approach. Thus, the work and changes that are being instigated may or may not come from those sources expected.

More and more organizations are changing personnel, strategic objectives are under more scrutiny, and the results are sometimes unexpected. All of this creates some degree of havoc and forces both internal and external business users to pursue and address these objectives beyond the tools, reports, views, and visuals already available. This pursuit ultimately allows organizations to reflexively return and improve their existing views, reports and tool solutions in a cyclical improvement process.

A PPM System that Missed the Initial Mark

An automotive support organization we once assisted was looking to select a project lifecycle tool using Microsoft Project Desktop 2010 and SharePoint. After a lengthy and deliberate planning process, the roll-out was compromised due to an undocumented customer (external force) that wound up playing more of a critical part in the system roll-out and structure. This unforeseen stakeholder represented a set of compliance and regulation requirements not initially defined or understood.

A new set of rules and requirements was created not only for the system but for the executive team. The Project Server PPM system had to be retooled quickly across the existing projects, views, and reports, and throughout the PPM structure. A lot of reevaluation and rerunning of the business intelligence reports was necessary to revalidate the original decisions and decision points. The key here is that up-front stakeholder identification and integration into the PPM lifecycle process enables a higher degree of satisfaction and communications through reporting.

While Project Server 2010 and its new features give the end user the ability to update and change the interface they are working with, the larger implementation has greater and more far-reaching impacts. In some organizations, this can be the kiss of death of whether a system is adopted.

We did not experience a failure of planning for scope definition with the automotive organization; we gained an opportunity to increase the realization of the potential capabilities of Project 2010 and SharePoint 2010. We had the option to replace more systems and shift additional business functions to Project 2010 and SharePoint 2010, which is what led to this dramatic change. The profile morphed due to the actual deployment of the system based on a wider set of stakeholder needs.

The Microsoft technology stack for PPM—Project Standard 2010, Project Professional 2010, Project 2010, and SharePoint Server 2010—has been available and supporting PPM campaigns for the majority of the twenty-first century. As the industry is realizing an uptick in specializations, such as Project Management Professional (PMP) and Microsoft Certified Professional (MCP), that have certified competencies in project environments, we have observed some new bad habits developing. Previous versions of Project Server have shown tremendous capabilities in key PPM initiatives, such as scheduling and resource forecasting; however, some technological weaknesses and a wide organizational immaturity of PPM have led to both the growth and the adoption of poor technology and processes as well as ill-founded, preconceived notions.

Project 2010 has addressed many critical requirements for effective PPM. Information relating to the robust features available in Project 2010 is lacking, which is leading to a gap in project organization's understanding of how massively the new product has been upgraded. We anticipate many Microsoft PPM supporters may build in too much buffer to compensate for past versions or previously troubled PPM roll-outs. This may lead to overestimation, overcompensation of time and effort, or misjudging the technology and process requirements.

In an established (or a maturing) PPM environment, two critical factors must be accounted for:

1. The acclamation and use of the technology by a wide variety of business users
2. The technology's functionality and ability to meet the needs of the business users

As discussed, adoption is part hands-on experience and part perception management. Change often is not viewed in a positive light, so demonstrating how the tool will make job functions easier and demonstrating the shallow learning curve is essential in ensuring organizational adoption.

Social and Personal Networks The Internet and Internet services offer a good portion of information channels for business users today, but information channels leverage multiple means, including word of mouth, third-party channels (representatives), research, ratings, search popularity, and the like. Microsoft Project has always ranked high in studies and research, but how does it fare in other information channels? One way to evaluate the situation is to look back a few versions, to one of the enterprise platforms in early 2000. Preliminary official studies were performed to evaluate Project users and their experiences. Decision makers and business users relied on a few formal and a number of informal information channels, mostly unproven and highly subjective, to get the data that would influence decisions.

The results of studies, comparative analysis from PPM implementing organizations, and the technologies, solutions, and training delivered with the technologies are being rated. PPM is more and more a vital part of the business culture, albeit still leveraged at a low maturity state. Processes are now leveraging PPM technologies (specifically Project) to meet not only corporate strategies but compliance issues, budget requirements, and regulations. PM practices, such as earned value (EV) in the construction and government industries, have long been used for litigation and tracking. Now a much wider set of organizations are leveraging these practices in scaled-down versions to gain more control over work and costs.

Moreover, as professional specialties increase, certifications for projects as well as industry certifications, such as project management, risk management, program management, training, and scheduling, are adding additional information channels, and communities of practitioners for business users to reference.

Sales professionals have an advantage when it comes to collecting feedback from their calls on prospective executives and decision makers. Sales experts in any industry have insight into the pulse of the market and what their prospects' competition may or not be doing. Many business professionals who serve as decision makers or influencers attend industry conferences and other sessions where they can network with colleagues. Another form of feedback is pure feedback, which is typically a direct, hands-on experience. But is it feasible to increase your contact with pure feedback while extending your ability to leverage peer feedback?

With information streaming in from multiple sources, often at lightning speed, it is important to maintain critical thinking when looking at feedback. A blend of sources should be evaluated, analyzed, and investigated before decisions are made.

Facts and Opinions (Blogs, Research, Professional Associations) In the age of Internet services, communication options are growing: texting, social media updates, blogs and video. We continually hear from customers and organizations that there is information overload. The issue is less one of information overload but more of inaccurate or unqualified information overload. Everyone has an opinion, not all opinions are fact, and people now have many more and farther-reaching channels with which to voice those opinions. When there are many sources of information about PPM tools and scheduling best practices, you need to find common watering holes and expert sources. When researching Project 2010, a good place is Microsoft's project sites or most valuable player groups (industry experts recognized by Microsoft who have a community-focused outreach).

Because the virtual world is always on, business users globally now have access to live people and current data 24/7. Many times this accessibility leads to discussions, responses, and additional opportunities to reconsider a course of action to address business problems. This pull in the market is leading to business users, such as consumers of PPM, being more knowledgeable about the depth and breadth of PPM technologies and processes. Additionally, social media channels, such as blogs, wikis, and videos, have content providers creating a push inertia that gets their thoughts and opinions out to a wide consumer base.

Common Challenges Challenges may arise when an organization assumes that each business unit has similar demands, objectives, and pains, and that it can use the same solution as a different organization or unit within the organization. In a situation similar to the halo effect, a portion or group within a company recognizes the value another part of the company is realizing from a solution, and they see that same solution relevant for their situation when it may not be.

Another common challenge is when an organization tries to create a solution that fits everyone. This has been referred to as solution crowdsourcing, or community-based design. In this case, each division or group defines the common solution to its specific needs; again, the organization is trying to create a one-size-fits-all design.

Mandated change is often the hardest for people to digest. It is important to understand that change initiated from the top does not automatically guarantee an easier adoption.

The assumption that the new system will make employees' jobs harder is prevalent with the introduction of new tools, technology, and processes. Employees do not want to feel that they will be forced to drink from a fire hose. People are very sensitive to learning curves, as they understand that work already on their plate still must move forward while they learn the new system.

Issues of control can create problems in the adoption of PPM. Employees worry, "I control my environment today; the new system is a threat to me and may make my job unnecessary."

Often new capital investments are viewed as wasteful; however, maintaining a status quo that is currently seeing little benefit rarely turns around into visible ROI. Sticking with the status quo can undermine future growth and increase or continue to promote the current inefficiencies, which could then outweigh the costs of a new system.

Key Steps to Creating the Win Analyze the bigger picture to get a better understanding of the current versus proposed TCO. Considering just the tangible costs of licenses and maintenance for the existing system is no longer sufficient to justify the new integrated approach. It is important to look at the increased visibility and decrease with human inefficiencies resulting from the nonintegrated and even broken systems should be factored into the decision making.

New generations of information workers are adopting the new social software solutions at an accelerated or rapid pace. Organizational rework must take into account this new way of doing business both internal and external to the organization through partner supply chains, social networking, and knowledge dissemination. A modern organization must ensure that the new PPM initiative and team is looking at the long-range goals of the enterprise.

Engage all key stakeholders in the organization on the initiative. The HR department can play a significant role in the area of resource management, skills inventory, and role definition. PPM is a multiplayer game, and the stakes are high.

Establish a positive change culture within the organization by ensuring that new PPM initiatives are successful and deliver on the promise of streamlined work, collaborative teams, and assistance in working smarter, not harder. Identify key initiatives where a win-win outcome is highly probable and ensure their success; then go and communicate that success. Viral adoption will take it from there.

Don't judge PPM merely by its past. A number of technologies had hiccups and burps when they first came out. Today most organizations already have one ERP system and one HR system; someday we will see the average enterprise develop a common platform for work and resource management.

PM is a relatively new discipline. Prevailing theory, such as the critical path method (CPM), has been around for only 50 years. Microsoft Project has been in the market for approximately 20 years, and there are already over 20 million users worldwide.

Perspective 6. Challenges and Critical Assumptions Related to Selecting Project 2010 as a Solution

This perspective addresses the various challenges and critical assumptions related to selecting Microsoft Project 2010 and sourcing expertise. Four major topics are covered:

1. Internal deployment versus vendor or partner involvement
2. What do the external information streams offer?
3. Advantages to sourcing a qualified partner
4. Critical assumptions: knowns and unknowns

Technological and social media capabilities have proven to be an effective channel for businesses to market, communicate, and interact, especially to manage projects as the global economy morphs from stovepipe industries into a virtualized economy.

Project 2010 offers the most comprehensive and collaborative PPM solution on the market. The new release has taken great strides in incorporating business processes, workflow, BI, and the most prevalent scheduling driver in the marketplace. But as mentioned earlier, the concept of PPM encompasses more than just the technology. The requirements for a successful, ROI-creating PPM implementation are varied and diverse, and involve multiple layers of the organization. Assumptions must be identified and validated. Expected challenges must be addressed to ensure that the organization is placing the best PPM tool for its needs into its environment. In today's sophisticated business environment, believing that decision makers are going to choose Project 2010 simply because of name recognition is a weak position. Organizations need to understand why the capabilities of Project 2010 make it the best tool to meet their needs now and in the future.

Business leaders are leveraging technological options. Corporate functions such as legal, HR, accounting, and IT are being subsidized through outsourcing or by tasking internal personnel to learn and manage the technologies using information available through the Web and social contacts. The internal competencies and capabilities of each person in the organization continue to rise as these changes are adopted. Similar to the process by which the personal computer changed the way data and information was created, shared, and stored, organizational business users are getting better at identifying and using technology to expose relevant information to make decisions. Unfortunately, these technological evolutions can instill a sense of confidence that organizations can do more with less, if not do it all with less. Indeed, technology is more business user friendly, but the need for solid planning and leveraging competent resources must not be overlooked.

Consumers and business decision makers are savvier than they have ever been. With organizations spending more than 50 cents of every dollar of their IT hardware budget on storage, and with the amount of data being stored and managed growing by double-digit percentages or more, savvy IT managers are exploring a variety of new strategies and technologies to reduce infrastructures costs. The pressure on IT professionals to store, preserve, and protect data while still making data accessible has intensified.

Organizations today continue to look for ways to reduce implementation and administration costs while driving value and profits directly to the bottom line. As information becomes more available and as people are more readily accessible, organizations are taking bold steps to select and implement change using technology and processes. Since PPM is intricately involved with corporate constraints, such as costs and resources, and with bottom-line results, such as actual work completed, there is a vested interest at the decision making level to do everything right.

Many times organizations mistakenly choose to install the technology themselves, internalize their process improvements, and leverage the growing access to information via Web and social channels. Although there is a strong business case for taking this approach in organizations that have employees who are knowledgeable in the technologies to be implemented, the risks are similar to those of laypeople self-diagnosing and self-medicating when they feel ill. Often people who feel sick are right: They have a cold. But other times they are mistaken and need prescriptions or perhaps more preventive measures. In hindsight, such people would have been better off had they sought assistance from experts sooner.

The next scenarios highlight the perspectives of an organization's option for going it alone and its option to secure assistance externally in support of a Project 2010 initiative.

Many organizations in today's economy are looking for more cost-effective measures and to utilize internal resources as much as possible. The challenge

Microsoft Project 2010 faces today is that it is technically easier to set up and deploy as an infrastructure than ever before, which makes it seem that it can just be "installed"; but the tool is not just a simple scheduling engine, it is a full-spectrum enterprise project/program and portfolio management tool. Therefore, many organizations are choosing to have their internal technical staff do the deployment. A good example would be similar to having IT install an ERP system like Systems Applications and Products (SAP) without involving accounting.

Prospective Options to Selecting, Deploying, and Supporting Project 2010 Project 2010 is capable of interacting and sending or retrieving information that may come from other systems of record (e.g., HR systems, time-sheeting, or actuals). Thus, often there are security concerns associated with that information and the ownership of system-of-record data. An example may be when integrating financial data, often from an ERP system, or product design system.

Organizations have established policies and effective processes around data compliance, financial, and product development environments. They have also dedicated employees to serve as internal experts to manage the data. Sourcing an external entity that may not have the depth and experience or qualifications to handle sensitive data to focus on Project 2010 poses a risk. In this instance, it may be beneficial for an organization to install, configure, and deploy Project 2010 internally to maintain control of the sensitive data.

Project 2010 leverages the SharePoint Enterprise platform and enables more integration of BI metadata to be threaded within the management structure for Project 2010. SharePoint has evolved as a business data management system faster than Project Server. Many organizations have already initiated document management and information portal campaigns prior to the consideration of a PPM system. Many instances have SharePoint supporting data from other legacy systems, including SAP in data processing, and project lifecycle management systems. Often SharePoint is already being used for continuous improvement, lifecycle management processes, and other information sharing and collaborative functions. Thus, organizations have acquired a significant level of knowledge working with the data in an enterprise environment. Adding Project 2010 may seem like a manageable step for internal resources.

The main business justifications for Microsoft Project 2010 are its ability to expose, track, and control resource work and costs related to actions that are tied to corporate objectives and goals. Because Project 2010 resides in an enterprise collaborative environment (namely SharePoint Server 2010), it is increasingly critical to business users that Project data are accessible and reportable. These requirements typically are core for effective delivery to the strategic objectives, which makes Project 2010 much more enticing to internal decision makers. As they know the business better than anyone else, they make assumptions that they know how to structure the technical business platform to meet their business needs.

What Does the External Information Stream Offer? As the Internet continues to grow and mature, so do the data. The U.S. judicial system is built on the premise that all individuals are assumed to be innocent until proven guilty in a court of law, and accused individuals have the right to receive a judgment regarding their guilt or innocence by a jury of peers. The Internet has followed a similar example, where data are true or valid until they have has been reviewed and rated by others. Decision makers now have access to more channels of scrutinized options that contain credible, validated information complements of social media and the virtual world.

In the past, research firms spent staggering amounts of time and capital to reach end users, rate products and services, and craft analyses of the findings that were meaningful to business decision makers. Today, search engines such as Google and Bing are advancing the search algorithms to mine data based on results criteria.

A huge array of Internet sources provide a broad range of resources to an organization that wishes to deploy a PPM solution without a partner. A number of these Internet sources either provide access to their resources via a subscription or act as storefronts for their partner offerings. The informational content provided often is only a part of the necessary knowledge required. Just as business decision makers have become savvier, marketplace providers have become more astute at crafting content to drive sales. Microsoft has created a focused portal for Project on its Web site (www.microsoft.com/project/en-us/project-server-2010.aspx). You can do a quick search on the Web site to find the central repository of information for Project Server 2010.

At the time this book was authored, key social and business networks that are directly involved with PPM include LinkedIn, the PMI Scheduling Community of Practice (formerly the College of Scheduling), and the Microsoft Project 2010 PPM Virtual Showcase. A new category of social collaboration software is breaking through the grid from multiple third parties. Virtual events sites, including the Microsoft Worldwide Events (WWE), offer a significant amount of self-paced, on-demand training.

A number of professional associations are primarily in the business of brokering knowledge, certifications, and people networking. These associations offer a multitude of complementary training, white papers, videos, and access to trial software and complementary learning products.

Advantages to Sourcing a Qualified Partner There is inherent risk in not ensuring that meeting the end users' needs remains critical for the success of implementation and adoption. Organizations have become more aware of end user needs and the critical relationship between user preparation and implementation success. However, old habits still remain: Organizations assume that once the solution has been installed and the training has been delivered, they will reach a level of competence that will remain constant.

An organization's ROI will improve only with a prolonged effort to ensure the ongoing competence and confidence of its users in the usage of the tool. According to a report from International Data Corporation: "Training on complex systems is a never-ending cycle. There are always new processes, new employees, and new locations that must be brought up to speed or brought online, and no group can be left out" (Anderson, 2006).

Many times a business system initiative is launched within an organization to meet its internal needs. For example, a product engineering team acquires PPM technologies to manage its project and programs. Demand inside the organization from another department (e.g., the sales department) forces the change with engineering. A very common example is when the sales department is selling work that has to be managed or built out by engineering, and what is promised to the customer by sales is proving difficult or impossible to build in engineering. This is a classic example where engineering departments have moved to standardize templates and folded sales organizations into that workflow so that cost estimates and proposals are based on some measure of scheduling and work reality. Mastering these insights requires experience with the inherent needs and capabilities of PPM; a qualified partner can help an organization identify and fold these needs and capabilities into a PPM implementation. This will save the organization time and heartache.

SharePoint is a game changer. "SharePoint, in a sense, is becoming an operating system," stated Steve Ballmer, chief executive officer for Microsoft, at the SharePoint conference in 2009. There is a new dynamic to the Enterprise Content Management space. It's being enabled and forever changed with the new release of SharePoint 2010. Many organizations already own SharePoint, and many partners have subject matter expertise across countless verticals and horizontals. Likely there will be an explosion of partners and solutions hitting the market in the coming years.

Qualified partners can help improve IT service excellence, lower development costs, and help organizations develop their competitive advantages through the innovative application of the Microsoft SharePoint platform.

PPM historically has had to compete for funding as an IT expense. As the PPM software has grown in size, scalability, and use, the infrastructure requirements have increased. Additionally, the collaboration elements with PPM technologies demand more from the organization's systems, processes, and end users. What normally starts in IT expands to other groups, leads to the need for additional training, possibly extensive consulting services, and ongoing system and process support. It truly is a culture change.

Project 2010 is a business application within a business platform. The approach for organizations and consultants to secure executive agreement requires a team effort and a shift in the approach used with earlier versions of Project Server implementations.

Critical Assumptions: Known and Unknown In previous versions of the Microsoft Project product, the technical architecture depended on leveraging and extracting metadata from environments. Typically, the SQL Server is a secured environment, with restrictions at the highest level due to the sensitivity of the information that has potential to be exposed. What typically happens is that Project Server data is made visible, bringing a dose of reality to the users and decision makers, who can (in some cases) have visibility for the first time and see that information in context—a PM context, which often requires a skilled "interpreter" of the message and impact of the data. We call this "telling the story." It is important to not just have metrics roll-up for reporting but to allow the addition of user-defined fields or comments. These metrics can be utilized, analyzed, and trended over time to see the business benefits and successes or failures of different activities, approaches, resources, or even individuals.

Another critical assumption is that organizations are constrained by the budget they are allocated. Individuals tasked with setting, managing and controlling the budget are the critical stakeholders. Many engagements, partners, and even internal stakeholders find themselves at the mercy of the financial controller, who must sign off on expenses.

There are times when an organization is more agile and the financial subject matter expert has assumed that the manager or requester of the project has the business case already approved. However, sometimes further explanation is required, and that justification tends to scale beyond a straightforward IT investment or training cost. PPM is not just a quality or continuous process initiative, although it affects and supports both by its very nature. A strong business case that is articulated in bottom-line facts and elements is required for the costs to be supported and approved. The stronger the business case, the easier it is to align the ROI for what the Project Server needs to report on.

As mentioned in the Project Server Demand Management white paper written and posted on Microsoft Web site (http://technet.microsoft.com/en-us/library/ff686781(office.14).aspx), understanding and quantification for the use of Project declines the farther away it is from the PMO construct. Many times pockets of activities and unique initiatives within organizations go unnoticed and remain undetected within the PMO.

Why is this an issue? Typically, these initiatives and activities require the contribution of resources—many times skilled resources. As organizations are utilizing shared resources on a majority of their initiatives, this lack of visibility can affect the bottom line when resource constraints affect project success.

Decision makers are of different profiles. According to a report, "Decision-making styles are behavior-based, so the key to correctly identifying an executive's tendency is to pay attention to what she does, not what she says" (Santosus, 2003, p. 124). Understanding how to present a well-articulated message is a *core* competency; delivering that message in a manner that resonates with its target audience is, however, the *critical* element that often is not taken into consideration.

Progressive Global Manufacturing Organization and the Challenges Often Faced in Choosing a PPM Solution

One organization we worked with was a global entity in the manufacturing industry that had been in existence for decades and was a leader in the space it served. Product lines were diligently managed by teams of experts in their respective fields, including certified project managers, certified engineers, and other industry experts.

The product teams used a third-party product for overall project visibility, but all the scheduling and forecasting was performed using Project 2003 Desktop or Microsoft Excel. Essentially, program management and key milestones were managed in the third-party application, and data were cascaded to the subgroups that report using Project or Excel.

One of the subgroups was looking to upgrade to Project Server 2010. Its goal was to secure better resource capacity planning and forecasting and a better ability to meet key program milestones.

This organization was making the next assumptions:

- Project 2010 would simply be a scheduling and resource planning tool it could use as a subset to the third-party application already in use.
- Since the staff of the subgroup was already using Project 2003 Standard, the organization believed that migrating all users to Project 2010 would be easy.
- The processes for managing resource load and related costs (which were currently done in Excel) would be easily adapted to Project 2010 by simply transferring data from legacy products to Project Server 2010.
- The third-party application already in use as the main PM software would drive the demand from the program level, and each subgroup would simply report its schedule status from Project 2010. The third-party program would still have better visibility.

Our analysis:

- Project 2010 is used to pull all groups together into a central Project platform to better align with the program level than the organization's current implementation of Project 2003. Project 2010 is an integrated system built for accessing data visibility at multiple PM levels.
- Project 2010 might initially work as a complement to the third-party application in use, but if the deployment was executed well, stakeholders should see more innate capabilities in Project 2010, and it will eventually replace the third-party application.
- Project 2010 data extends across all Microsoft Office applications. It offers collaboration and other Web capabilities that the third-party application does not. This allows for easier end user adoption of usage from familiar interfaces. This deep integration removes the need for third-party add-ons, reducing potential points of failure and the requirement for customized, costly programming.
- Project 2010 will serve as a business and BI platform, including earned value and strategic planning for all programs.
- Selecting a qualified consulting vendor that specializes in Microsoft Project and PPM will ensure a smooth integration of all systems in use.

Sometimes there are opportunities for POEs by capturing a departmental need and scaling the larger solution to address that one need. Over time, more capabilities can be introduced to the organization to illustrate that Project 2010 accomplishes in itself what multiple systems are currently used for. Organizations are looking for cost savings within their IT architecture, and this is a perfect example of how the multiple capabilities of Project 2010 can address vital issues that might be considered non-PPM issues.

Common Challenges Simply upgrading from Project 2003 Standard to Project 2010 and dozens of Project 2010 clients internally can be tricky. An experience and qualified partner can assist in guiding this migration.

Synchronizing data via integration between Project 2010 and LOB applications requires proper architecture and professional partner support.

Taking existing Excel and other reports and transferring the data directly into Project 2010 requires skilled IT configuration that internal sources may not possess.

In the last scenario with the global manufacturing industry entity, the organization did not factor in user training for Project Server 2010. A tool is only as good as its use. If the end user cannot function efficiently within the tool, fragmented workarounds will begin to appear.

Especially when blending multiple technologies into one system, licensing can be intricate and confusing. Organizations must ensure that adequate licensing is obtained for all components. Use of Microsoft Developers Network (MSDN) licenses to manage the Project 2010 production environment may not be sufficient.

Failure to take the appropriate technical and PPM-related training prior to system configuration and deployment is a common flaw in deployments. These complex systems require different role-based training tailored to job responsibilities and functions.

Inexperience with the system architecture or lack of appropriate hardware can result in poor system performance, as when an organization tries to run everything on one server.

PPM, whether utilizing one main technology platform or multiple system components, remains process driven. Having no staging system or change control processes in place will decrease efficiencies in execution and bottom-line dollars as projects spiral out of control and are not traceable in a meaningful way for the organization.

Technologies are always subject to tweaks and adjustments. If the organization does not apply cumulative updates and service packs as they become available, the system will break or have chasms that are prime territory for hackers.

Attempting to "boil the ocean" or "leap into the future," often as a result of a senior management directive (e.g., "We need this new system online next month!"), can do more harm than good. Bells and whistles are shiny and attractive, but if the solution is not scaled to the organization's current status, end users will become overwhelmed and strong ROI is not achieved.

Key Steps to Creating the Win All organizations should consider engaging a qualified partner (i.e., partners that have obtained the PPM competency through Microsoft), but this is especially true of Level 0 or Level 1 organizations, as they are just beginning to form their PPM foundation. Work done at this critical stage will set the scene for successful PPM adoption across the enterprise.

Work with your Microsoft licensing specialists and value add reseller or large account reseller to ensure adequate licensing is in place concurrent with the deployment.

Ensure that adequate Microsoft Premier Support coverage is in place to enable assistance from Project Server technical support resources prior to the deployment.

Define a transition plan and then follow that plan. Watch for scope creep in this critical area. You will be much further ahead of the game if you do first things first. Again, a certified Microsoft Solution Partner has the experience (and battle scars) to ensure success.

Ensure that end users are adequately trained and have had hands-on experience with scenarios they are going to encounter on a daily basis. Making end users comfortable with the new tool will accelerate adoption.

Create end user documentation that describes the normal process flow. Detailed documentation remains a data-driven point of reference for future resources to learn from or for organizations to engage in continuous improvement.

Not everyone will know what to do next, especially if something goes wrong. Ensure that there are trained and identified support personnel to answer key questions as the system goes live. It will take a while to build up internal institutional knowledge

End users often will be able to learn from each other and provide valuable feedback to the system administrators and key PMO personnel about common issues and needs. An easy way to capture this information is to provide for brown-bag lunch training sessions.

Build a safe and friendly community around the new PPM solution. Very often new systems die before they generate any real value due to unresolved issues, data reentry and rework, and puzzling situations that no one seems to know how to resolve. Have a regular communications meeting about new capabilities and changes, so that end users are not surprised with new things that are coming at them. Such meetings also provide one-stop shops for vetting issues encountered by end users and will lessen the chance of ad hoc and disjointed fixes (kludges) being implemented.

Reward teams for successes. All too often we single out individuals for their efforts instead of sharing tribal wins. This can result in too many Level 0 heroics and hold back organizational maturity. There is also the risk that these people move on, taking valuable how-to knowledge with them when they leave.

IMPORTANT CONCEPTS COVERED IN THIS CHAPTER

It's not just about the technology, the processes, or the people; it is all about ensuring that we have the best possible impact on the business. Therefore, we must address PPM from a business perspective.

Business leaders are now seeking information and knowledge management, BI, and analytic capabilities that are far beyond the current capabilities of their IT infrastructure. With the recent downturn in the economy, businesses are also being forced to reevaluate their current investments in IT, methods, and processes with which they will run their business in the coming decade. Information workers across the board are being asked to take on more and make do with a lot less.

These are only a couple of the key factors driving the need for better integrated work and resource management solutions. An ever-increasing competitive landscape and the associated time-to-market considerations are also driving the need for better end-to-end information and knowledge management.

With the addition of the SharePoint Server 2010 platform, integration and collaboration have been extended from within a single tool to an integral piece that is easily accessed throughout the organization. This portal for information and document collection creates organizational assets that will benefit at all levels of PPM but also aid in an organization's growth and development.

Microsoft Project 2010, built on SharePoint 2010, introduces a new level of PPM technology to organizations and business users that far surpasses anything on the market today. In one connected package that keeps business at the core, Project 2010 enables efficient project execution and measurable ROI that will drive business forward on a strong platform into the next decade.

Key Summary Points

Key summary points are highlighted here to remind the reader of some of the vital considerations presented in this chapter.

- As PPM software has grown in size, scalability, and use, the infrastructure requirements have increased. Additionally, the collaborative elements with PPM technologies demand more from the organization's systems, processes, and end users. What normally starts in IT expands to other groups and leads to additional of training, extensive consulting services, and ongoing system and process support. It truly is a culture change.
- Organizations are constantly working to improve the value of their investments. Regardless of the economic state or health of the particular organization, discretionary projects and initiatives receive significant scrutiny. How well money is spent and managed within initiatives determines not only the economic success of an overall strategy but also the personal success of those who put those initiatives into motion.
- Organizations are more volatile than previously anticipated. Due to the global economic changes of 2008 to 2012, a number of external factors have impacted the ability of organizations to select and adopt updated technology. This makes the choice of the appropriate technology essential in getting the most from the organization's investment.

- Change occurs only when the environment is turbulent. Emerging markets globally are realizing growth trends because of an imbalance. This flux pushes people out of their comfort zones. When the environment moves away from status quo innovation, updating and eventually change takes root.
- Much of the controversy around organizational maturity models stems from a mismatch between the amount and type of processes applied at each level and the organization's basic cultural mental model (regardless if it is a volatile, stable, or turbulent market).
- If organizations attempt to push such changes too fast or too far, they often encounter cultural conflicts and wholesale rejection by all the concerned parties. The PPM maturity model enables senior management to avoid such problems. It provides a framework that can help facilitate communication with executive management by comparing their organization's PPM processes and attributes to those in the Gartner model.

REFERENCES

Anderson, Cushing. 2006. *Process Training, Enabled by Tools and Processes from SAP Education, Is Critical for Technology Adoption at Kimberly-Clark*. Accessed September 30, 2011. www.sap.com/cis/services/education/russia/infopak/customersuccess/Kimberly.pdf

Ballmer, Steve. 2009. *Keynote Remarks*. Presentation, 2009 SharePoint Conference, Las Vegas, Nevada, October 19, 2009.

Cooperrider, David L., and Diana Whitney. *Appreciative Inquiry: A Positive Revolution in Change*. San Francisco: Berrett-Koehler, 2005.

Hammer, Michael, and James Champy. 1993. *Reengineering the Corporation: A Manifesto for Business Revolution*. New York: HarperBusiness.

Morgan, Mark, Willam A. Malek, and Raymond E. Levitt. 2007. *Executing Your Strategy: How to Break It Down and Get It Done*. Boston: Harvard Business School Press.

Santosus, Megan. 2003. "Who's the Boss: CIOs Need to Learn to Tailor Their Message to Different Kinds of Decision-Makers." *CIO* (October 15), pp. 122–124; available at www.cio.com/article/29850/How_to_Tailor_Your_Message_to_Different_Kinds_of_Decision_Makers_?page=2&taxonomyId=3164 (accessed September 30, 2011).

Wikipedia. (n.d.). "Application Lifecycle Management." Accessed September 25, 2011. http://en.wikipedia.org/wiki/Application_lifecycle_management

Wikipedia. (n.d.). "Technology Acceptance Model." Accessed September 23, 2011.

CHAPTER 3

MEETING CFO NEEDS WITH PROJECT/SERVER 2010

IN THIS CHAPTER

This chapter helps any individual who is involved with work management understand the importance and capabilities of Project Server in tying financial and economic metrics to the needs of the chief financial officer (CFO) and other financial stakeholders.

We explore the capabilities of Project Server, its newer features, and its ability to connect to other systems. We look at how this results not only in higher success for projects but also in better visibility and forecasting for the projects/programs to their original planned return on investment (ROI).

What You Will Learn

- How to address CFO and organizational financial reporting needs with Project Server
- How a project management office (PMO) can assist in delivering project cost accounting and forecasts
- How to ensure good financial buy-in for decreased risk to a project
- The benefits of tying project reporting to measured ROI and outcomes
- How work management can be critical to organizational success
- How to enable strategic planning to bottom-up reporting and actuals

HOW THE CFO GETS THE ATTENTION OF THE PMO

Attention, CFOs: Imagine being able to increase the health of your portfolio by leveraging portfolio management. There are many factors to consider and put into play, but having the proper type of technology is critical. Running a company with fiscal prudency requires adhesion to compliances, processes, and discipline to protecting the profitability.

A project/program management office (PMO; sometimes called a project controls or project standards office) can play a key role in validating, collecting, and reporting uniform financial and work forecast metrics that are vital to a CFO's ability to review trends and look at future work, cost, and resource information related to capital and expense projects or programs.

Financial environments are now, more than ever extending to project management systems and include the tracking of spending and associating multiple channels of revenue and capital gains against projects. Decisions are made and initiatives are supported that are aligned with either strategic objectives or legal requirements. Some of the supported initiatives are the costs of doing business; others are needed for the future health of the company. From a financial manager's perspective, not all projects in flight make sense for the bottom line: Some are needed for business growth, some for market expansion, and in many cases some are needed to keep the lights on. For example: A project that retools a production line or updates products for the emerging mobile phone or apps market is just about immediate financial gratification.

The CFO and those responsible for managing the corporate bottom line and strategic financial plans typically manage portfolios (diversified centers of monetization). Financial portfolio managers or fund managers are experts in specific investment areas and strategies. Participants with capital "invest" in each manager with the expectation that they will gain an ROI. For the individual investor, these investment vehicles include such things as mutual funds, stocks, and others.

In the investment world, portfolio/fund managers are held accountable for ROI, performance of their strategies and decisions over time, and other fiscal metrics. In order for financial investment managers (fund or portfolio managers) to be successful, they need high-performing technology specifically tailored to their environment. The technology needs to be current, enable communication via multiple mediums, and have business users' tools for executing business intelligence (BI), analysis, and modeling and for allowing access to real-time data.

So let's go one step further and add corporate portfolio/fund manager to the CFO's list of titles/responsibilities.

Now the CFO is responsible for monitoring the health and performance of each of the investment centers (e.g., departments/divisions, channels, programs, etc.) He or she needs technology and processes that are suited not only for financial management but also for business and work management, including dynamic communication capabilities, tracking, forecasting, and modeling to determine trends, changes, and opportunities.

In order for CFOs to realize effective management of the company from a portfolio perspective, they need to have technologies and processes that capture all business data and information paths and that work against common constraints (i.e., budget, resources, and time) across the entire enterprise. This chapter highlights the benefits, proposed actions, and gaps that a CFO should consider for strategic fiscal planning.

In today's world, the relationship between the financial organization and the PMO is more important than ever. With Project, organizations are responsible for the implementation of projects that support organizational strategy and in many cases help to quantify, define, and track the strategic value proposition of proposed projects. It is increasingly more important to associate these value propositions with the financial and values that a CFO will need to evaluate.

What the PMO Is Expected to Deliver to the Bottom Line

Every executive, manager, and project manager knows that financial management is an important—and sometimes critical—element of their job. Executives are held responsible for the overall profitability of the business, middle managers are concerned about department budgets and meeting revenue targets, and project managers are accountable for making sure that their projects are accomplishing their objectives within a set of financial constraints. Successful business execution is dependent on having timely and accurate financial information available for all of these roles.

There is an important link between profitability, project costs, and assets, as managed by the PMO, and the reporting and financial summary that CFOs are looking for. Far too often, important financial information is distributed via spreadsheets with little thought about presentation. The data may be inaccurate or incomplete. Without a clear, proper presentation of that data, it is difficult to determine its validity. Even complete, accurate data sets can appear questionable when poorly presented. Unclear data presentations are thus unsuitable for good, informed decision making.

Additionally, financial data must be timely. It cannot be so old that decisions are no longer actionable or so new that decisions are based on incomplete or inaccurate data.

While a full discussion of financial management is too broad to cover here, we can look at financial management from a project portfolio management (PPM) perspective. In other words, from a project or portfolio perspective, what does a high-level executive expect to see from the PMO for actionable decision making?

As mentioned, we cannot ignore the tight relationship that exists between the PMO and the financial organization, since the good or bad financial performance of the strategy implementation has a direct impact on the finances of the organization.

The PMO and those responsible for the area of finance must work together as a team to effectively manage the organization's financial performance. The technology that is available today lets users monitor financial performance information practically in real time during strategy execution. This real-time information enables the project area and the finance area to make much better decisions about the path to be taken by the organization, based on the financial impacts.

Managing Financials: Start with the Project Business Case

In the context of either an individual project or a portfolio of projects, any discussion of financial information needs to start with the business case. From the CFO's perspective, what is a business case? The business case, or purpose behind the decision, contains the "what," the "why," and the "how much" compared to the dual constraints of budget (amount of capital available) and resources (amount of work capacity available). The business case should detail the expected benefits or returns over a period of time as well as the expected timing or roll-out and the cost/effort investment required.

Organizations can leverage the business case approach to determine justification, prioritization, and relevancy of projects against the constraint of both delivery and strategic alignment with future organizational goals and direction. This relationship gives executive stakeholders the ability to measure the project's financial performance.

Creating a business case for a project may require content and focus as well detailed solution outcomes that will differ across the corporate landscape based on needs of the different corporate stakeholders. The business case should include relevance to vertical industries, department needs, product lines impacted or supported, drawing upon the resources of different functional disciplines of the business group, departments and divisions and the overall resources capabilities to deliver. Essentially tying the business case to strategic drivers needs to not only look at the corporate goals, but also now takes into account the ability of any part of the organization tasked with that delivery, whether they have the capacity, skills to deliver.

For example, take three different potential projects or initiatives, a new product, a new business system, or a new process needing to be implemented for compliance. A business case for a new product or service may focus on generating incremental revenue or increasing market share. The business case for a new business system may focus on increasing efficiency or reducing costs. The business case for implementing processes required for compliance may focus on avoiding costs resulting from noncompliance (fines and business impacts associated with loss of reputation or public image).

These different projects created and approved based on the strengths of their comparative business cases typically share denominators residing on both sides of an investment equation: benefits and costs. An organization that is going to leverage PPM technology must understand that quantifying or weighing and deciding on the approval of projects cannot just be measured in revenue generated.

CFOs functioning in support of or parallel to a PMO may have different or unique business value propositions that should be used to rate, rank, and approve and prioritize projects. To a CFO, these different or unique business elements should be captured by the project business case, which means that in defining a project up front, these metrics or measurement vehicles should be defined ahead of time for the business to present to senior management. The methods by which the project business case details are captured may also vary widely throughout different parts of an organization. In larger companies, the different business units or divisions may have similar mini-portfolios of projects (compliance, optimization, revenue), but these initiatives may need to be rated, ranked, and then combined with other departments so that a CFO and other organizational leaders can review them across the entire corporate landscape.

In some organizations, the business case is a stand-alone document or part of a proposal. In others, it is embedded in a statement of work or project charter. Regardless of the differences, there are some common questions that should be answered in the business case:

- Does this project and the level of investment it requires support the organization's strategic objectives? If yes, how?
- What is the benefit analysis and expected ROI? How soon can we expect to see an ROI? How will the return be measured?
- What types of investment are required, and what type of project will the investment dictate? Capital or incremental? Division specific or enterprise wide?
- Are there any interdependencies among projects, programs, and their related costs versus benefits? For example, to receive the benefits, we can't do Project X without also doing Project Y.

These questions are just a few examples of what CFOs and financial executives track as they manage projects, programs, or any work-related initiative.

Benefits: Designing a Project Process to Measure Outcomes

The past decade has seen an increasing level of corporate project maturity as well as an increasing level of awareness of the importance of tracking and measuring project costs and benefits. Organizations associated with project management like the Project Management Institute, have a larger than ever member base and the prevalence of understanding of project management and certified project management individuals continues to grow. Business and financial stakeholder groups have adopted a rationale for requiring that projects are linked to expected benefits associated with project deliverables or outcomes. For instance, when evaluating scope change that impacts a deliverable, the additional cost or time would be justified against the potential change or increases to the expected financial benefit to be realized by the organization.

As CFOs become more integrated into the business of defining and impacting projects from inception to delivery, they find they are practicing project benefit planning (similar to financial planning actions). Project benefit planning identifies

Table 3.1 Examples of Foreseen Benefits when Project Planning

	Example Project 1	Example Project 2	Example Project 3
What are the annual or year-over-year (YOY) benefits?	Increase in sales (e.g., %, change YOY, etc.)	Increase in customer satisfaction	Improved governance, regulatory compliance
What tracking and measurements/metrics will be used?	Reports/dashboards/key performance indicators	Surveys, case studies	Assessments/ certifications/third-party audits
How are the timing and milestones defined?	1/3/5 year forecasting	YOY/tracking against a baseline	Gateway reviews/ maintenance
How will the metrics be reported?	U.S. dollars/intellectual properties/assets, financial revenue generated	Positive versus negative ratings	Maintain certifications

what the expected outcomes or benefits of the project will be and establishes how those benefits can or will be measured and when each benefit is expected to occur. Additionally, financial stakeholders in companies today are finding value in measuring the level of performance for project types within their portfolio and using those measurements to anticipate the performance of future investment. Table 3.1 shows some examples of the different ways financial executives and stakeholders might characterize benefits for project planning. (Note: These examples may also be characterized as business drivers in a portfolio assessment.)

Financial executives take note: Work and resource assignments across an enterprise (classified as projects) may not be in play for strategic or financial benefits. In essence, not all projects have financially quantifiable results.

Projects that directly impact the financial bottom-line need to have metrics established early in the project lifecycle in order to track time and effort and to quantify this against the overall costs versus the benefits realized. Tracking project ROI is sometimes included as part of project scope, because there may be projects with intangible benefits or benefits that are difficult to define. CFOs have advantages when using enterprise project/portfolio management systems like Microsoft Project 2010 to define and prioritize project types (capital projects where a tangible asset will remain on the books versus other projects that are more difficult to quantify, such as projects that optimize processes, systems, or people). There is always a way to track intangible project ROI, and these fields or key metrics would be needed to establish comparison analysis, reports, and/or portfolio baseline tracking analysis for the project from one period to the next.

Effective CFOs will develop and manage strategic and investment benefit plans. From a portfolio management perspective, having a benefit plan provides essential information to focus organizational resources on projects based on return and strategic alignment. It also aids portfolio managers in the process of selecting a diverse set of projects that will provide long-, short-, and mid-term returns.

From a project perspective, the benefit plan provides guidelines for managing project changes so that costs and benefits at least stay in alignment. It can also help to provide a basis for evaluating potential approaches to accelerating project returns, such as iterative development or incremental releases of products or services.

Investments: How to Derive Quantifiable Value from Project Costs

PPM is a powerful mechanism to plan, track, manage, and deliver financial results and an easy place to embed the needed fields and reporting requirements to manage and track such information. Managing financial deliverables by leveraging a PPM ecosystem is a cultural and executive shift, but those organizations that commit to doing so will realize success and provide better visibility and responsiveness to their financial systems and stakeholders.

Let's break down the scenario of tracking financials and tracking deliverables. We may start by asking questions like: What activities and costs are being identified and managed? How much work package decomposition (task-level detail) is sufficient to expose the details required to gain accurate metrics? Not every organization tracks project costs in the same way. Resource costs are captured in a variety of ways depending on the organization, business environment, and industry. One example of resource costs that an organization may track focuses on the human side of things: the time or effort expended by its employees. Alternately, an organization could track material change through its depreciation of assets, use of materials for production, and so on. Some projects are required to track and report capital expenses separately from operating expenses; in other projects, the differentiation and tracking of capital expense is handled by accounting. For example, a construction project may be building or improving a capital asset, hence a capital project, whereas a software implementation would be considered an expense category project.

Ideally, the business case describes both the total budget (expected cost) for the project and a breakdown by category of actual cost. Leveraging PPM technology like Microsoft Project 2010, a CFO has the ability to closely control the tracking of expenditures and use that data to analyze and model the project's progression. In this way, comparing the current fiscal state of a project to its planned costs becomes both easier to manage and more powerful. This comparison allows the project sponsor to dynamically adjust the funding plan for the project, whether that means switching to a partial rolling wave budget (where the budget is detailed out only in the areas known and the rest of the project is estimated at a high level), as needing a capital budget, or an expense budget. The stakeholders receive reports on how much is being spent in each of these categories, which is more meaningful than merely reporting against a lump-sum amount. A project business case review, in partnership with the finance department, is now possible.

As a CFO, you will have improved visibility into what the finance department and related PMO division are managing. By using Project Server 2010, you can define the target's projects and manage the baseline of project costs across the existing work portfolio. Scope changes, changes in material or supply costs, and unanticipated risks all may impact the costs called out in the business case. In some cases, these changes will result in changes to the baseline budget. In others, the project may be canceled so that resources can be focused on efforts more closely aligned with the organization's strategy and objectives.

The financial and strategic plans need the support and involvement of both the financial stakeholders and the business user stakeholder classes. The project manager is responsible for tracking both the costs and the use of approved resources to complete work packages and deliver expected project outcomes. The project manager is also responsible for making financial decisions regarding the project. To meet these responsibilities effectively, the project manager must establish and maintain processes to ensure that information regarding the project's finances is accurate and timely. More often than not this requires the project manager to maintain records of financial commitments (such as purchase order amounts) and actual expenditures. This record keeping may be in addition to those maintained in the financial or enterprise resource planning (ERP) system, which, while accurate, may not reflect expenditures until months after the purchases have been made. The intent is not to duplicate the accounting function but instead to provide the basic reporting needed to control and report project costs.

For tracking project costs, the simplest approach is to use the built-in features of Microsoft Project Professional 2010 to maintain a project financial account (a record where expenditures are adjusted against the budget and the project manager can, at any point in time, determine how much of the budgeted amount remains). While this is effective in tracking expenditures, it assumes that the amount of budget that remains is adequate to cover the remaining costs of the project. Alternatively, the project manager may examine both the actual cost and the planned cost of the work remaining in the project. From these data, reports and dashboard views such as a budget variance can be reported.

Earned value (EV) is a capability of Microsoft Project 2010 and a project tracking method that is often used by organizations. EV is considered a reliable financial performance and forecast indicator by the project management community (Project Management Institute standards). Determining EV is an advanced PM competency, and its effectiveness is dependent on the organization's commitment to processes.

One of the benefits of EV when using Microsoft Project 2010 is that it ties the cost to the work performed. Rather than looking just at the costs incurred to date, EV also looks at how much work was accomplished for cost.

For example, let's assume that you have a project with a budget of $150,000. Halfway to the project's promised completion date, the work is estimated to be 50 percent complete, and your costs are $100,000. According to EV, the "value" of the work planned is $150,000. At 50 percent complete, you've earned half of it ($75,000). The difference between what was spent and what was earned is the cost variance ($25,000). Additionally, if you divide the earned value by the actual cost, the result is the cost performance index (CPI), a measure of how efficiently money is being spent to accomplish the work. In this example, the CPI is .83, meaning that we are earning $.75 of value for each $1.00 spent. More simply put, it's costing more to do the work than originally planned. Furthermore, if it is taking more costs to accomplish the work for the first half of the project, it is highly likely that we will need that much more to accomplish the remainder. By dividing the original budget ($150,000) by our CPI, we can forecast that the total amount needed for the project is $200,000.

Ultimately, CFOs are looking for better fiscal tracking tools, specifically related to planning work activities, budgets across the enterprise, and forecasting. Effective financial management and reporting must be meaningful, accurate, and timely. It must be meaningful in that it reflects the key financial indicators tied to the project's business case, allowing decision makers to examine both benefits and costs of a project and compare them to the expectations. Developing a reporting framework around the business case and focusing on the most important metrics for the benefits and costs is the best approach to ensure financial reporting for executives that meets these goals.

WHAT AND WHY IS WORK MANAGEMENT CRITICAL TO ORGANIZATIONAL SUCCESS?

Work management is the practice of focusing on the specific activities, tools, processes, and reporting that helps an individual get work done. Work management is not reserved just for project work; it can include activities such as time spent on issues (not in a schedule), trouble tickets, project, program and portfolio work, and any type of to-do list that is associated with a resource's daily work.

The key to understanding work management is that the effort spent in this discipline is designed to reduce waste and optimize a resource's output and maximize its value to the organization.

As organizations or companies get larger, they tend to struggle with siloed activities and projects that, in many cases, are duplicated in other parts of the organization. Or they find that resources doing the work approach managing their work in different ways, using different tools and methods that don't lead to an easy way to achieve visibility.

A good PPM software product should enable some flexibility for tracking, managing, and reporting work for resources, and centralize that information into easy and cohesive views to report progress or expose overallocation, underallocation, or duplicate efforts. Not only does Project Server 2010 deliver this flexibility, but Microsoft is continuing to expand and emphasize the ability to bring work, and work activities, past, present and future into view of the project organization through its different products and collaboration portal.

The better visibility from which to work includes not only an individual level but also those who are responsible for that work, creating a higher level of value to the organization and a more rapid ability to address issues, slippages, or risks before they derail a project or a resource's ability to deliver.

The approach that Project Server 2010 brings to the project management organization enables managers to evaluate the resource capabilities of both potential and current projects and to establish the right strategic and tactical priorities and objectives. This information in turn validates the corporate initiatives to focus on and execute those projects that provide the greatest business value.

As we know, all strategy requires financial, human, and material resources in order to be successful and to be carried out. The human resources tend to be more complex to administer and manage, and in most cases they have competing priorities. Work management is where the rubber hits the road and where good PM delivers its highest value. However, this is also where the highest likelihood of strategic value disconnects or misdirection can happen. It is important that even at the lowest tactical level those doing the work understand their connection to the strategic outcomes and key deliverables required by a project.

With all the work being managed across an organization and resources (especially those in a matrixed organization) working on projects with competing priorities and activities, it can be a daunting task to align both the value and the work. For planners, managers, and even the individual resources, it can be difficult to see progress and status and to address resource allocation. However, thanks to technology, we now can clearly identify the real capacity of our resources according to their roles and skills within the organization. Doing this allows us to make a much more realistic plan of our ability to execute strategy and the time frame in which the components of the strategy can be executed, based on the organization's financial and material resources.

Managing the capacity of the organization's human resources is not an easy task. The principal challenge facing organizations today is the need to integrate and understand the effort required of our resources to execute projects while fulfilling all of the activities of normal day-to-day operations and then capturing and consolidating that data in a way that will allow us to leverage the information to plan efficiently for completing both projects and operations. This is where leveraging a PPM system like Project Server 2010 bridges that gap by centralizing resource and strategic information in one central database for reporting, tracking, and review by CFOs and other senior management.

Managing Independent Departmental Initiatives

All organizations have different departments, and each one has an important contribution to make toward achieving strategic objectives. All departments must work in a synchronized fashion under the leadership of that organization. Some organizations utilize a PMO to obtain and consolidate status updates from each of the various departments to ensure that strategy execution is on target.

While utilizing a PMO makes the process easier, different types of information may need to be tracked along with different workflows, depending on which part of the organization a project is being worked under.

In the past, organizations would stand up, or install multiple instances of PPM technologies to support and simplify the managing and tracking of projects related to different business units (departments) and their associated fields and workflows.

Project Server 2010 contains a new feature—the "Department" field—to helping split apart and manage different department initiatives. This field is a multi-value lookup that allows both strategic planning of portfolio business drivers and prioritization initiatives to be created and applied to different departments while also allowing projects, workflows, project detail pages, and almost every part of the enterprise PPM system to be differentiated by this field. (See Figures 3.1, 3.2, 3.3, 3.4, and 3.5.)

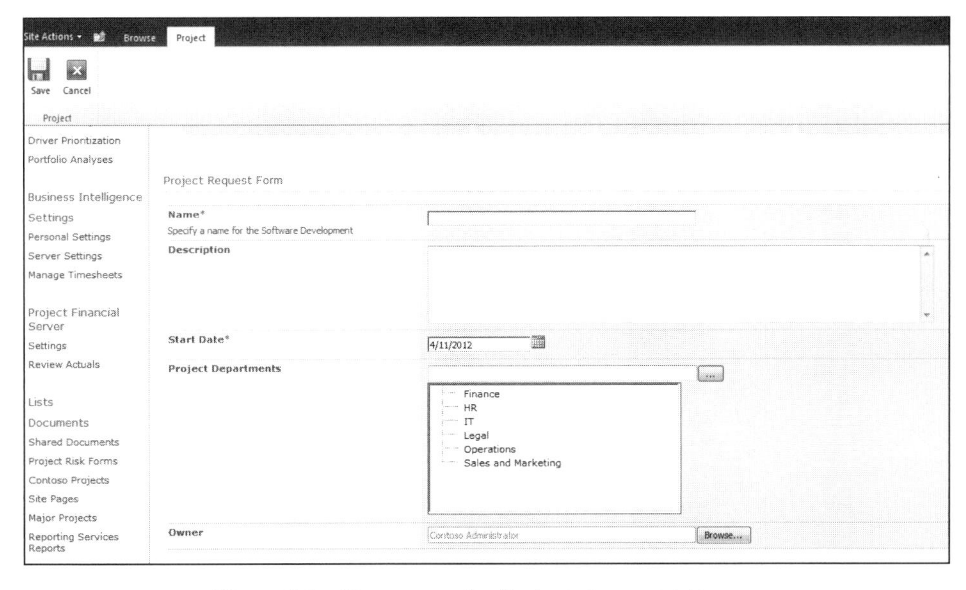

Figure 3.1 Department for Projects Source: Advisicon

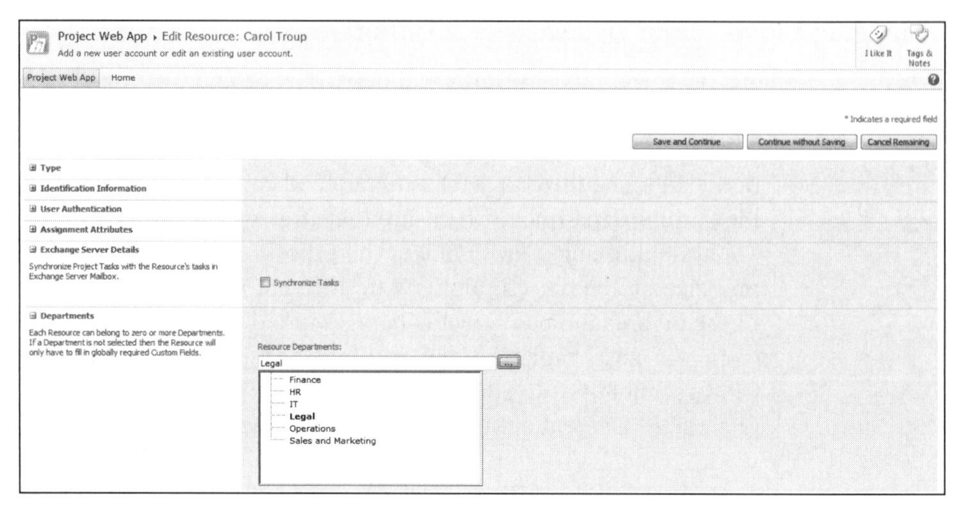

Figure 3.2 Department for Resources Source: Advisicon

Managing Constraints to Your Advantage: Addressing the Planning and Execution Bottlenecks

The constraints most commonly identified in the execution of strategy are tied to the availability of an organization's financial and human resources. The skill of personnel, not just the number of warm bodies, within the organization is

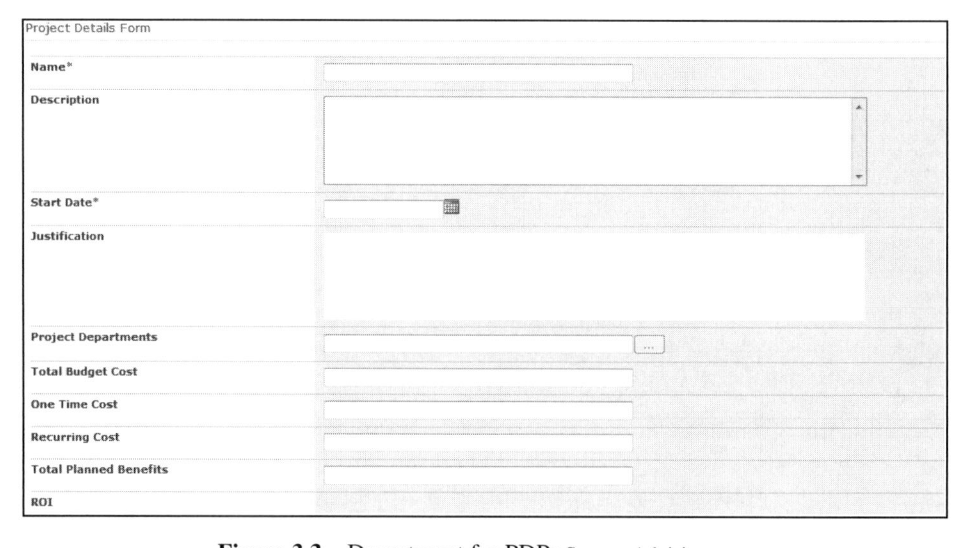

Figure 3.3 Department for PDPs Source: Advisicon

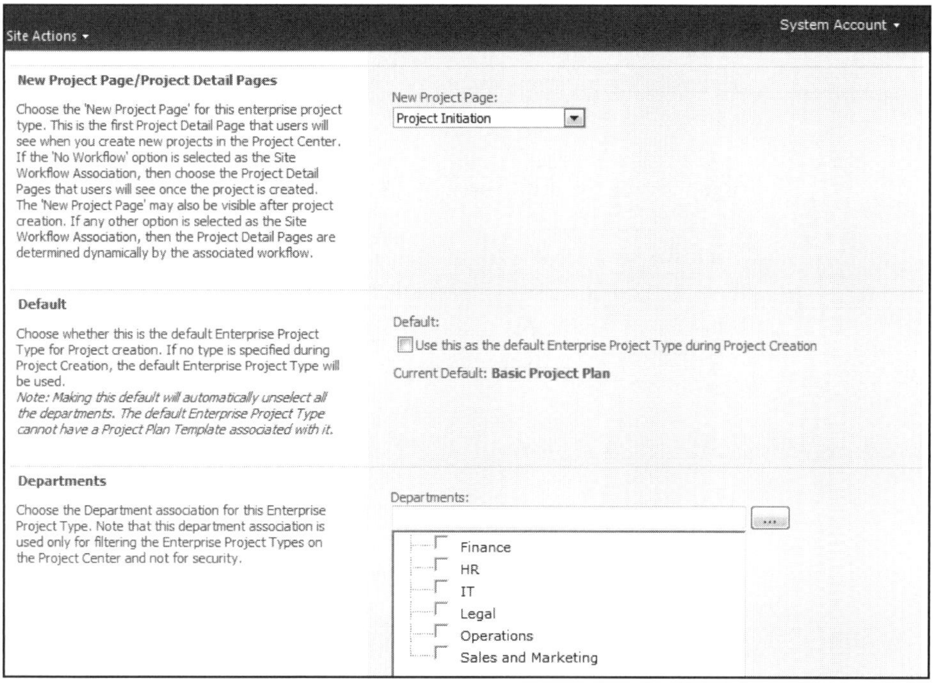

Site Actions ▾ System Account ▾

New Project Page/Project Detail Pages

Choose the 'New Project Page' for this enterprise project type. This is the first Project Detail Page that users will see when you create new projects in the Project Center. If the 'No Workflow' option is selected as the Site Workflow Association, then choose the Project Detail Pages that users will see once the project is created. The 'New Project Page' may also be visible after project creation. If any other option is selected as the Site Workflow Association, then the Project Detail Pages are determined dynamically by the associated workflow.

New Project Page:
Project Initiation

Default

Choose whether this is the default Enterprise Project Type for Project creation. If no type is specified during Project Creation, the default Enterprise Project Type will be used.
Note: Making this default will automatically unselect all the departments. The default Enterprise Project Type cannot have a Project Plan Template associated with it.

Default:
Use this as the default Enterprise Project Type during Project Creation
Current Default: **Basic Project Plan**

Departments

Choose the Department association for this Enterprise Project Type. Note that this department association is used only for filtering the Enterprise Project Types on the Project Center and not for security.

Departments:
Finance
HR
IT
Legal
Operations
Sales and Marketing

Figure 3.4 Department for EPTs Source: Advisicon

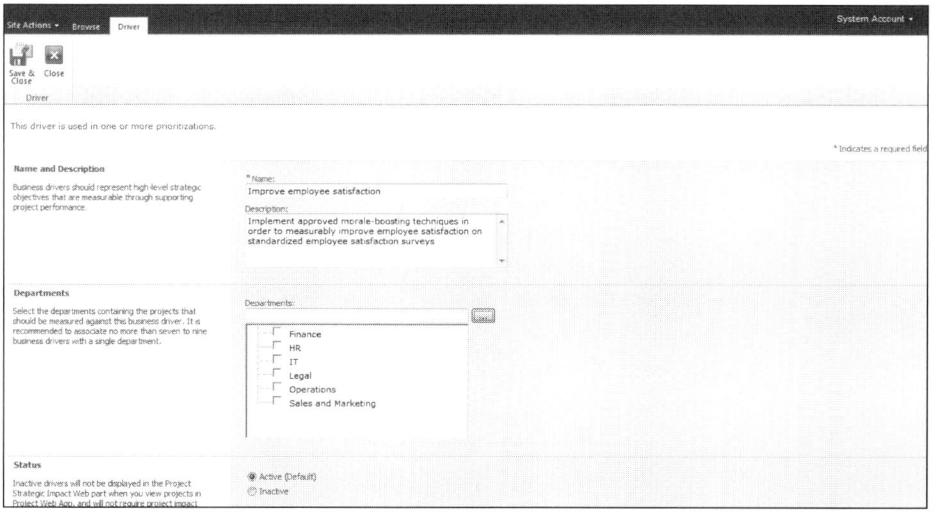

Site Actions ▾ Browse Driver System Account ▾

Save & Close Close
Driver

This driver is used in one or more prioritizations.

* Indicates a required field

Name and Description

Business drivers should represent high-level strategic objectives that are measurable through supporting project performance.

* Name:
Improve employee satisfaction
Description:
Implement approved morale-boosting techniques in order to measurably improve employee satisfaction on standardized employee satisfaction surveys

Departments

Select the departments containing the projects that should be measured against this business driver. It is recommended to associate no more than seven to nine business drivers with a single department.

Departments:
Finance
HR
IT
Legal
Operations
Sales and Marketing

Status

Inactive drivers will not be displayed in the Project Strategic Impact Web part when you view projects in Project Web App, and will not require project impact

Active (Default)
Inactive

Figure 3.5 Department for Portfolio and Business Drivers Source: Advisicon

the most critical factor that can become a bottleneck and constraint for a project organization. Not having the necessary skills or the lack of their availability because of time constraints will surely prevent the execution strategy the organization needs.

During the stages of project portfolio planning, the main challenge facing organizations is to clearly identify the real availability of resources to assign to strategic initiatives.

During implementation, the main challenge is to stay on course with the activities as planned. This is difficult because daily changes in the strategy and operations cause resources to divert their attention from project assignments that may impact project delivery dates and the results the projects should deliver.

Validating Proposals and Holding the Business Accountable

A largely new and sometimes overlooked part of Project Server 2010 is its ability to create proposals and to present a high-level impact to both future and existing workload, without spending too much time building detailed information that may not be approved.

The proposal may go through its project lifecycle of phases and stages, with the project manager or proposer filling in key fields as required, in fact even building a resource plan that showcases high-level resource needs across weeks, months, or years. This resource plan allows the proposer to give an estimate that is used in providing demand information to executive management without necessarily drilling into a tremendous amount of detail.

By enabling an organization to create key fields through a proposal that can be approved, tied to workflow, and filled in just in time, through the stages of a project, an organization never loses the key metrics associated with it. Those metrics can be used to validate assumptions and even compare the original plan with actual values (time, costs, work, and the value proposition) without forcing proposers to spend a great deal of time on a project that may never be approved.

In the authors' experience, we have seen cases where 60 percent of projects proposed were never approved. In many cases, when the projects were ready to be reviewed at the next planning cycle, the information had been lost or the resources had left the organization so the exercise of preparing a proposal had to be started all over again.

Some organizations we have worked with launch 30,000 projects or more each year. Years of human effort are wasted when a lot of time is spent building proposals and a high percentage of projects are never approved. By leveraging the Proposal and Resource Plan features of Project Server these organizations saved tens of thousands of hours.

Although your organization may not initiate thousands of projects, imagine the time savings by having the ability to review, validate, and archive good projects that may not be approved during a planning cycle but may be initiated at a later time.

From a financial standpoint, comparing a proposal-turned-project to its original estimates and business case provides a closed-loop analysis for CFOs and other executives to begin understanding and better managing the strategic initiatives across the organization.

SYNCHRONIZATION OF STRATEGIC OBJECTIVES TO ACTUAL EFFORT

Effectively aligning strategy with the tactical operations of the organization is no easy task, but it is vital to establish a very close link between the definition of the strategy and the daily operations delivering the project and functional tasks of the prioritized work activities by the organization.

It is easy to define how to achieve this synchronization, but it is difficult to implement. Remember that the strategic benefit or value in some cases is an after-the-fact metric; in some cases it is measured a full year after a project is completed.

In this section, we explore the need to integrate the process of strategic planning with the key metrics and deliverables of the tasks laid out in delivering the project. We also discuss the relevance of helping team members see the impact of their work and activities, that leads to strategic metrics or values. These metrics can be quantified and rolled up and then reviewed by senior management.

Imagining Structuring Strategies that Can Be Delivered

One key area that CFOs and other leaders in organizations strive to achieve is executing on a strategy and validating its success. In the book *Good to Great*, author Jim Collins addresses the fact that executives are held accountable not for how brilliant a strategy is but for how well they execute against that strategy. One value provided by Project Server is its organization of initiatives associated with programs or portfolios. By embedding at the earliest stage the strategic value proposition in an enterprise project field and capturing this metric, an organization can group the project center views and automatically have the ability to roll up planned, actuals, and any variance of these key metrics on the screen.

In almost every organization we have worked with in deploying PPM technology, these fields are turned into dashboards, and senior management begins reviewing "exceptions" rather than trying to drill into all projects in the hope of exposing issues or risks to their portfolio of projects.

Fluency of Actuals Against Estimates

By default, the tracking or progressing of tasks in projects or work defined in Project or Project Server automatically allows anyone in Project Server to review the actuals associated with the baseline, estimates, or planned work.

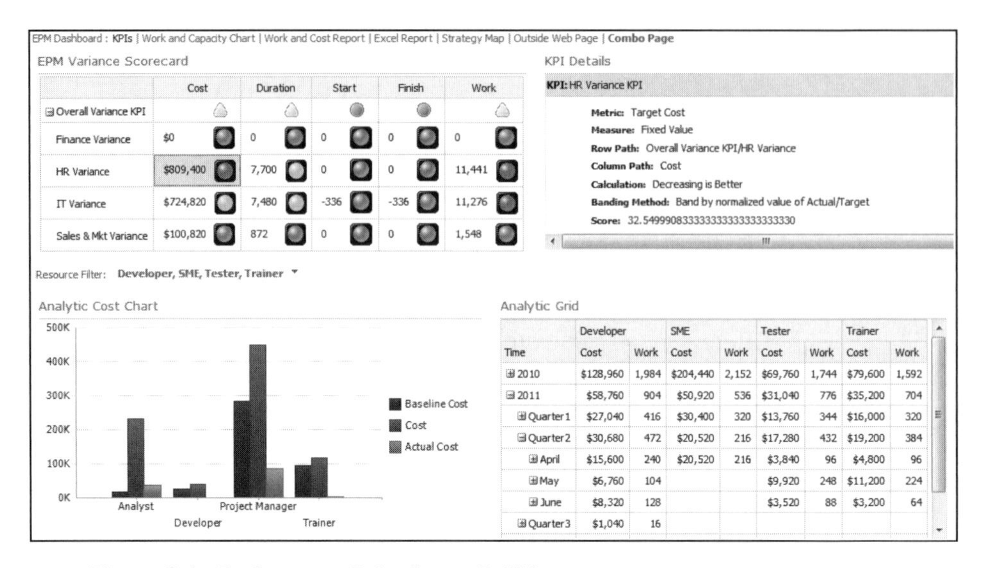

Figure 3.6 Performance Point Server Drilldown Source: Advisicon Figure 3.7 is an out-of-the-box KPI that points to Project Server information.

In Project, there are a series of EV fields and variance fields that do the calculation of planned versus actuals for work, cost, duration, and start and finish dates. These fields are native and require no additional work to add to a view or table.

The amazing part of Project Server 2010 is its rich reporting tools that come along with SharePoint. For example, Performance Point Server allows for a Web drilldown directly against Project Server data. Excel Services and SharePoint KPI

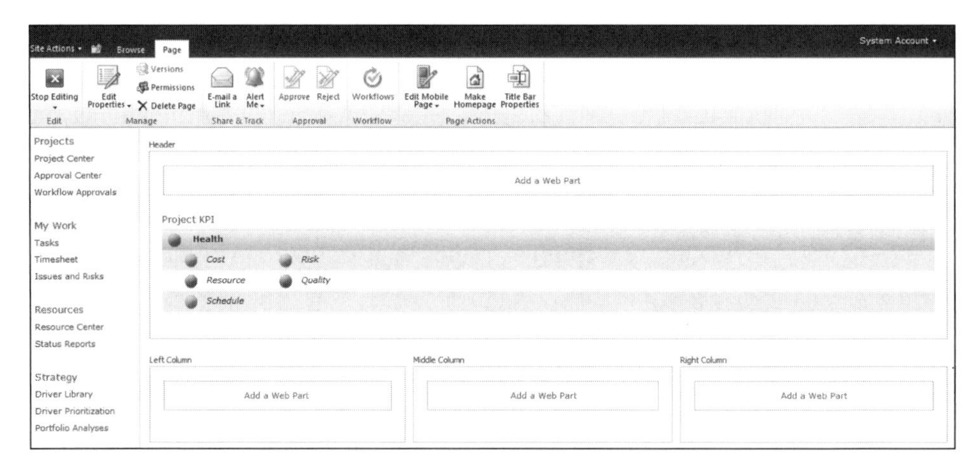

Figure 3.7 KPI Web Part Source: Advisicon

(key performance indicator) Web Parts are designed to help present graphical BI regarding status and the reporting of actuals against estimates.

Figure 3.6 is an example of the reporting capability of Performance Point server displayed in SharePoint directly against Project Server data.

In Chapter 10, we explore more of the dashboard reporting capabilities and options associated with Project Server 2010.

The Big Picture: Approaching Top-Down and Bottom-Up Planning and Control Options

In this book, we have discussed and will continue to explore both the capabilities of detailed planning and rolling up information and key metrics to the big picture and business goals and objectives. We also detail the strategic planning capabilities of Project Server 2010. In Chapter ,we compare the idea of strategic planning to the existing work portfolio.

In a number of cases, the big initiatives and strategic projects launched by organizations lose momentum and connectedness with the vision and intent of company leaders; as a result, the project is watered down from the original intended goals of these initiatives. Often, by the time a CFO or senior company leader finds out what the end result was or would be, there was no time to respond, redirect, or refocus the initiative to hit key objects.

This may be hard for the reader to believe, but many times senior management feels frustrated and powerless to shepherd core or key projects once they have begun because the strategic controls, metrics, and information relating to the project are fractured or difficult to assemble and report against. This is one of the reasons that Microsoft has invested in connecting its ERP software (Dynamics) products to Project Server, to enable the tracking and direct reporting of the financial systems to strategic project tasks and activities.

Project Server 2010 puts the top-down tools and controls directly inside each and every project and establishes role-based views that allow a quick review of any key metric, whether task or rolled-up project-level information. Thus, drill-down connectedness is available not just with schedule and task details but directly with the collaboration portal (Project Workspace) for all project information associated with a project.

Defining a Long-Term Solution that Enables Stakeholder-Class Scalability

An important part of a PPM implementation is to fold in the different stakeholders and their classes to the information, reporting, and metrics that will help them be successful. The more stakeholders benefit from a PPM tool, the more likely that they will support and reinforce the processes and activities that allow that system to succeed.

We have seen cases where PPM implementations have helped to bridge siloed reporting, tracking, and to help eliminate turfing activities leading to galvanized and centralized organizational behavior. These implementations helped to ensure that resources from all parts of a company followed the processes and updated information around their part of the project, thus leading to better unified maturity in Project, Program and Portfolio Management practices and reporting. This integration of stakeholders ensures the highest success rate for a large PPM implementation and also creates a synergy of support for PMOs and the information technology organization that may be responsible for the PPM implementation.

By starting with best practices around requirements gathering, first identifying stakeholder groups, classes, and representatives and then folding them into the requirements process, you can ensure that the scalability (upward or downward) of details and reporting metrics are identified. CFOs and the financial organization have a significant stake in being a part of this. By avoiding making the ERP or accounting system become a PM system or the inverse, by ensuring that the PPM system does not become the project cost accounting system of record, you can combine the best of both worlds, enabling the least amount of effort with the maximum amount of reporting, scalability, and drill-down reporting.

Integrating ERP and Other Financial Data

As mentioned earlier, ERP integration with Project Server 2010 is a highly sought after feature. Microsoft has included it in its ERP products (Dynamics), and other third-party products have been created to assist in the combining of systems.

The benefits and scenarios for integrating Microsoft Project Server 2010 with ERP, in particular, integrating the actual time reporting from ERP systems with the timephased planning and resource forecasting power of Project Server 2010, create one of the best solutions for seeing earned value with minimal effort. It effectively links actuals from the source of record (whether financial or time related) for organizations that want to maximize their work management and establish more fluid and effective reporting without duplicating time or cost entries.

Some companies and partners have worked directly with organizations to use the richness of the application programming interface with SharePoint, Project Server, and the SQL server database to integrate bidirectional reporting and updates. This integration is cost effective and enables the source of record actuals (both cost and time reporting) to bring actuals to the planning activities that project/program offices or scheduling groups use Project Server 2010 for.

This connection and the processes around validation and viewing the ERP information in a Project environment enables companies to drastically reduce the time spent in planning and trying to map the actuals directly with the way that the organization schedules work. Enabling the feature-rich reporting capabilities of Microsoft Project Professional, Project Server and SharePoint puts that planning and reporting directly in the hands of the end users.

At the time of this writing, several companies have created integration modules or have done significant work bringing these systems together. These companies and brief descriptions are listed next.

- **Advisicon Inc.** An international project, program, and portfolio management company, teaching, authoring, and consulting in the establishment of PMOs and the technologies to optimize and automate their processes and methodology best practices.
- **Campana & Schott.** An international consulting company for PM and process optimization. With the combination of management and technology consulting, Campana & Schott improve business processes and automate them by using innovative information technology.
- **The Project Group (TPG).** An international full-service provider of consulting, implementation, hosting, products, and training for PM and Microsoft Project.

IMPORTANT CONCEPTS COVERED IN THIS CHAPTER

In this chapter we covered the importance of involving and engaging the financial management of an organization early in the implementation of Microsoft's PPM product. Project Server 2010 was designed with top-down scalability in mind and can fold both bottom-up tracking and work management with strategic alignment and business value reporting in one centralized environment. The list below are key topics covered in this chapter and reinforce the value of Project Server 2010.

- Project Server empowers the PMO to support and work with the CFO and for a CFO to help ensure good reporting by basic up-front business case tie-in with projects and proposals.
- The key to financial reporting and strategic value proposition evaluation reporting is found through adding these key metrics to the phases and stages associated with a project's lifecycle.
- Understanding work management is important to an organization's critical success in ensuring good alignment with strategic goals and objectives for a project.
- Actuals and estimates are designed to be native and easily exposed for financial and other progress reporting at a glance.

- ERP and financial data can be linked and integrated with Project Server through both native features and third-party products/modules.
- Scalability of growing project maturity can be top down or bottom up. Organizations can start at either end easily and meet in the middle, without having to do all at the same time.

REFERENCE

Collins, Jim. 2001. *Good to Great: Why Some Companies Make the Leap … and Others Don't.* New York: HarperBusiness.

CHAPTER 4

THE BUSINESS SHAKES HANDS WITH THE MICROSOFT PROJECT 2010 PLATFORM

IN THIS CHAPTER

This chapter provides you with an understanding of how the Project Server 2010 platform is more like an enterprise resource planning (ERP) system and should be treated as such.

A large number of companies have simply missed the value of enterprise project management (EPM), have not done a complete implementation, or have participated in a failed implementation. Business and information technology (IT) managers need to learn the value of EPM and understand how the communication and workflow capabilities native to Project Server 2010 can address the project, program, and portfolio management requirements of small and large businesses alike.

What You Will Learn

- Project Server 2010 is built on SharePoint Server 2010 technology. It integrates with Microsoft Office 2010 and Exchange Server 2010 to provide a powerful and familiar work management platform.
- SharePoint Server 2010 has the fastest adoption rate by business users (with over 150 million users in 2011 alone). It now serves as the business platform for enterprise project management.
- The project portfolio management (PPM) lifecycle requires communication and workflow capabilities that are inherent in Project Server 2010.
- Users are no longer limited to a desktop tool. Working in the Project Server 2010 environment enables end users also to access information via the Web.
- Line of business integration has a well-established history with previous versions of Project Server.

LOGICAL ARCHITECTURE IS MORE NATURAL FOR BUSINESS USERS

Small and large organizations are seeking capabilities that empower both business users and IT professionals to create information management solutions quickly. In order to facilitate the increasing demands on available and skilled resources, it is essential to be able to manage, monitor, and assess the status of all work and projects across the enterprise.

With the general adoption of organizational project management (PM) and associated governance practices over the past decade, growth in portfolio, program, and project management processes, methods, and application packages has erupted. In the 1990s, the focus was generally on the management of individual projects. Today's software includes portfolios of projects and integrated cross organizational project and resource tracking and reporting.

EPM is today's solution for handling multiple projects efficiently and is fast becoming a necessary set of tools and methods for any company involved in managing work or projects. EPM helps organizations gain visibility and control across all types of projects. Its strategy enhances the decision-making process, improves alignment with business and priorities, maximizes resource utilization, and enhances project execution to optimize the return on investment.

As Project Server is gaining visibility as an enterprise solution for managing the work and resources of an organization, it needs to integrate with existing systems, starting typically with ERP systems (Fiessinger 2010).

Figure 4.1 illustrates a framework that describes the key interrelationships of EPM, including the organization's enterprise architecture (EA), business strategy, project management office (PMO), the portfolio management system, and of course the internal and external stakeholders.

An EPM framework provides a basis for successfully managing and maximizing the value an organization achieves from its project investments. The key elements include:

- **Business strategy.** Provides the business drivers for both the development of the EA and the IT strategic plan.
- **Enterprise architecture.** The enterprise's key business, information, application, and technology strategies, and their impact on business functions and processes
- **Portfolio management.** The strategic management of an organization's investments and projects to align project investments with associated business initiatives
- **PMO.** Responsible for executing the changes determined by the PM function
- **Internal and external customers.** Department, individual, or process within an organization that supplies another; external customers are those persons who come from the outside for the enterprise to fulfill their needs

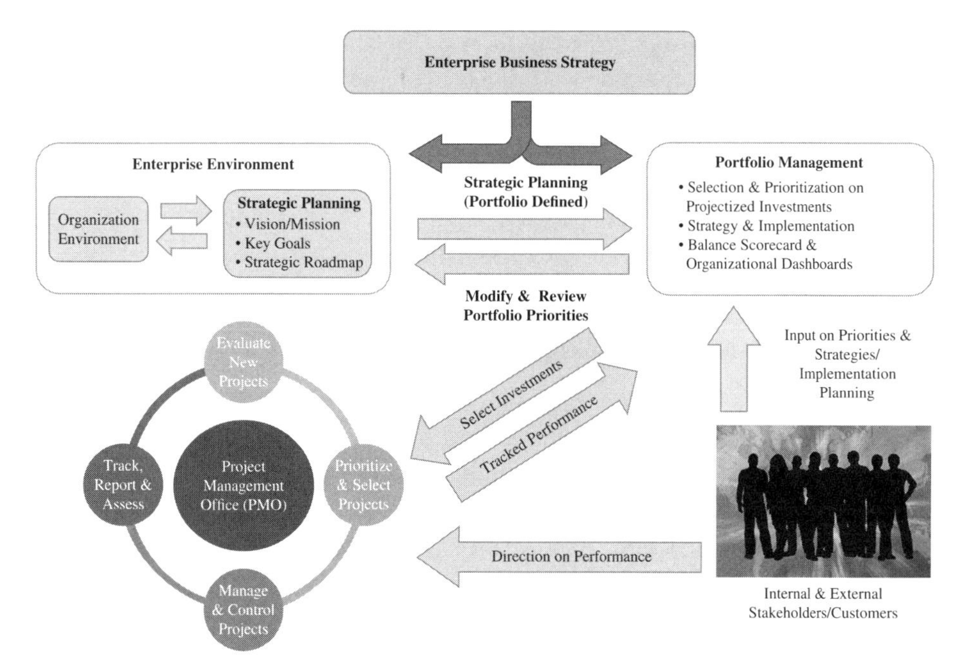

Figure 4.1 Enterprise Project Management Framework Source: Advisicon

This chapter focuses on the portfolio management aspects of Microsoft EPM and describes how the solution provides the centralized business application platform to help assess, prioritize, and manage the work and resources of the organization.

Business Platform Designed for the Business User

The logical architecture of Microsoft's EPM 2010 solution was designed around business users' needs to collaborate, prioritize, and manage the work and resources of the enterprise. The focus is more holistic therefore, so is the deployment platform.

Traditional approaches used over the last couple of decades (selecting best-of-breed business applications and attempting to integrate them into a single logical structure) have largely failed and are now yielding to organizations seeking platform choices (e.g., SAP, Oracle Fusion, Microsoft SharePoint Solutions). We are currently in a buy versus build era.

With over 100 million users (Henschen 2010) at the time this book was written, SharePoint is not only considered to be the most successful application platform that Microsoft has ever built, it is also providing the best opportunity for business users to begin to standardize on an EA that will have a significant number of solutions built on top of SharePoint 2010.

Microsoft EPM Platform

Let's examine one the first significant business solutions to be built on top of SharePoint 2010, Microsoft EPM 2010.

The Microsoft EPM 2010 solution is a flexible, end-to-end PPM platform used by organizations across many industries to automate work and PM processes. The Microsoft EPM 2010 solution is comprised of Microsoft Project Server 2010, Microsoft Project Professional 2010, Microsoft Office 2010, and of course Microsoft SharePoint 2010.

The Microsoft EPM 2010 solution provides these key PPM capabilities in a single integrated system:

- **Demand management.** Captures all work and resource requests centrally, then manages according to a governance workflow
- **Portfolio selection and analytics.** Prioritizes, optimizes, and selects project portfolios that align with the organization's business strategy.
- **Resource capacity planning.** Proactively manages resources throughout the project lifecycle
- **Schedule management.** Plans and communicates both simple and complex project schedules
- **Program management.** Initiates, plans, and delivers strategic programs
- **Financial management.** Measures and controls project and portfolio financial performance
- **Time and task management.** Captures time and task status updates from team members
- **Team collaboration.** Connects teams to share information and drives project collaboration
- **Issues and risk management.** Identifies, reduces, and communicates issues and risks that might impact project success
- **Business intelligence (BI) and reporting.** Measures project performance and gains visibility and control across all portfolios

Figure 4.2 illustrates the components of the Microsoft EPM solution.

Workflows

A governance workflow is about creating the lifecycle for any new proposal or initiative. It includes defining the various stages a project goes through during its lifecycle.

Project Server 2010 workflows are designed to model the organizational governance processes and provide a structured way for projects to proceed through the various phases and stages of the project lifecycle. Example stages include proposal creation, proposal initial approval, final budget approval, and so on.

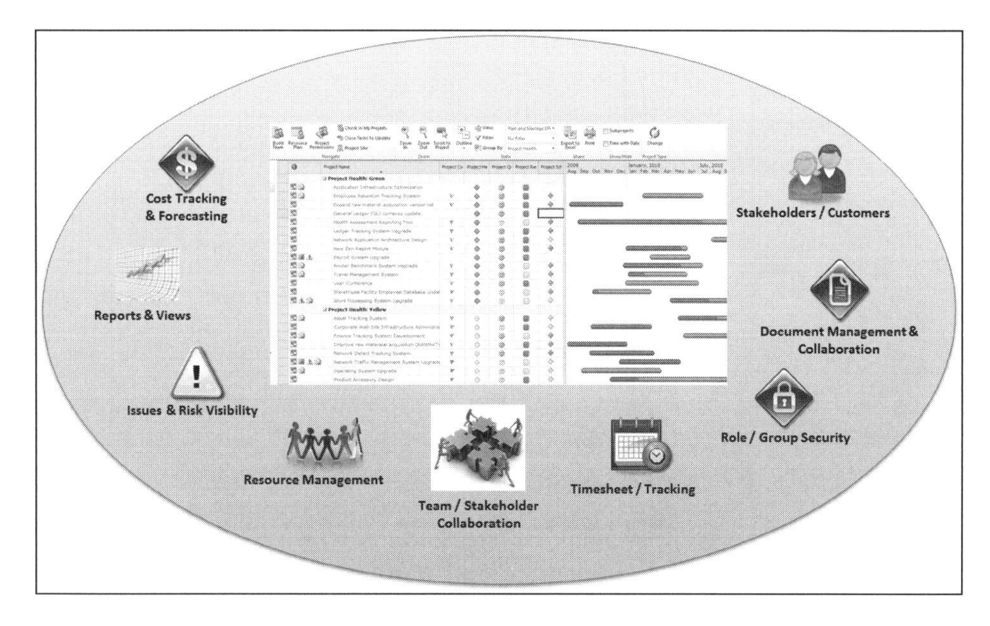

Figure 4.2 Microsoft EPM Solution Components Source: Advisicon

The Project Server 2010 workflow platform is based on the Windows SharePoint Services 2010 workflow platform, which in turn is based on the Windows workflow foundation capabilities of the Windows Server.

Enterprise Project Types

The basic workhorse behind the Project Server 2010 design was the introduction of the new enterprise project type (EPT) and the built-in workflow capability. Every project is associated with an EPT, which governs it through its project lifecycle.

EPTs are project templates that represent various types of projects and nonproject work within the portfolio. Normally, project types are aligned with individual departments—for example, engineering projects, software development projects, or human resources projects. Using project types helps categorize projects that have the same approval steps or share a similar project lifecycle. Think of the EPT as a project definition or template.

As a project goes through its lifecycle or workflow (i.e., phases and stages), the workflow will determine which project detail pages (PDPs) will be presented depending on a predefined process for a given project type.

Native to Project Server 2010, EPTs provide support for a combination of phases, stages, a single workflow, and any number of PDPs.

Business User Benefits

From a business perspective, the use of EPTs helps guide projects through their respective lifecycles using workflows that can define stage gates, approval processes, and distribute project progress information to stakeholders.

The application of EPTs and PDPs can help provide a standardized approach to proposal definitions, including work, cost, and timeline details.

Other business benefits of this approach include:

- A standardized approach to work and project demand capture
- A uniform process-driven approach to assessing work and project selection to the business priorities
- Common criteria for reporting project progress (cost, schedule, and project deliverables)
- Common operating platform (single source of project information including total demand, work and resource forecast, and project actuals)

You Don't Need to Be a Technologist to Be an Effective Practitioner

Defining projects in Microsoft's PPM 2010 environment doesn't require that you know the deep scheduling algorithm for the Project Professional tool. In fact, the building and migrating of work, planning, and key metrics around projects is a matter of filling in forms. Depending on your particular role, you may be looking only at high-level information rather than detailed planning.

In Project Server 2010, the ability to work from a Web-based environment to link, update, progress, and manage a schedule has removed much of the fear and angst of working in Project's desktop version. Don't get us wrong; Project Professional is still called the rich client for a reason. It has some extremely powerful capabilities, and with a little training in the dos and don'ts in building, working, and managing with Project, you can be extremely effective and spend less time in the tool and more time in and around the projects you are working on. However, the key idea with working in Project Server 2010 is that you are not tied to being in a desktop tool. Instead, the Web-based environment enables end users to access the information from the Web.

Proposing a Project

Let's now take a look at a lifecycle for a typical project. The aim here is to illustrate by example that you don't need to be a technologist to effectively use the Microsoft EPM 2010 solution.

As shown in Figure 4.3, Project Server provides us with a drop-down list of the EPTs that are available to us (e.g., a basic project plan or a marketing campaign). EPTs are highly configurable, as we detail in chapter 9.

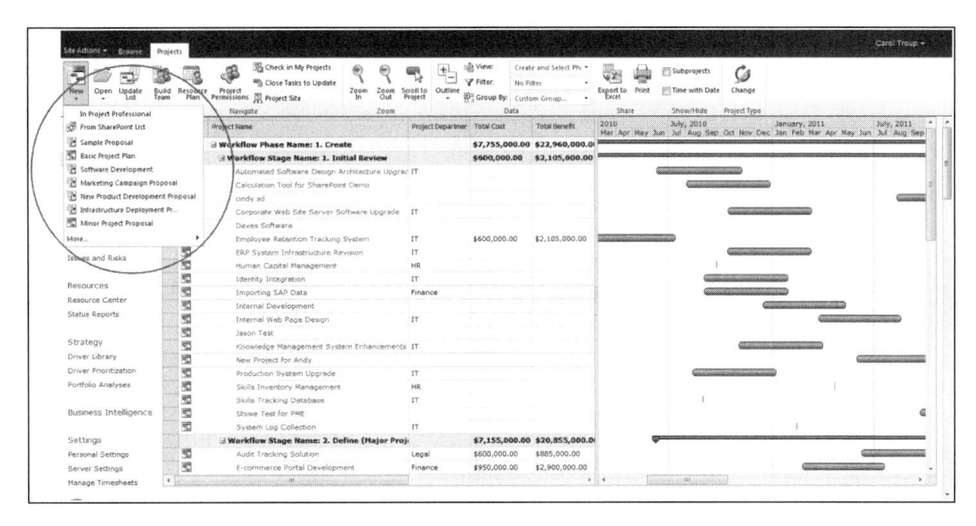

Figure 4.3 Selecting an EPTs Source: Advisicon

Based on our assigned department, role, and security level (all of which are fully configurable for each user within the EPM system), we will be presented with those EPTs that we are authorized to see. This security design provides a powerful combination of flexibility and access to information while maintaining control over system functions and data, essentially making information available to users on a need-to-know basis.

For this example, we will choose the Sample Proposal project type.

Defining the Project Proposal

In the first stage, we must enter summary information about the project proposal needs. Proposals are created in the Project Web Application (PWA). Anyone who has access to PWA can view proposals.

To create a project proposal, however, a user must be granted permission. EPM can be set up to allow certain authorized users to propose new projects or ideas while allowing all users to view what others are proposing.

Figure 4.4 illustrates a project proposal, which is essentially a configurable form defined to the Project Server system. This form provides a straightforward, minimal approach to gathering basic information about the project (e.g., the project's description, start date, etc.).

This proposal form can be configured to enter basic project data:

- Project name
- Project description
- Proposed start date and end date

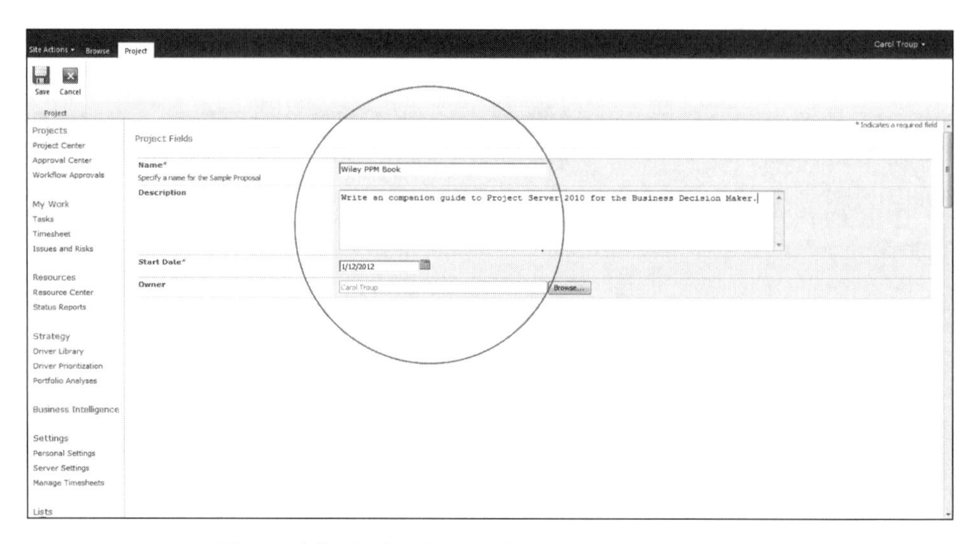

Figure 4.4 Project Proposal Form Source: Advisicon

- Proposed cost
- Proposer name
- Proposer department

When the proposal is saved, the Project Workflow defined for this project type is started. The workflow status is then presented to the user, indicating in this example that we have now entered the Initiate stage. From here, we can decide to accept the proposal, and if so, move it to the Define stage. (See Figure 4.5.)

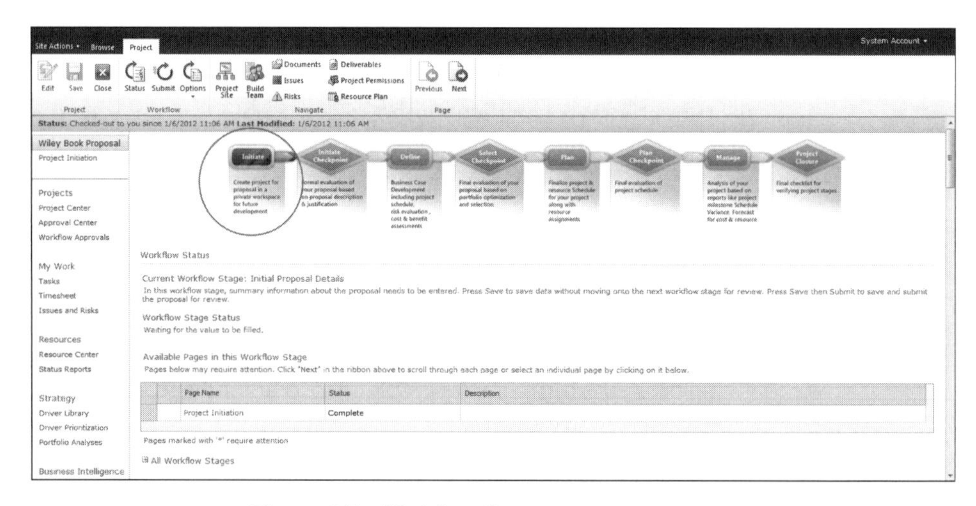

Figure 4.5 Workflow Status Source: Advisicon

Workflows are fully configurable and provide the ability to configure different workflows for different project types, departments, and so on.

Reviewing and Approving the Proposal

As we progress through the governance workflow, if the proposal is approved, we will continue to collect project metadata. Eventually this proposal will become a full-scale project.

Proposals should contain sufficient information to allow a business decision maker to approve or reject them. Chapter on portfolio management provides additional details on project selection and optimization (see Figure 4.6).

Figure 4.7 illustrates the review and approval functions provided to the end user by the EPM 2010 solution. Once again, this is a highly configurable component of Project Server 2010 and even supports auto-approval, based on definable workflow criteria.

Developing the Business Case

As we can see in Figure 4.8, the workflow has progressed to the "define" stage, where we will detail the business case development (potentially including the project schedule, risk evaluation, and cost and benefits assessments).

Estimating Required Resources

We now need to develop a resource plan that will be used during portfolio analysis to determine the requirements and impact that this project will have on the overall portfolio.

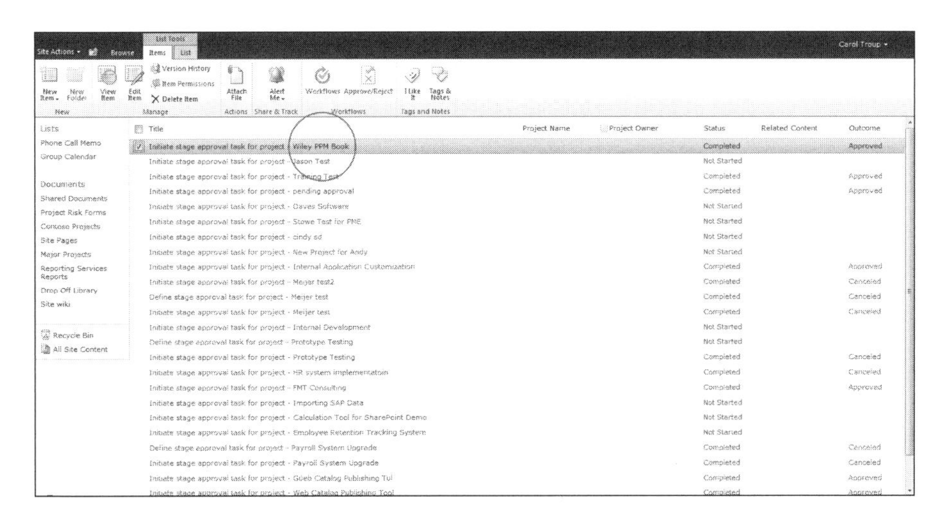

Figure 4.6 Approvals Source: Advisicon

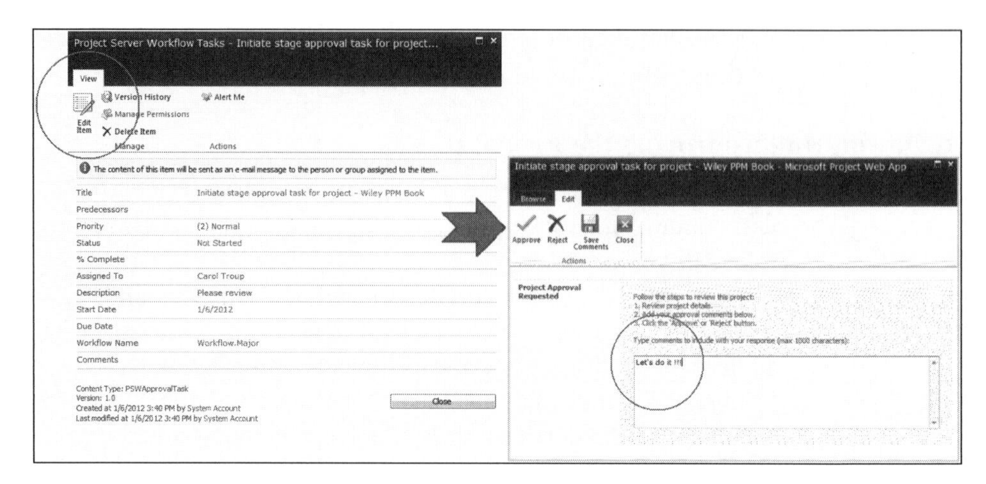

Figure 4.7 Workflow Stage Approval Source: Advisicon

Figure 4.9 illustrates the process of associating generic resources (placeholders) to the proposal. This process will provide an overall level of effort (capacity) that will be required to complete the project being proposed.

At this stage, we need not concern ourselves with the named individuals, so we assign generic resource names and estimate their full-time equivalent (FTE) requirements for the term of the project. Figure 4.10 illustrates the definition of the resource plan that gets associated with the individual project.

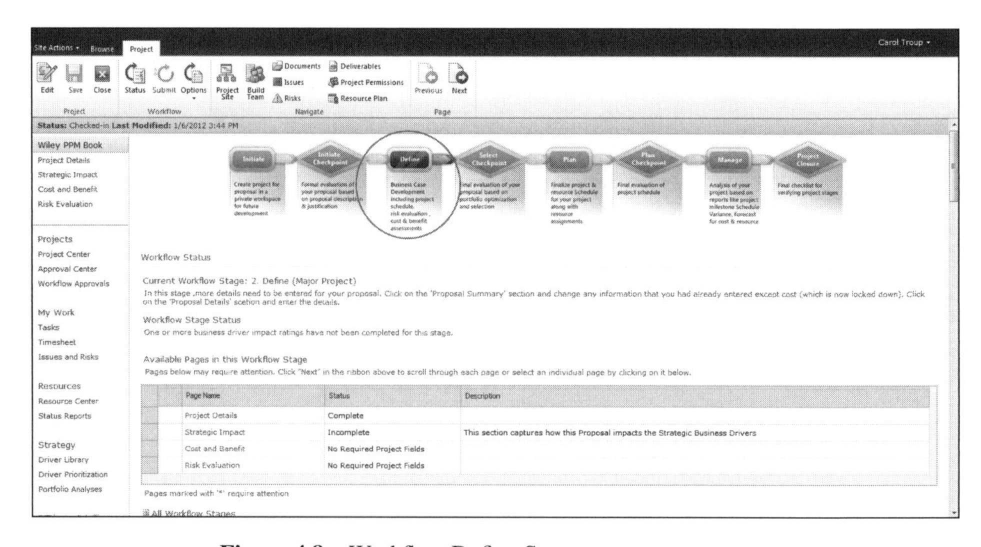

Figure 4.8 Workflow Define Stage Source: Advisicon

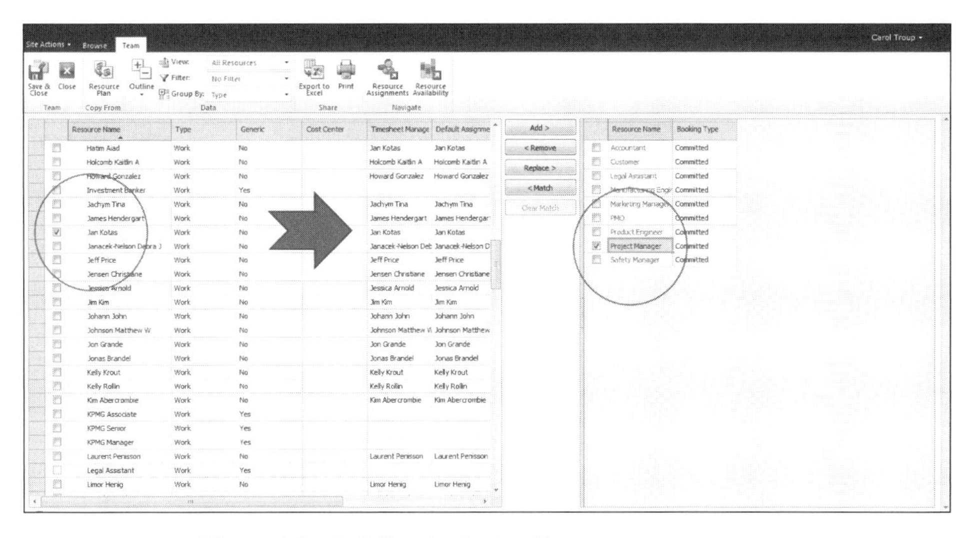

Figure 4.9 Building the Project Team Source: Advisicon

This capacity planning capability is part of Project Server 2010 and provides end users the ability to display resource data by time scale (i.e., days, weeks, months, quarters, or years) and calculate the resources based on the type of work unit they wish work in (i.e., hours, days, FTEs).

The source of the resource planning data can also be selected from either the resource plan or the project plan. Utilization can be calculated from the project plan up to a specified point in the project plan (i.e., rolling wave) (Shaker 2010).

Figure 4.10 Estimating the Resource Plan Source: Advisicon

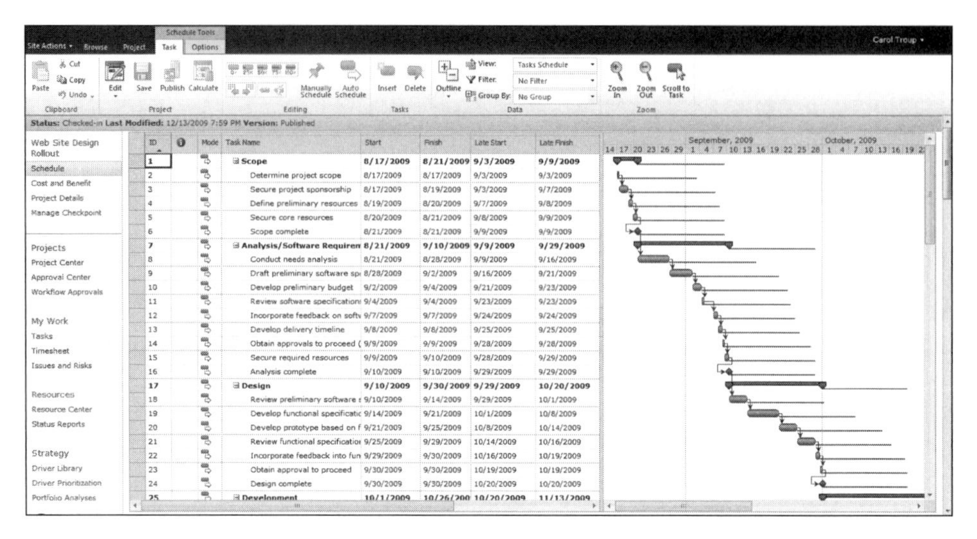

Figure 4.11 Planning the Schedule Source: Advisicon

Planning the Project

If the business proposal has merit, an organization will initiate a project in order to implement an aspect of corporate strategy, to realize a business case, and to create a set of deliverables. Projects should exist for very clear reasons (Hillson 2009).

If the proposal is approved, the project moves into the detailed planning stage of the project lifecycle. Tasks that need to be performed to complete the deliverables of the project are detailed as illustrated in Figure 4.11.

Chapter provides additional details on how to perform good schedule development and some best practices for defining dynamic schedules that will derive forecast completion dates and help determine when the deliverables (and the project) will be complete.

Planning and scheduling of project tasks, along with placing their estimates and dependencies into a sequence, is a critical step to ensure successful project outcomes. According to Jim Snyder, one of the founders of the Project Management Institute, "If we spent more time Planning and Scheduling, we wouldn't have to do so much Project Management."[1]

Doing an initial schedule based on the work breakdown structure and resource assignments typically shows the project team the time and resource requirements and provides a starting point for an iterative schedule management process.

[1]As quoted on the PMI website: www.pmi.org/About-Us/Fellows/James-R-Snyder.aspx (accessed January 13, 2012).

Managing the Project(s)

Once we have a fully authorized proposal, planned our resources, and finalized our schedule, our focus can turn to managing the project (i.e., managing and tracking execution of the schedule, monitoring and reporting progress, and delivering project products).

The real key here is to manage all the projects of the enterprise in a manner that provides visibility across all factors that can impact the successful delivery of each project. This needs to be done in a manner that is consistent, accurate, and timely in regard to:

- Scope
- Schedule
- Resources
- Costs
- Issues and risks
- Deliverables
- Quality
- Benefits
- Overall health

This is a real challenge for most organizations because there is no central location where all these data can reside. Typically this information can be found in several nonintegrated sources (e.g., project schedules, PowerPoint presentations, and spreadsheets).

According to extensive research by the European Spreadsheet Risk Interest Group (2012, p. 1) on the risks of using spreadsheets within business, the majority of spreadsheets (>90%) contain errors. The research goes on to say that "because spreadsheets are rarely tested, these errors remain. Recent research has shown that about 50% of spreadsheet models used operationally in large businesses [have] material defects."

The Microsoft EPM solution centralizes all of this project operational/performance data into a centralized SQL database where data integrity can be properly managed. In addition, because the data are located in a common "shared" repository, they can be viewed (in real time) and grouped, filtered, and provided on a secure need-to-know basis using views.

Figure 4.12 illustrates how projects can be managed centrally, providing the right level of detail that is appropriate and sufficient to the applicable end user role (i.e., not inundating users with all of the gory details of the project).

Information can also be extracted at any time, from any view, to Excel, Visio, and other data tools, to be manipulated, formatted, e-mailed, or even printed, should a report be required for a meeting (where, for some strange reason, no computers or projection screens are allowed).

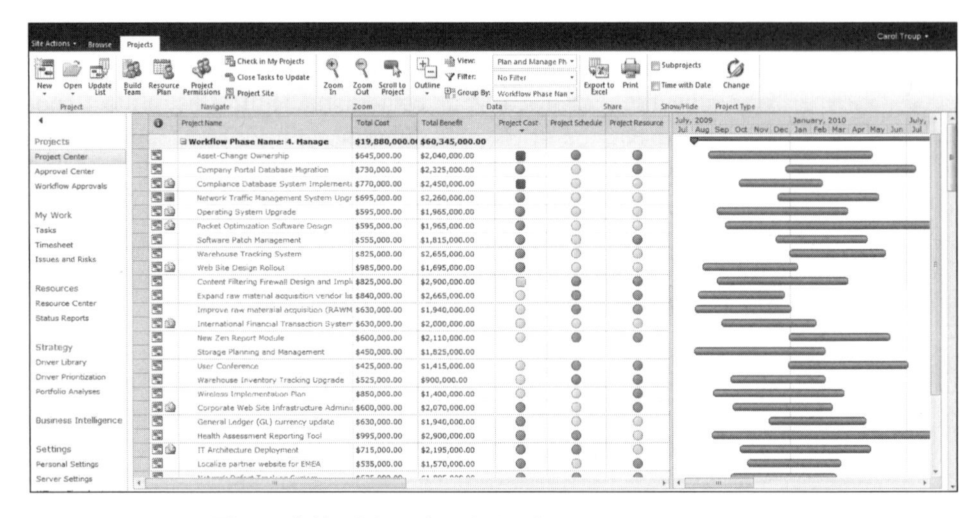

Figure 4.12 Managing the Projects Source: Advisicon

Figure 4.13 illustrates an example of a project that has reached the Manage stage of the project lifecycle. Here, for example, is where analysis of the project is performed, based on reports such as milestones, schedule and cost variance, and forecast of key target dates and deliverables.

Business Intelligence/Reporting

By now, you are probably realizing that it makes sense to track project performance in a manner similar to how we track financial performance. What are the

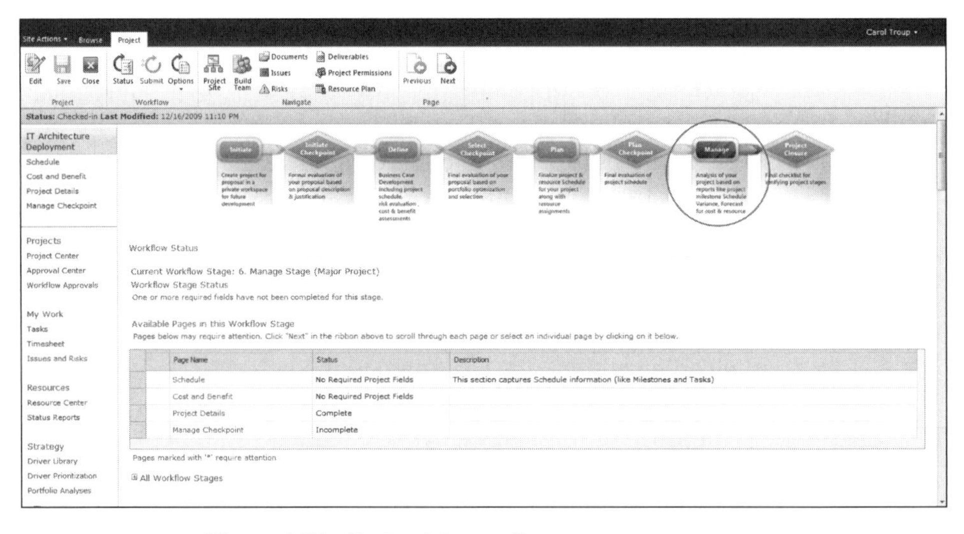

Figure 4.13 Project Manage Stage Source: Advisicon

actuals relative to the plan? Are we on budget? On schedule? While we are tracking, we also need to be producing quality deliverables that meet the needs of the stakeholders.

One of the key challenges that organizations of all sizes struggle with is how to make sense of all the data. Spreadsheets and PowerPoint presentations are commonplace; however, so are the eleventh-hour fire drills that are associated with their preparation.

BI is critical for decision makers in order for them to make good business directions. Project portfolio intelligence must also become a critical approach to planning and managing the enterprise investments in the organization's critical resources; and BI and reporting must become pervasive throughout all phases of the project portfolio lifecycle. This list breaks out the key components that BI can be used to clarify and provide good analysis.

- **Planning.** Ensuring that we are collecting and analyzing total demand, aligning to business priorities, detailing the work breakdown, estimating the resources
- **Managing.** Monitoring and controlling project execution
- **BI and reporting.** To support better decision making

The definition of business intelligence is "a set of methodologies, processes, architectures, and technologies that transform raw data into meaningful and useful information used to enable more effective strategic, tactical, and operational insights and decision making" (Everson and Nicolson 2008).

With the large amount of data located in a variety of tools and formats, one of the challenges that organizations currently face is how to collect, categorize, understand, and make decisions about project data.

The Microsoft EPM 2010 solution offers integrated BI that provides visibility into the project portfolio facilitating decision support for proactive project and work management. The EPM 2010 solution includes a variety of online views and powerful BI and reporting services to help organizations gain insight, visibility, and control across their project portfolios.

EPM 2010 is built on top of SharePoint Server 2010, which provides end users with the capability to take advantage of all the tools included in the Microsoft BI platform, including Excel Services, PerformancePoint Services, Visio Services, PowerPivot for Excel, and SQL Reporting Services.

Figure 4.14 illustrates an integrated project dashboard comprised of textual and graphical project data:

- **EPM Variance Scorecard.** Key performance indicators (KPIs) that change colors based on formulas that are measuring cost, duration, start, finish, and work variances

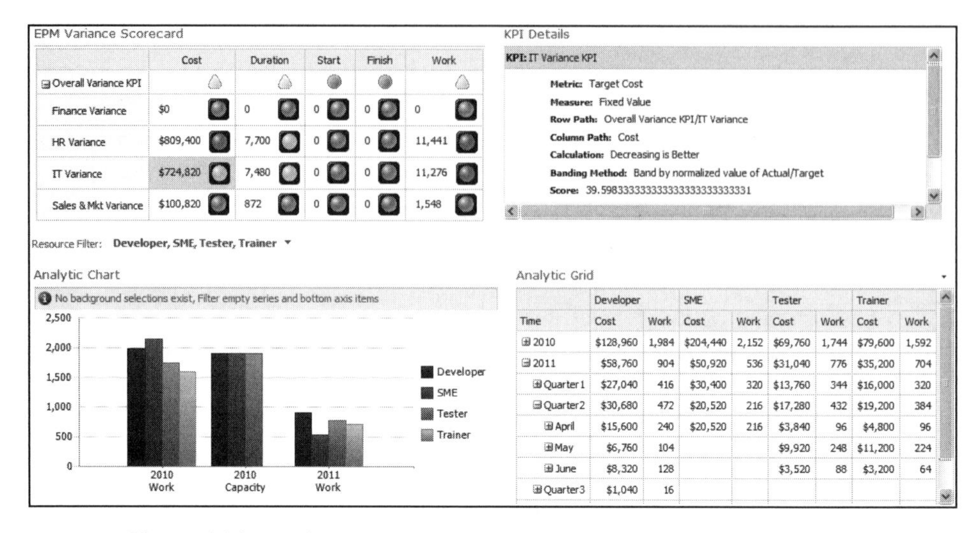

Figure 4.14 Business Intelligence Project Dashboard Source: Advisicon

- **KPI Details.** Text attributes to support the KPI chart
- **Analytic Chart.** Example bar chart illustrating hours of work over time
- **Analytic Grid.** Text summary of cost and work for each project role over time

This example dashboard provides summary PM performance data to end users and also acts as a menu to drill down into the details. We discuss this in more detail next.

Dashboards are essentially a mashup (Web application hybrid) created from Web Parts that are configured to meet the needs of an end user role. Dashboards present data, initially in a high-level roll-up format, then allow drilldown into details that provide insight into specifics of a project and its related components.

PMOs can also create dashboards to ensure that end users receive relevant, useful information that meet the needs of the roles they perform. More technical users can employ more sophisticated tools, such as SQL Server Reporting Services, to create more involved reports.

Project Server 2010 also includes a library of preconfigured template reports that end users can customize, using familiar tools like Excel. These reports can then be published using Excel Services and incorporated into dashboard views.

Figure 4.15 illustrates a consolidated project summary view where KPIs, cost summary information, and other information can be displayed using different types of presentation formats (e.g., tabular data, traffic light indicators, dials, etc.).

The Microsoft EPM 2010 solution further provides end users with the ability to drill down into data using a decomposition tree. Decomposition trees are a new feature in Microsoft SharePoint 2010 that allows end users to drill down on

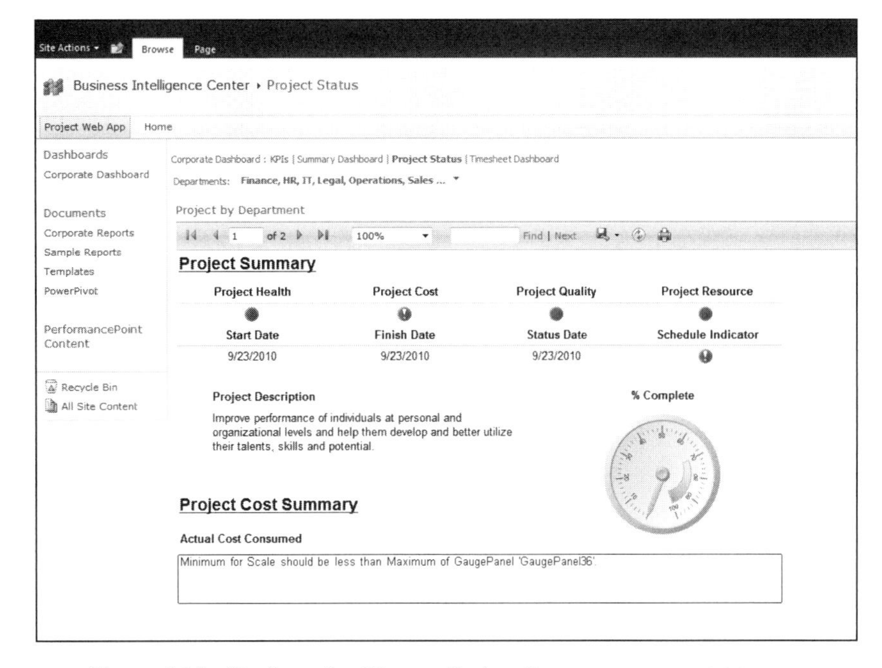

Figure 4.15 Business Intelligence Project Summary Source: Advisicon

reports generated from analysis services where the data can then be displayed in a consolidated dashboard format.

The PerformancePoint analytics tool produces a decomposition tree that is used to perform root-cause analysis by viewing how individual members in a group contribute to the whole (Microsoft, 2012a). Decomposition trees can be used to examine how an individual value in a dashboard can be broken down into its contributing elements.

Figure 4.16 shows a PerformancePoint query in a pop-up window that allows a nontechnical end user to navigate project data, by pointing, clicking, and using drop-down menus to filter and display specific data (e.g., departmental costs broken down by projects and then by how each role contributes to those costs).

PerformancePoint Services included with the Microsoft EPM 2010 solution provides several new and improved features to help monitor and analyze an organization's project and work performance. Dashboards can also be developed to include more sophisticated KPIs in Web-based scorecards.

A True Enterprise Platform

The Microsoft EPM 2010 Solution is a fully integrated, end-to-end PPM solution used to automate PPM processes (TechNet 2011). The EPM solution includes these products to provide a comprehensive desktop and Web-based PPM solution:

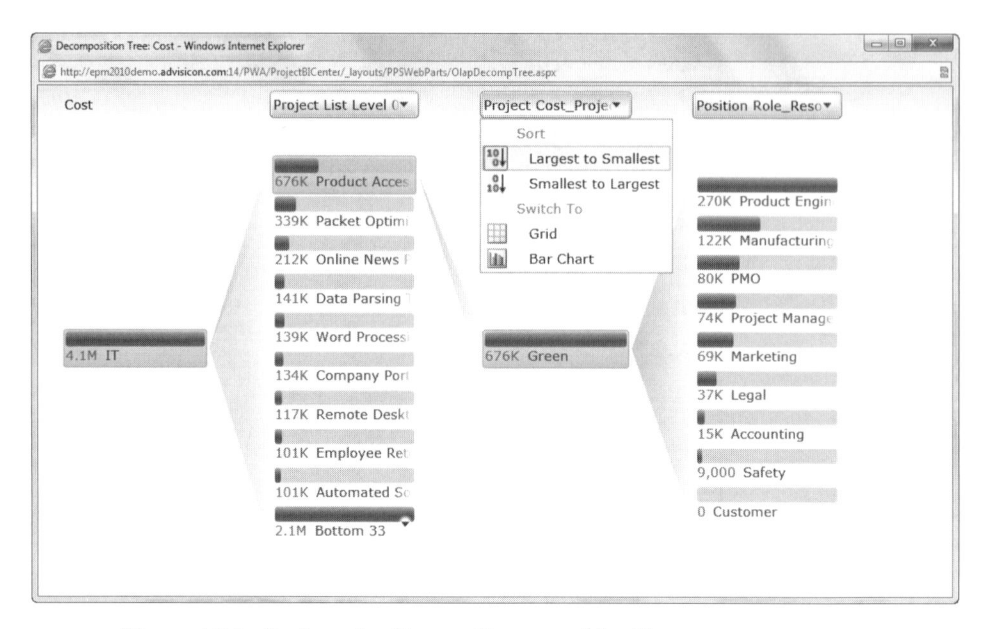

Figure 4.16 Business Intelligence Decomposition Tree Source: Advisicon

- **Microsoft Project Professional 2010.** Provides connectivity between the desktop and Project Server 2010, to ensure that organizations can achieve the added business benefits of unified PPM.
- **Microsoft Project Server 2010.** Brings together the business collaboration platform services of SharePoint Server 2010 with structured execution capabilities to provide flexible work management solutions.
- **Microsoft SharePoint 2010.** Project Server 2010 is built on SharePoint Server 2010 to provide true multitier architecture by using the new service application model.

Project Server 2010 on SharePoint 2010 is the only truly integrated portfolio, program, and project management solution on the market today. Organizational attempts to integrate of best-of-breed products with enterprise work and resource planning and reporting have largely failed up to now.

Collaboration

With the more traditional management of individual projects, focus has always been on project schedule, budget, work, scope, and the deliverables of these projects. Team collaboration, however, is the backbone that supports and drives success with planning and execution effectiveness.

Businesses need to approach collaboration strategically and seek to align people, processes, and technology with the organization's project investment goals.

In the case of collaboration, that alignment must also take into account the level of trust required to improve the chances of success for each type of collaboration (The Economist Intelligence Unit 2008).

Microsoft Project Server 2010 is built on SharePoint Server 2010 and combines powerful business collaboration capabilities with structured execution functionality to provide a flexible and secure work and PM solution. Core capabilities include:

- BI platform to easily create reports and powerful dashboards
- Custom project site templates for each type of project to provide a one-stop collaboration workspace for the team
- Review and approval tracking throughout the workflow
- Enterprise search to find people and effectively mine project data
- Enhanced teams communication with wikis, blogs, discussion forums, and My Sites

Team collaboration has evolved over the past decade from informal techniques into a recognized discipline that helps organizations more effectively find and share information. Tools that support collaboration have evolved to sophisticated solutions, such as Microsoft SharePoint Server 2010.

Team Sites

SharePoint Server 2010 team sites provide a place where the team can communicate with each other, share documents, and work collaboratively on a project. Separate project sites can be created for each project, or one site can be shared by several project teams (e.g., for a specific program).

Project teams can add information to the site, such as events, contact names, and phone numbers of people that a project team communicates with. The team site is really the one-stop-shop for all project information; it includes:

- **Announcements.** Where end users can post information for the team.
- **Calendar.** Create and attach a project calendar to a team's Web site.
- **Contacts list.** Stores information (including names, telephone numbers, e-mail addresses, and street addresses) for people who are part of the team.
- **Issues and risks.** Can be associated with projects, tasks, and documents to permit team members to keep track of their status.
- **Links list.** Displays hyperlinks to Web pages of interest to team members. By default, a view of the Links list appears on the home page.
- **Project document library.** Stores documents that are related to a specific project. Access to documents in the library is based on permissions that can be set for project managers, team members, and other stakeholders.
- **Tasks list.** Provides a to-do list for team members.

This list describes how the core team context is established using SharePoint Team Sites, social networking capabilities, and e-mail to provide alerts and notifications.

Document Management

SharePoint Server 2010 includes document management capabilities that the project manager and team members can use to control the lifecycle of documents for all projects in the organization (i.e., how they are created, reviewed, and published and ultimately how they are disposed of or retained).

An effective document management solution specifies:

- What types of documents can be created in an organization.
- Which templates are available for each type of document.
- What metadata can be associated with each type of document.
- Where a document should be stored at each stage of its lifecycle.
- How document access is controlled at each stage of its lifecycle.
- How documents move within the enterprise, as team members contribute to document creation, review, approval, publication, and retention/disposition.
- What document policies are applied to ensure that document-related actions are audited, documents are retained or disposed of appropriately, and how important content is protected.
- How documents have to be converted from one format to another as they move through the various stages of their lifecycle.
- How documents are handled as corporate records (plan of record) and how they need to be retained/disposed per legal requirements and corporate guidelines.

SharePoint Server 2010 includes capabilities that support all of these aspects of document management. Applications within the Microsoft Office system also include features that support each stage in a document's lifecycle (e.g. review, approve, publish, etc.).

Organizations should take time to thoroughly understand the requirements of the enterprise document management solution and carefully plan the system based on the capabilities of Microsoft SharePoint Server 2010 (TechNet 2010).

Governing Project Components Defining the Work Management

In this section we present the key elements of enterprise governance and provide an overview of the PPM lifecycle. (We will expand on this further in Chapter 5 and in Chapter 8) We describe the importance of both to the business.

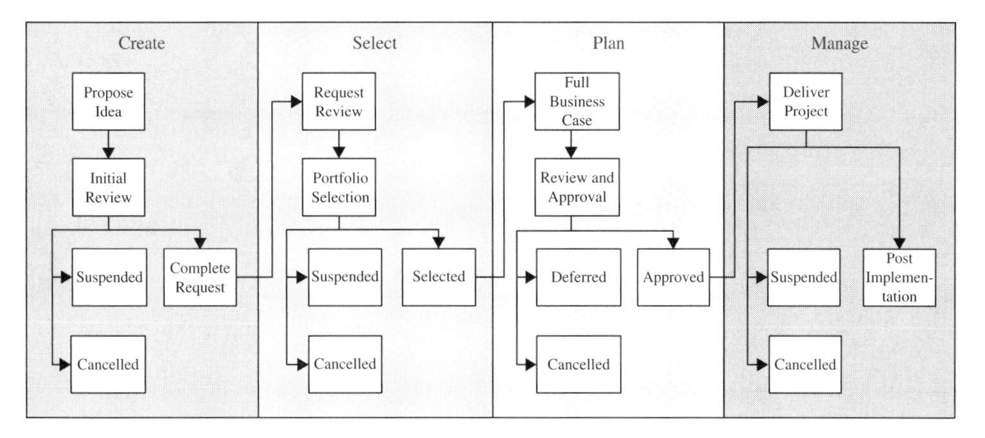

Figure 4.17 Example Governance Workflow Source: O'Cull 2009

Enterprise Governance

Portfolio governance and lifecycle management enable organizations to define processes that synchronize the efforts of distributed teams to consistently create the best possible products, capture greater market share, and increase customer satisfaction.

Figure 4.17 illustrates an example governance workflow that includes four key phases (create, select, plan, and manage). A phase represents a collection of stages, grouped to identify a common set of activities in the project lifecycle. A stage represents one step within a project lifecycle (e.g., propose idea, initial review . . . deliver project).

Phases and stages are managed in Project Server 2010 through the use of enterprise project templates (ETPs; described in the next section) that help guide projects through each stage through the use of workflows.

Project Portfolio Management Lifecycle

According to Wikipedia, project portfolio management (PPM) is a term used by project managers and project management (PM) organizations, (or PMOs), to describe methods for analyzing and collectively managing a group of current or proposed projects based on numerous key characteristics.[2]

The Microsoft EPM 2010 solution enables organizations to manage the continuous flow of projects from concept (demand) to completion (closure).

A number of the key elements of the PPM lifecycle include:

- **Demand management.** Provides a consolidated view of the total work and resource demand picture across the entire organization.

[2]*Wikipedia*, s.v. "Project Portfolio Management," accessed January 23, 2012, http://en.wikipedia.org/wiki/Project_portfolio_management.

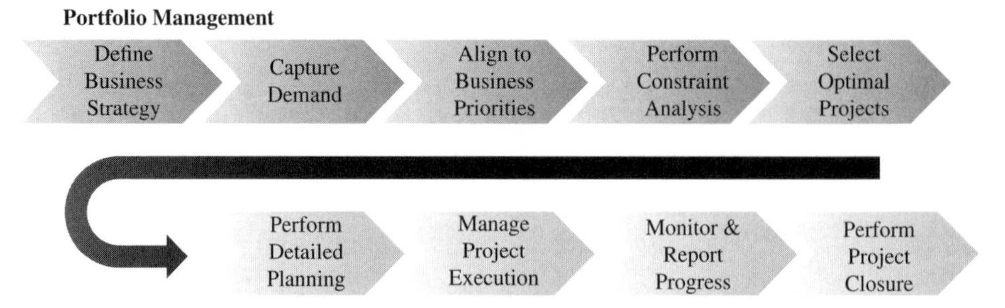

Figure 4.18 Business Driver Relationship to Projects Source: Advisicon

- **Capacity planning.** Proposes an initial assessment of resources (i.e., people, money, and time) when determining organizational capability of resources (human, financial, material).
- **Project prioritization.** Determines how each project will be prioritized to ultimately affect project selection.
- **Project selection.** Defines which projects will be selected as successful candidates for detailed planning and execution. Figure 4.18 illustrates an organization's relationship from the business mission/goals/strategy(s) to project selection. Business drivers are specific to the EPM 2010 system and help align projects to the priorities of the business.
- **Detailed project planning.** The scope of work is broken down into work packages, the network of activities is developed, and tasks are assigned to named resources so that a work schedule can be prepared.
- **Project execution.** The products (deliverables) of the project are performed.
- **Monitoring and reporting.** A key aspect in the management of a project, where the work, schedule, and financial performance are measured and reported to all stakeholders of the project. The EPM solution has significant capabilities in this regard, as we discussed throughout this text.
- **Project closure.** Where the contractual and administrative completion processes of the project are performed. A lessons learned step may also be included at this point (Ivanenko 2009).

Figure 4.19 provides an overview of a project portfolio lifecycle.

PPM Is a Business Imperative

PPM is the art and science of balancing an organization's product and PM skills and resources to achieve optimum strategic, financial, and operational impact across all product lines in all lifecycle phases.

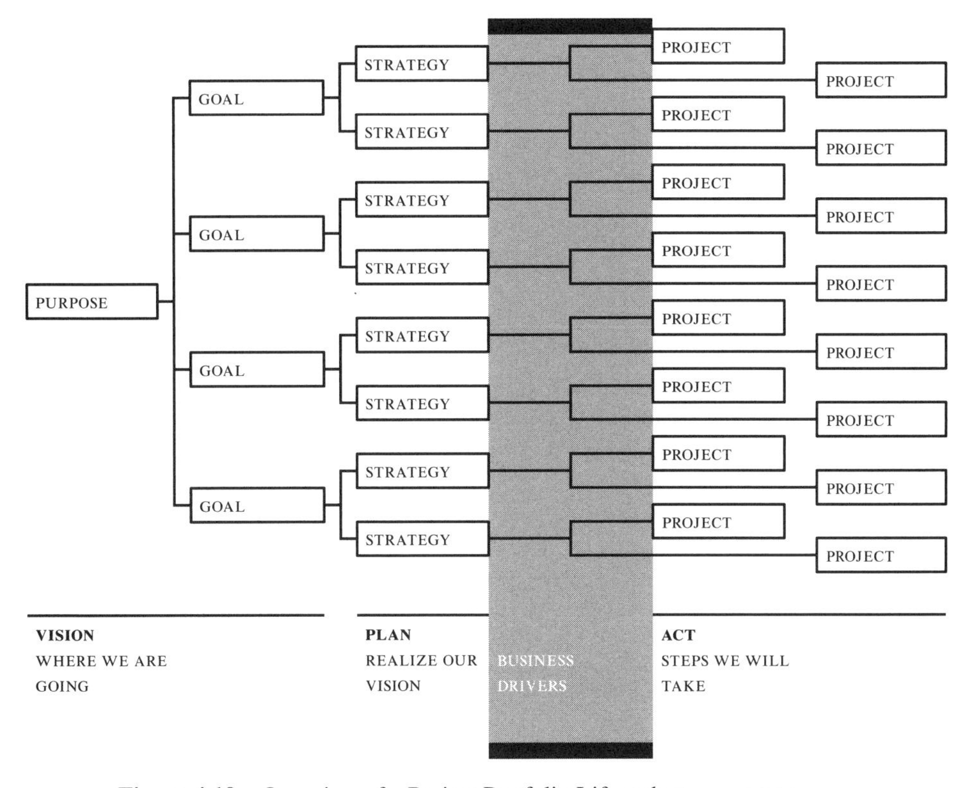

Figure 4.19 Overview of a Project Portfolio Lifecycle Source: Advisicon

Once the business executive management has defined the mission, goals, and strategies, we can derive a set of business drivers that help us align specific projects that will address the business priorities of the organization. Figure 4.18 helps us visualize this relationship.

Business drivers are the people, information, and tasks that support the fulfillment of a business objective. Drivers can include the people, knowledge, and conditions (e.g., market forces) that initiate and support activities for which the business was designed.

Understanding and properly defining business drivers is a key step in ensuring the success of an enterprise PPM system. Business drivers are an effective way to ensure alignment between strategy and execution as they:

- Provide the linkage between the business strategy and the portfolio of projects.
- Ensure a consistent way for key stakeholders to agree on cross-organization business objectives.
- Establish a basis for mapping projects back to business priorities.

Even if specific projects support business objectives, they are not necessarily guaranteed to be selected for execution. All projects in the portfolio compete for limited organizational resources (Levine 2005).

MICROSOFT PROJECT 2010 PLATFORM IS HIGHLY EXTENSIBLE

The EPM 2010 solution provides the fundamental components to automate the governance business processes and provides a scalable, connected, and extensible platform to meet the needs of the business and align with the organization's EA standards. This includes configurable business drivers (that help align and prioritize the projects/initiatives) along with custom workflows (to enforce business rules) that can be defined using Windows Workflow Foundation, using Project Server events and Web services available through the Project Server Interface (PSI).

In Chapter 8 and Chapter 8 we review in more detail the workflows and programmability of Project Server, SharePoint, and the development tools (SharePoint Designer and Visual Studio). In the next section, we discuss how the PPM lifecycle can be realized by the highly extensible Project Server 2010 platform.

Extensive Work Flows, Forms, or Approvals? No Problem!

In PPM, a project lifecycle is a process that spans key governance phases. Example demand management phases are create, select, plan, and manage.

The Planning and Management phases are accomplished by the more familiar PM processes and tools, using Project Professional and PWA.

Workflow models the governance processes and provides a structured way for projects to proceed through the various phases. Workflows, along with other proposal data in project detail pages (PDPs), are captured and integrated within Project Server 2010.

Relationship of Project Server 2010 Elements

To better understand how EPM manages a project lifecycle, it is important to understandthe relationships of the key Project Server 2010 elements, along with the role of each element. These relationships are illustrated in Figure 4.20.

The elements are defined next.

- EPTs encapsulate phases, stages, a single workflow, and PDPs.
- Phases represent a collection of stages grouped to identify a common set of activities in the project lifecycle (e.g., create, select, plan, and manage).
- Stages represent one step within a project lifecycle. A stage is composed of one or more PDPs linked by a common theme. Stages at a user level appear as steps within a project. At each step, data must be entered, modified, reviewed, or processed.

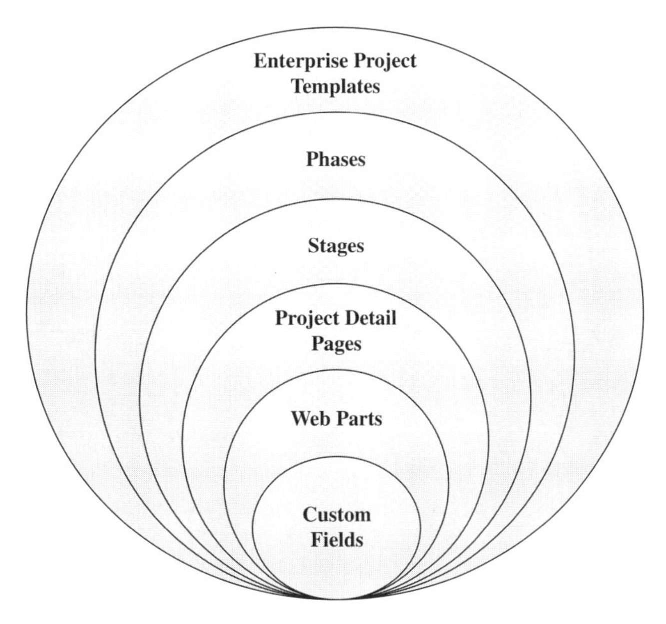

Figure 4.20 Relationships of Project Server 2010 Elements Source: Advisicon

- PDPs are used to display or collect information from the end user.
- Web Parts are located on PWA pages. They communicate with the PPSI and also use standard SharePoint Server 2010 Web Parts.
- Custom fields extend the attributes of tasks, resources, or projects in Microsoft Project 2010.

The EPM 2010 architecture provides flexible design and configuration for all aspects of the end user interface experience. Coupled with the powerful and integrated workflow capability and integrated reporting capabilities of SharePoint Server 2010, Microsoft EPM 2010 is the most extensible platform to meet the needs of the business and align with the organization's EA standards.

Workflow Integration

Workflows are another core feature of the Microsoft EPM 2010 solution. A project lifecycle can include long-running processes that span many phases. Governance phases include project proposals, analysis of business impact, selection, creation, planning, managing, and tracking of work/projects.

Workflow integration of portfolio and PM in Project Server 2010 provides a rich and extensible platform for building workflows, based on the SharePoint Server 2010 workflow platform.

From a business perspective, this capability provides total flexibility to configure the enterprise system, from work/project demand capture, to how requests are assessed and approved, through to the sequence of planning steps that need to be completed before the start of a project, and ending up with work will be executed, monitored, and controlled as it moves through the execution phase of the lifecycle.

Extensible Development Platform

With the 2010 release, Microsoft Project Server is also increasingly becoming a compelling development platform. Microsoft Project Server 2010 is designed to support high levels of programmer productivity by building on Microsoft SharePoint Server 2010 and the Microsoft.NET Framework.

There is also a Project Development site that provides resources specifically to support the Project Server 2010 developer community. This site includes a Software Develop Kit (SDK) that contains documentation, code samples, how-to articles, and programming references to help customize and integrate the Project 2010 clients and Microsoft Project Server 2010 with a wide variety of other desktop and business applications for enterprise PM (MSDN 2011).

Developers can use the SDK to extend the out-of-the-box PWA user interface by acquiring or creating new Web Parts, developed with Microsoft ASP.NET.

Open Source Platform

To assist with the rapid ramp-up of customers being able to achieve custom results quickly when customizing Project Server 2010, a number of solution starters and code samples are available to download for free from Microsoft. These solution starters include deployment packages, source code, documentation, and webcasts (MSDN 2011).

There is also an open source site provided by Microsoft. CodePlex is Microsoft's open source project hosting Web site that can be used by developers to create new projects to share with the world. Through CodePlex, you can join others who have started their own projects, download open source software, and even provide feedback (CodePlex 2012).

OK, It Manages the Work, But What About the Financials?

The International Federation of Accountants (2004) commissioned a task force on rebuilding confidence in financial reporting to look at ways of restoring the credibility of financial reporting. The report was published in 2004 and set out 10 recommendations, one of which was: "Corporate management must place greater emphasis on the effectiveness of financial management and controls."

Chapter provides additional information on the management of financial controls, approaches and the role and responsibilities of the chief financial officer.

This section discusses cost management from a project perspective and the roll-up capabilities of the Microsoft Project 2010 desktop and the EPM 2010 solutions.

Project Cost Types

Microsoft Project 2010 calculates the costs for resources based on a number of factors and resource types. A wide variety of resource costs need to be tracked throughout the lifecycle of a project, including:

- **Regular and overtime rates,** which are calculated based on the pay rates that are specified for a resource and the amount of work that is performed by that resource. The standard and overtime rates are calculated separately, based on their specific pay rate, and then rolled up to reflect the total labor costs for a given task (Microsoft, 2012a).
- **Per-use costs,** which can be applied to a work request to track a professional fee for example. Per-use costs can also be allocated to material resource when costs must be tracked each time some item is used. You can enter more than one per-use cost for each resource.
- **Fixed costs,** which are assigned to tasks and are useful for planning and capturing task costs that arise in addition to those arising from the assigned resources. Fixed costs can be applied to a task only, not to resources. You can also enter fixed costs for the entire project, to track overall project costs for example.
- **Cost resources that are assigned to tasks,** which might include airfare and lodging. This is typically a one-time cost per task, but the cost resource assignment also can be contoured across the duration of the task (i.e., start of task, prorated over the duration of the task, or at the finish of the task).

Rate Tables

You can model more complex billing schemes using the rate table features provided in Project. Figure 4.21 shows how a resource standard rate can be changed starting on a specified date.

Be careful when changing the standard rate for a resource, as this can affect the cost of tasks that are already 100 percent complete.

Budget Resource

Microsoft Project 2010 introduced a budget resource capability that provides the ability to track budget at the project level. Budget resources cannot be assigned to individual tasks in a project.

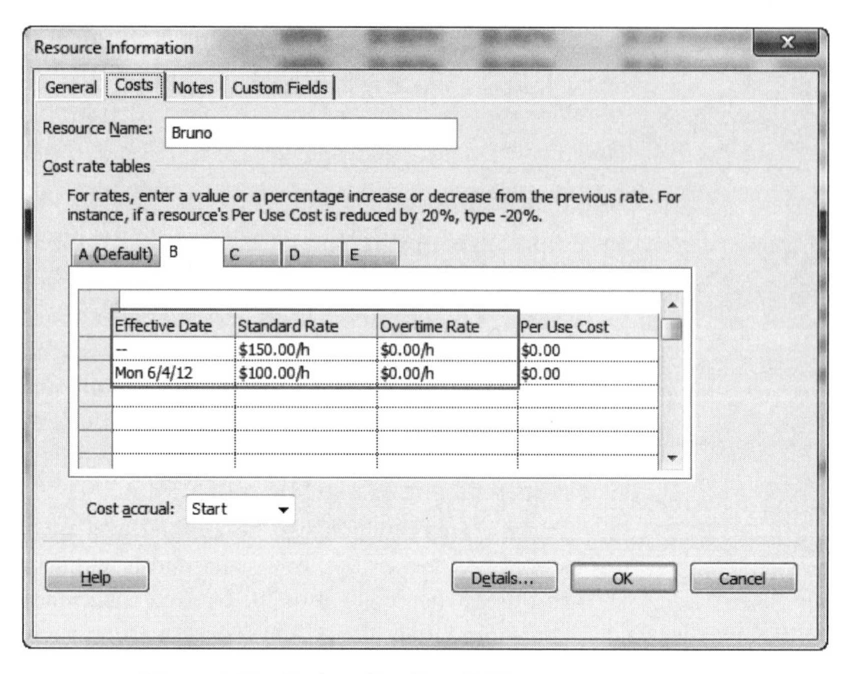

Figure 4.21 Project Cost Rate Table Source: Advisicon

Figure 4.22 illustrates setting up a budget resource to track project costs. Other budget resources can be defined to track work and material budgets.

Figure 4.23 provides a summary view of the project including the budget and planned costs and work. Notice that budget resources can be assigned only to the project summary level.

EPM 2010

Although cost tracking and management for individual projects can be managed effectively with Project 2010 desktop, the enterprise view requires the robust capabilities of the Microsoft EPM solution offered by Project Server 2010.

The Microsoft EPM 2010 solution provides a robust platform that can aid an organization's management of resources (people, materials, and expenses). EPM can greatly enhance an organization's ability to define, align, plan, manage and monitor finances at a project, program, or portfolio level (Arbutus Solutions 2011).

EPM 2010 provides a variety of online views (out of the box) that easily facilitate enterprise financial reporting. Cost fields from individual projects can be published in Project Server (e.g., cost, baseline cost, actual cost, cost variance, remaining cost, etc.).

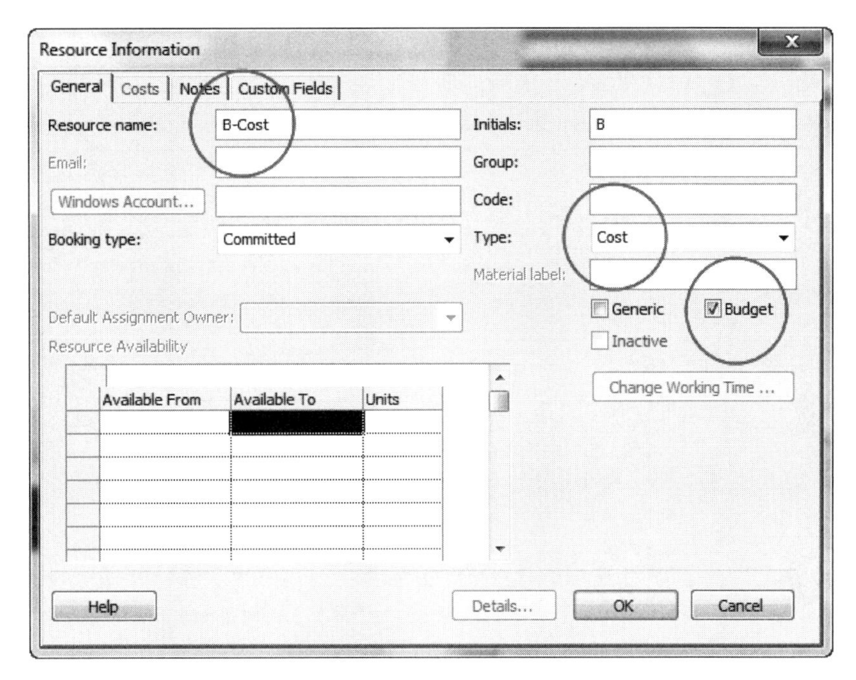

Figure 4.22 Project Budget Resource Source: Advisicon

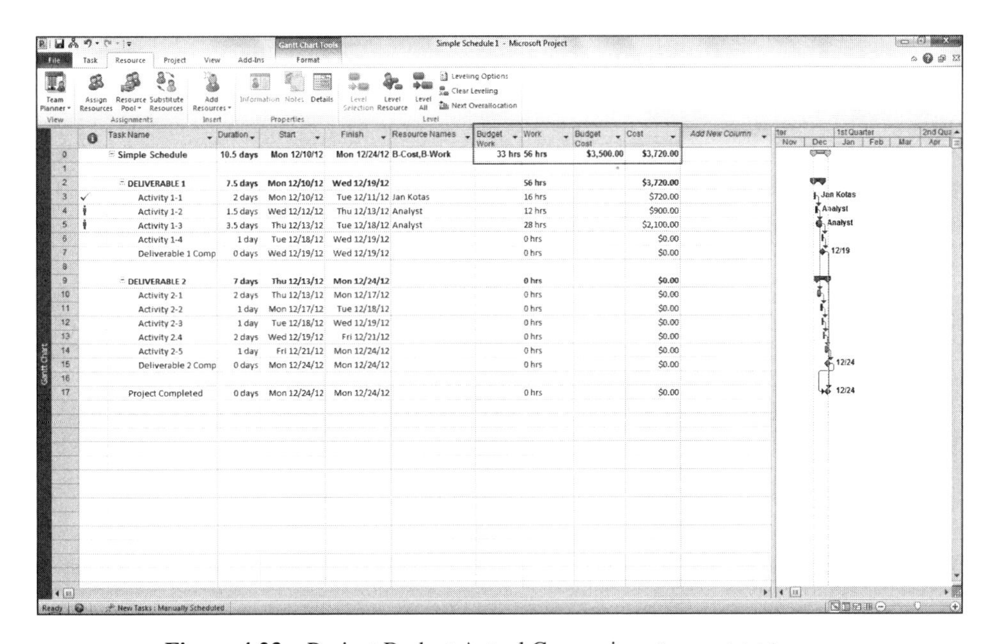

Figure 4.23 Project Budget Actual Comparison Source: Advisicon

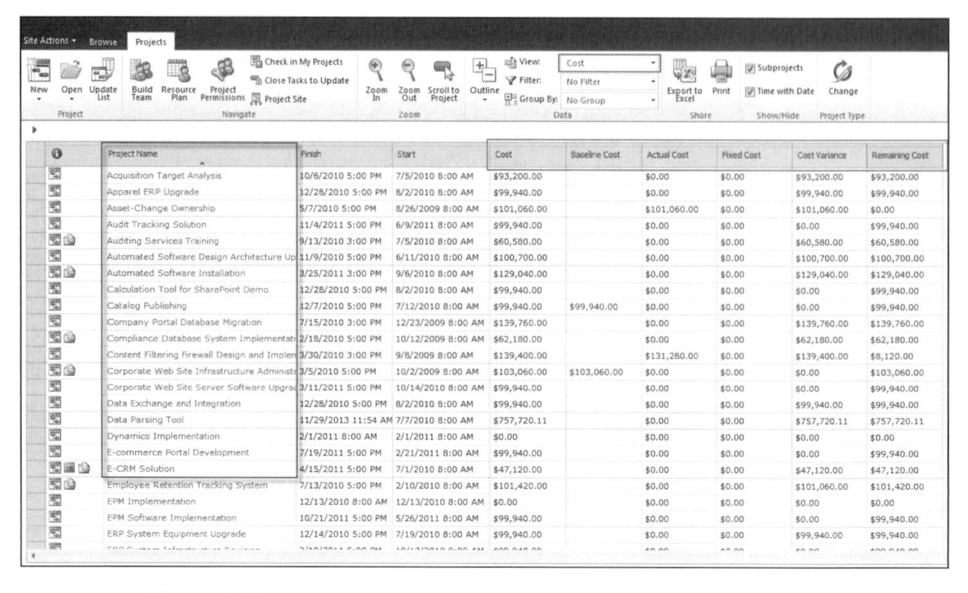

Project Name	Finish	Start	Cost	Baseline Cost	Actual Cost	Fixed Cost	Cost Variance	Remaining Cost
Acquisition Target Analysis	10/6/2010 5:00 PM	7/5/2010 8:00 AM	$93,200.00		$0.00	$0.00	$93,200.00	$93,200.00
Apparel ERP Upgrade	12/28/2010 5:00 PM	8/2/2010 8:00 AM	$99,940.00		$0.00	$0.00	$99,940.00	$99,940.00
Asset-Change Ownership	5/7/2010 5:00 PM	8/26/2009 8:00 AM	$101,060.00		$101,060.00	$0.00	$101,060.00	$0.00
Audit Tracking Solution	11/4/2011 5:00 PM	6/9/2011 8:00 AM	$99,940.00		$0.00	$0.00	$0.00	$99,940.00
Auditing Services Training	9/13/2010 3:00 PM	7/5/2010 8:00 AM	$60,580.00		$0.00	$0.00	$60,580.00	$60,580.00
Automated Software Design Architecture Up	11/9/2010 5:00 PM	6/11/2010 8:00 AM	$100,700.00		$0.00	$0.00	$100,700.00	$100,700.00
Automated Software Installation	3/25/2011 3:00 PM	9/6/2010 8:00 AM	$129,040.00		$0.00	$0.00	$129,040.00	$129,040.00
Calculation Tool for SharePoint Demo	12/28/2010 5:00 PM	8/2/2010 8:00 AM	$99,940.00		$0.00	$0.00	$0.00	$99,940.00
Catalog Publishing	12/7/2010 5:00 PM	7/12/2010 8:00 AM	$99,940.00	$99,940.00	$0.00	$0.00	$0.00	$99,940.00
Company Portal Database Migration	7/15/2010 3:00 PM	12/23/2009 8:00 AM	$139,760.00		$0.00	$0.00	$139,760.00	$139,760.00
Compliance Database System Implementati	2/18/2010 5:00 PM	10/12/2009 8:00 AM	$62,180.00		$0.00	$0.00	$62,180.00	$62,180.00
Content Filtering Firewall Design and Implen	3/30/2010 3:00 PM	9/8/2009 8:00 AM	$139,400.00		$131,280.00	$0.00	$139,400.00	$8,120.00
Corporate Web Site Infrastructure Administ	3/5/2010 5:00 PM	10/2/2009 8:00 AM	$103,060.00	$103,060.00	$0.00	$0.00	$0.00	$103,060.00
Corporate Web Site Server Software Upgrad	3/11/2011 5:00 PM	10/14/2010 8:00 AM	$99,940.00		$0.00	$0.00	$0.00	$99,940.00
Data Exchange and Integration	12/28/2010 5:00 PM	8/2/2010 8:00 AM	$99,940.00		$0.00	$0.00	$99,940.00	$99,940.00
Data Parsing Tool	11/29/2013 11:54 AM	7/7/2010 8:00 AM	$757,720.11		$0.00	$0.00	$757,720.11	$757,720.11
Dynamics Implementation	2/1/2011 8:00 AM	2/1/2011 8:00 AM	$0.00		$0.00	$0.00	$0.00	$0.00
E-commerce Portal Development	7/19/2011 5:00 PM	2/21/2011 8:00 AM	$99,940.00		$0.00	$0.00	$0.00	$99,940.00
E-CRM Solution	4/15/2011 5:00 PM	7/1/2010 8:00 AM	$47,120.00		$0.00	$0.00	$47,120.00	$47,120.00
Employee Retention Tracking System	7/13/2010 5:00 PM	2/10/2010 8:00 AM	$101,420.00		$0.00	$0.00	$101,060.00	$101,420.00
EPM Implementation	12/13/2010 8:00 AM	12/13/2010 8:00 AM	$0.00		$0.00	$0.00	$0.00	$0.00
EPM Software Implementation	10/21/2011 5:00 PM	5/26/2011 8:00 AM	$99,940.00		$0.00	$0.00	$0.00	$99,940.00
ERP System Equipment Upgrade	12/14/2010 5:00 PM	7/19/2010 8:00 AM	$99,940.00		$0.00	$0.00	$99,940.00	$99,940.00

Figure 4.24 EPM Project Costs Summary view Source: Advisicon

Figure 4.24 outlines an EPM project costs summary view of all the projects from across the enterprise. This summary facilitates roll-ups of individual project performance data, providing more timely access to financial, resource, and schedule information. Additional custom views can be configured easily.

End users can also take advantage of all the reporting tools included in the Microsoft BI platform, including Excel Services, PerformancePoint Services, Visio Services, PowerPivot for Excel, and SQL Reporting Services (TechNet 2011).

Options to Integrate with Microsoft Dynamics

The next-generation ERP architecture will be provided by Microsoft Dynamics AX 2012, which promises to deliver:

- **A model driven layered architecture** that will accelerate the application development process, providing the ability to build solutions more quickly, with less coding, and deliver solutions more quickly.
- **Unified natural models.** A key focus of the Microsoft Dynamics AX 2012 architecture is to facilitate flexible organization and business process models. These models can be set up when implemented for an organization and, more important, can be kept up to date and support changes in business requirements without having to modify the software source code or even implement different ERP systems (MSDN 2011).

EPM and ERP Interoperability

Interoperability between ERP and other applications, such as EPM, should be a key requirement for an organization's future-state ERP architecture.

Microsoft Dynamics AX 2012 offers productivity and familiarity in its design through interoperability with applications such as Project Server and SharePoint, thus extending the reach of the ERP solution within an organization.

The value of using an integrated approach is that:

- Most financial management systems offer minimal PM capabilities (i.e., they are often milestone based with little capacity planning or resource management functionality). Concepts such as critical path, earned value, or resource leveling are foreign concepts in the financial management arena.
- PPM solutions focus on work and resource management and approach financial management from an individual project costing perspective. PM capabilities typically compare a hierarchy of projects to a cost center, bill of materials, or other classification.
- This integrated solution allows the organization to realize benefits from both the PM capabilities in Project Server 2010 and the financial management capabilities in Microsoft Dynamics.
- Integration partners and independent software vendors can create customized business solutions using both Project Server and Microsoft Dynamics AX.
- The integration of Microsoft Dynamics AX and SL 2012 with Microsoft Project Server 2010 allows the user to create projects in either system and synchronize select data across both applications. This synchronization provides the ability to draw on both the core PM capabilities of Project Server and the financial management capabilities in Microsoft Dynamics AX.
- The architecture of the Project Server 2010 and Microsoft Dynamics AX integration works on the basis of a synchronization service that bridges the two and synchronizes projects and workers between the two servers.
- Multiple ways can be used to deploy the necessary components with Project Server 2010 and Microsoft Dynamics AX integration. Figure 4.25 presents a high-level view of the integration between Microsoft Dynamics AX 2012 and Project Server 2010 using a three-server topology (Ashtiani, Aggag, and Buenafe 2011).

Third-Party Data Exchange Solutions

In addition to integrating Project Server with Dynamics AX/SL (Microsoft ERP/Accounting Systems), third-party provider solutions also enable the bidirectional data exchange between Microsoft Project Server and other ERP solutions. These gateway software interfaces ensure the complete consistency of all mapped data in each of the participating systems.

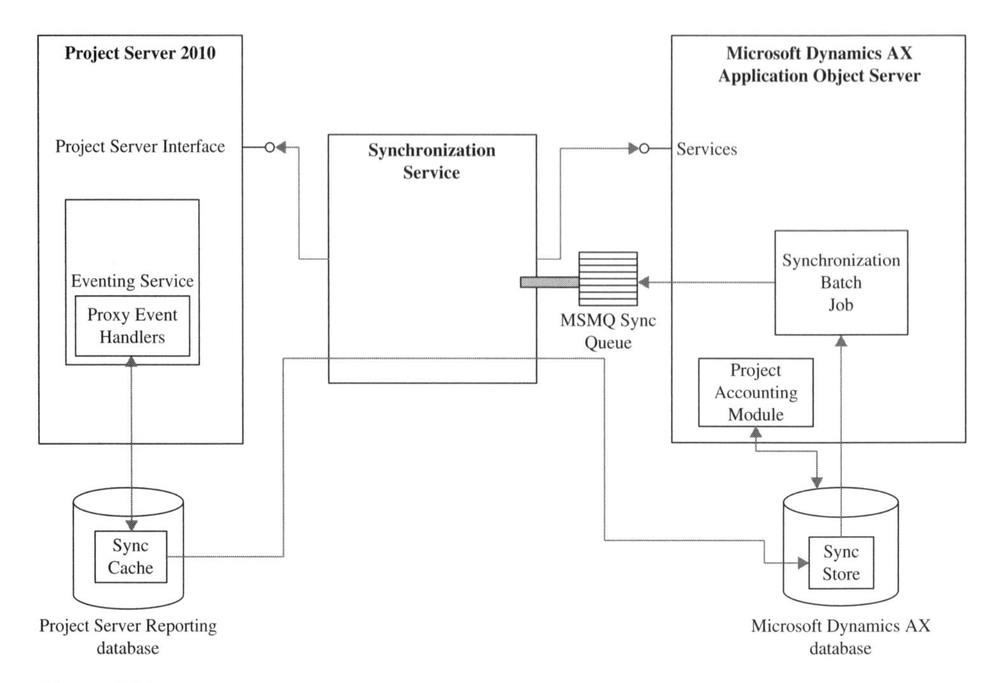

Figure 4.25 Integration between Microsoft Dynamics AX 2012 and Project Server 2010
Source: Ashtiani, Aggag, and Buenafe, 2011

As Project Server is gaining visibility as an enterprise application and a key PPM solution within companies, customers are realizing that Project Server must integrate with existing systems, starting typically with ERPERP systems, such as SAP, Oracle, and PeopleSoft.

IMPORTANT CONCEPTS COVERED IN THIS CHAPTER

This chapter discussed how the Project Server 2010 platform is more like an ERP system and should be treated as such.

Key Summary Points

Key summary points are highlighted here to remind the reader of some of the vital points covered in this chapter.

- The Project Server 2010 platform is more like an ERP system and should be treated as such.
- Project Server 2010 is built on SharePoint Server 2010 technology and is integrated with Microsoft Office 2010 and Exchange Server 2010.

- Project Server 2010 provides a powerful and familiar work management platform.
- Organizations must learn the value of EPM.
- IT organizations, PMOs, and other organizations need to learn how the communication and workflow capabilities native within Project Server 2010 can address the PPM requirements of small and large businesses.
- The PPM lifecycles requires communication and workflow capabilities that are inherent in Project Server 2010.
- Working in the Project Server 2010 environment enables end users to access information via the Web. With Project, you are not limited to a stand-alone desktop tool.
- An integrated tool set is required to cover the complete PPM lifecycle.

REFERENCES

Arbutus Solutions. 2011. "White Paper: Microsoft Project Server 2010—A Financial Management Solution for Projects, Programs, and Portfolios." Accessed January 20, 2012, at http://technet.microsoft.com/en-us/library/hh144780.aspx.

Ashtiani, Babak, Khalid Aggag, and Catcat Buenafe. 2011. "Microsoft Dynamics® AX 2012 White Paper: Microsoft Project Server 2010 Integration." Accessed January 16, 2012, at www.microsoft.com/download/en/details.aspx?id=15671.

CodePlex. 2012. "Find Projects and Downloads—Project Server." Accessed January 20, 2012, at www.codeplex.com/site/search?query=%22project%20server%22&sortBy=Relevance&licenses=|&ac=3.

The Economist Intelligence Unit. 2008. "The Role of Trust in Business Collaboration—An Economist Intelligence Unit Briefing Paper Sponsored by Cisco Systems." Accessed January 17, 2012, from http://graphics.eiu.com/upload/cisco_trust.pdf.

European Spreadsheet Risk Interest Group. 2012. "Basic Research." Accessed January 13, 2012, at www.eusprig.org/basic-research.htm.

Everson, Boris and Norman Nicolson. 2008. "Topic Overview: Business Intelligence—An Information Workplace Report." Available at www.forrester.com/Topic+Overview+Business+Intelligence/−/E-RES39218?objectid=RES39218.

Fiessinger, Christophe. 2010. "Integrating EPM with Line of Business Solutions Such as ERP Systems." Accessed January 13, 2012 at http://blogs.msdn.com/b/chrisfie/archive/2007/04/20/integrating-epm-2007-with-line-of-business-solutions-such-as-erp-systems.aspx.

Henschen, Doug. 2010. "Review: SharePoint 2010 Gets Overdue Upgrades." *Information Week*, May 10, 2010. Accessed January 13, 2012, at http://informationweek.com/news/windows/reviews/224701321.

Hillson D. A. 2009. *Managing Risk in Projects*. Farnham, UK: Gower.

International Federation of Accountants. 2004. "Enterprise Governance Getting the Balance Right." Accessed January 20, 2012 at www.cimaglobal.com/Documents/ImportedDocuments/tech_execrep_enterprise_governance_getting_the_balance_right_feb_2004.pdf.

Ivanenko, Dmitri. 2009. "Lessons Learned: Voices on Project Management. Independent Ideas and Insights by and for Project Practitioners," Accessed January 23, 2012, at http://blogs.pmi.org/blog/voices_on_project_management/2009/03/lessons-learned.html.

Levine, Harvey A. 2005. *Project Portfolio Management: A Practical Guide to Selecting Projects, Managing Portfolios, and Maximizing Benefits*. San Francisco, CA: Jossey-Bass.

MSDN. 2011. "Project 2010 SDK Documentation," Accessed January 4, 2012, from http://msdn.microsoft.com/en-us/library/office/ms512767.aspx.

O'Cull, Heather. 2009. "Project 2010: Introducing Demand Management." Official blog of the Microsoft Project product team. Accessed January 16, 2011, from http://blogs.msdn.com/b/project/archive/2009/11/13/project-2010-introducing-demand-management.aspx.

Shaker, Kareem. "Rolling Wave Planning Using Microsoft Project 2010," 2010. Kareem's Blog. Accessed January 16, 2012, from http://kareemshaker.com/project-management/rolling-wave-planning-using-microsoft-project-2010/.

TechNet. "Document Management Planning (SharePoint Server 2010)," 2010. Accessed January 18, 2012, from http://technet.microsoft.com/en-us/library/cc263266.aspx.

———. "Project Server 2010 with SharePoint Server 2010 Architecture (Overview)," 2011. Accessed January 5, 2012, from http://technet.microsoft.com/en-us/library/ff686783.aspx.

CHAPTER 5

END USERS' CRITICAL SUCCESS FACTORS: USING MS PROJECT 2010

IN THIS CHAPTER

Here we examine, from an end user's perspective, those critical success factors (CSFs) that can have a significant impact on the appropriate and effective use of Microsoft Project. CSFs are used in project management (PM) to identify key areas that need to be considered to make a project succeed.

We explore key factors associated with the effective use of PM tools in small businesses and large enterprises, from the individual desktop client through to the departmental and enterprise server. Scalability, configurability, ease of use, integration, and collaboration are five key factors that will positively impact organizational effectiveness and end user satisfaction.

What You Will Learn

- How organizations learn how to get things done, using an effective set of processes, tools, and technologies
- How PM is effective in small business and the enterprise
- The impact of organizational maturity on PM success
- How to initiate and manage projects using the Microsoft Project client
- The key steps for effective scheduling and resource planning using the Microsoft Project client
- Ways to perform collaborative work management through integration with SharePoint
- How agile work management and Project Server worlds are coming together
- How to integrate project-specific information using Project Professional
- Why the moving target is now feasible leveraging Project Server
- How customizing is as easy as right click/left click using the new ribbon
- About Project Frontstage and Backstage

PROJECT MANAGEMENT IN SMALL BUSINESS AND THE ENTERPRISE

Organizations of all sizes utilize some form of work, resource, or PM methodology and tool set. PM methods and disciplines vary widely: from ad hoc, reactive approaches to much more advanced PM or product line management (PLM) approaches. PM tools also range widely from individual desktop clients to centralized server-based software solutions.

PM tools can also be stand-alone or highly integrated with organizations' enterprise resource planning, human resources, customer relationship management, or other line of business systems. No one size fits all and some organizations may have a little bit of everything with various solutions and approaches also being acquired through organizational acquisition and merger.

For the purposes of brevity and inclusivity, any reference to PM is meant to include all three management practices: project, program. and portfolio management. These are sometimes referred to as project portfolio management (PPM).

Project Management and Organizational Maturity

As organizations improve in their understanding of how to get things done, an effective set of processes, tools, and technologies are typically implemented, usually resulting in improved organizational efficiency. This move from a task-oriented (ad hoc) approach to a higher level of work management rigor requires a different approach to managing the work, resources, and other project metadata of the organization.

A wide array of various work and PM methods and best practices are available: from a more agile or scrum-based software development technique to the more methodologically driven approach, such as PM or system development life cycle (SDLC).

Portfolio governance and lifecycle management enable organizations to define processes that synchronize the efforts of distributed teams to consistently create the best possible products (Cooper, Edgett, and Kleinschmidt 1998), capture greater market share, and increase customer satisfaction.

Typically there are five key phases to initiate and then manage a project through its lifecycle. You may recognize these as the Project Management Institute PMBOK® Process Areas (PMI 2008) illustrated in Figure 5.1.

1. Initiation phase
2. Planning phase
3. Executing phase
4. Monitoring and controlling phase
5. Closing phase

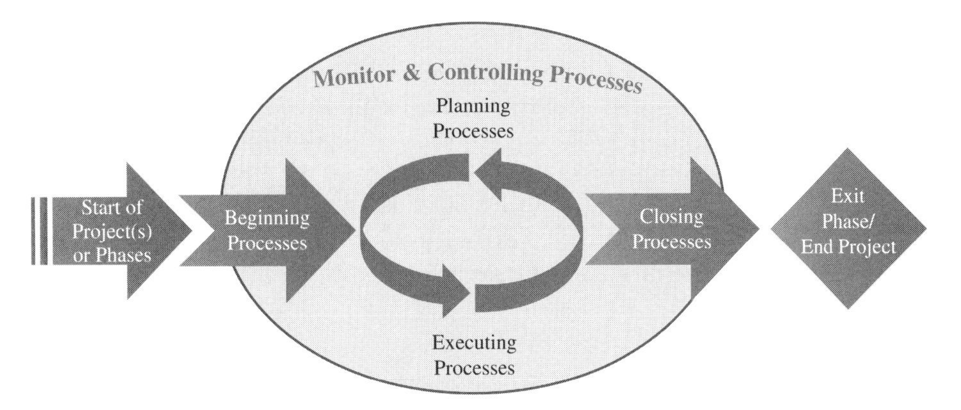

Figure 5.1 Project Management Process Areas Source: Advisicon

Initiation Phase

The initiation phase is performed when defining a new project (or a new phase of an existing project) by creating an outline of the work to be performed and obtaining authorization to initiate the project.

Because Project Server actually can start in the pre-sales or pre-initiation phase via the activities, documents collected, and information tracked even before a project is "approved," understanding the key artifacts (i.e., documents) and metadata associated with taking a project from pre-planning or a proposal phase to a full-blown project is now more important than ever.

For example, a manager may not want to invest too much time in creating an elaborate schedule yet needs a high-level resource proposal or a resource plan to allocate possible resource needs that could be weighed against the overall existing work in the project portfolio. The example continues with a person still planning and organizing information and documentation associated with what the project will be. Doing this requires collecting, capturing, and storing all related artifacts in a central repository or possibly a SharePoint site. Microsoft Project Professional and Microsoft's full PPM solution can provide all of these capabilities to a project manager (or a person planning a project). The great thing is that every action can be scalable, based on the level of information and details needed at that point in time (phase or stage).

One of the key artifacts in successful PM during the initiation phase is the project charter. One of the primary elements of the project charter is a high-level schedule of the project. At this stage of the project lifecycle, there may be little detailed information in terms of the work packages or tasks to be performed. As a result, this initial schedule is likely to contain high-level information just to sufficiently describe the key deliverables and target dates or milestones and to provide some statement of the resource capacity required to produce the deliverables of the project within the allocated time frame.

Other key outputs of the initiation phase typically include a high-level (milestone) schedule, a high-level cost estimate, an outline of the project team, and role definitions for each project team member.

A project manager is also assigned at this time and provided with the authority to deliver the project's product on time and on budget, according to a set of requirements that outline specific functionality and deliverables.

Planning Phase

Detailed planning is required to establish the scope of the project, refine the objectives, and define the course of action required to attain the objectives that the project was undertaken to achieve. The planning phase includes activities both to do detailed planning of activities and to perform resource assignment.

Two perspectives must be considered: a portfolio perspective and a project perspective. From a portfolio perspective, activities revolve around overall capacity planning and maintenance of the project portfolio delivery schedule. From a project perspective, planning involves detailed project planning and assignment of named resources to a project.

Project Server provides a fully integrated Portfolio PM solution that supports both perspectives:

1. **Portfolio management planning and analysis tools.** These include demand capture, capacity planning, business alignment and prioritization, constraint analysis, and reporting.
2. **Program and PM planning tools.** These include detailed scheduling, resource assignment, task status and timesheets, issue and risk management, collaboration, business intelligence, and reporting.

Key outputs from phe Planning stage include the project plan and the schedule, which are two distinct elements. The schedule is really a subcomponent of the project plan detailing the timeline for the deliverables and is (hopefully) resource and cost loaded to ensure an accurate depiction of who is going to deliver what by when. Schedules do not necessarily need to be resourced at this point; however, without resources, they are only a depiction of work elements and their associated milestones. Those milestones may not be attainable due to resource constraints.

As an organization's business environment fluctuates, the delivery of the project portfolio will also be affected, resulting in potential increases and decreases in scope, schedule, and budget. Changes in market forces can also result in new priorities. Ongoing reoptimization of the project (or the portfolio of projects) is necessary to ensure alignment with the organization's strategy. Replanning likely will become a necessary factor to consider.

Execution Phase

This is the stage where tasks and activities, which are defined in the detailed project plan, are executed to satisfy the project specifications (i.e., where the "product" of the project is produced).

The primary outputs from the execution phase are the deliverables (i.e., the products of the project). Results from the project's execution phase may also require replanning and rebaselining of the project schedule. Changes may be introduced due to resource productivity and availability, unanticipated risks, or changes in scope.

During the execution phase, project managers and team members progress their work using either the Microsoft Project Web Application (PWA) or the Project Professional client, via the My Tasks view or time-sheeting capabilities, or directly to the project schedule, depending on the approach used to manage the project.

Issues might be encountered during this phase that will need to be resolved. Potential risks identified during the planning stage might occur and require the innovation of a planned mitigation. The dynamic nature of the project execution phase needs to be carefully monitored and controlled to ensure that all of this uncertainty will be managed.

Monitoring and Controlling Phase

Successful projects require that we track, review, and regulate the progress and performance of the project; identify any areas in which changes to the plan are required; and initiate the corresponding changes.

This phase includes key activities to track the overall progress of projects. During this phase, there are three primary areas of focus:

1. Schedule tracking and schedule forecasting take place on an ongoing basis throughout the project. Schedule tracking is performed to ensure that the deliverables (i.e., products) of the project are occurring in a timely manner. Schedule analysis techniques are often utilized to determine corrective action to the schedule (e.g., in the event of schedule slippage). In some industries (e.g., construction), in litigation and damages might result when projects don't go according to schedule.

   The Project desktop client, both as a stand-alone product and when connected to Project Server, provides a powerful and dynamic means of tracking team progress to allow project managers to forecast timeliness of the deliverables and ability to deliver the required functionality and to project an estimated cost.

2. Resource management primarily occurs at the project level and involves the onboarding and offboarding of project team members and the assignment of resources to tasks.

   Project Server provides a centralized resource pool for managing the assignment and tracking critical team resources.
3. Determining actual project costs and forecasting future project costs is required to ensure that a project is working within its financial boundaries. At the portfolio level, cost analysis can contribute to tracking overall performance of a group of projects.

By now it should be clear that one of the critical capabilities of Project Server is its ability to continuously forecast the project end date and cost. The uncertainty of some estimates in project planning, coupled with the dynamic nature of projects, establishes a critical case for a solution that can track people resources and hard costs according to the agreed-on plan. Federal and state governments are now demanding the use of earned value management (*Federal Register* 2011), a performance-based tool that gives agency managers an early warning of potential cost overruns and schedule delays during the execution of their projects.

The Microsoft Project Server solution makes it much easier to comply with Office of Management and Budget Circular A-11 (Executive Office of the President Office of Management and Budget 2011) and American National Standards Institute/Enterprise Application Integration (ANSI/EAI) Standard 748-A (National Defense Industrial Association 2005).

Closing Phase

Once the project has completed, the closing stage is where we finalize all activities across all PM process areas to formally close the project. Key outputs of the closing stage typically include administrative and contractual closure, along with lessons learned and all other project related artifacts (e.g., schedules, status reports, risks and issues logs, etc.).

Microsoft Project schedules are updated throughout the project lifecycle, so there is little to do other than archiving a read-only copy for project audit purposes. Completed project schedules also make great templates for estimating projects that are similar in scope or approach.

Project Management in Small Business Using Microsoft Project Desktop Client

Successful PM requires tools that are easy to learn and use. This requirement presents a challenge to information technology (IT) departments responsible for selecting PM tools that can support better management of small projects.

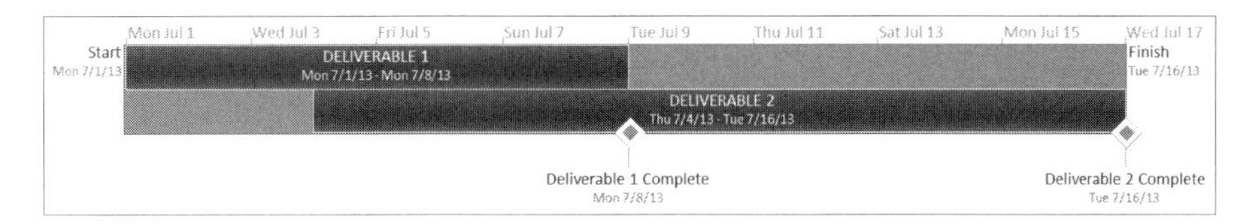

Figure 5.2 Milestones Chart with Microsoft Project Timeline View Source: Advisicon

The capabilities of the Microsoft Project Desktop client may meet the majority of PM needs of a small business.

Project Desktop

Now that we understand the PM lifecycle, let's take a look at how we can effectively use the Microsoft project client in small business.

First, we need to break the scope of the project into specific deliverables or work packages to provide a more outcomes-based approach to the project (i.e., an approach that considers the key deliverables or outcomes that need to be delivered to adequately meet the requirements of the project).

A milestone (Level 1) chart is a good first step in developing any project or program, regardless of the size or complexity of the initiative. Figure 5.2 is an example of a milestone chart produced using the Microsoft Project Timeline View.

A properly defined and maintained schedule (e.g., all tasks are connected into an activity network, there is a minimal use of constraints, milestones have been defined, etc.) can then be used as an effective "Forecast" (Uyttewaal 2010) to assist the project manager and other project stakeholders with a forward-looking view of the remaining work for a project. Figure 5.3 illustrates an example activity network using the Microsoft Project Gantt Chart.

The forecast schedule can then be used to determine if the work packages and deliverables are most likely to land according to schedule.

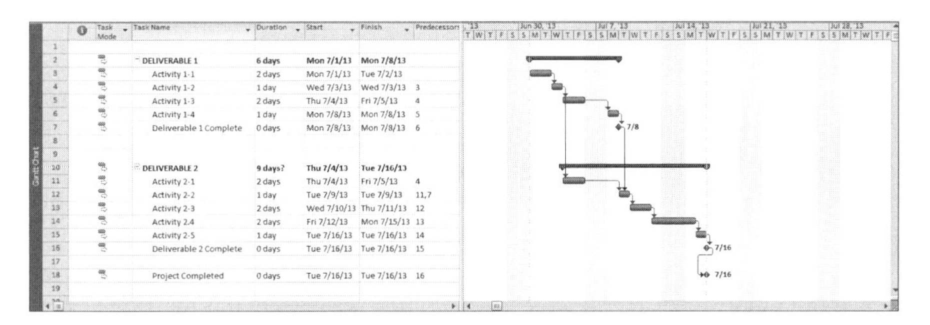

	Task Name	Duration	Start	Finish	Predecessors
1					
2	**DELIVERABLE 1**	**6 days**	**Mon 7/1/13**	**Mon 7/8/13**	
3	Activity 1-1	2 days	Mon 7/1/13	Tue 7/2/13	
4	Activity 1-2	1 day	Wed 7/3/13	Wed 7/3/13	3
5	Activity 1-3	2 days	Thu 7/4/13	Fri 7/5/13	4
6	Activity 1-4	1 day	Mon 7/8/13	Mon 7/8/13	5
7	Deliverable 1 Complete	0 days	Mon 7/8/13	Mon 7/8/13	6
8					
9					
10	**DELIVERABLE 2**	**9 days?**	**Thu 7/4/13**	**Tue 7/16/13**	
11	Activity 2-1	2 days	Thu 7/4/13	Fri 7/5/13	4
12	Activity 2-2	1 day	Tue 7/9/13	Tue 7/9/13	11,7
13	Activity 2-3	2 days	Wed 7/10/13	Thu 7/11/13	12
14	Activity 2.4	2 days	Fri 7/12/13	Mon 7/15/13	13
15	Activity 2-5	1 day	Tue 7/16/13	Tue 7/16/13	14
16	Deliverable 2 Complete	0 days	Tue 7/16/13	Tue 7/16/13	15
17					
18	Project Completed	0 days	Tue 7/16/13	Tue 7/16/13	16
19					

Figure 5.3 Activity Network with Microsoft Project Gantt Chart Source: Advisicon

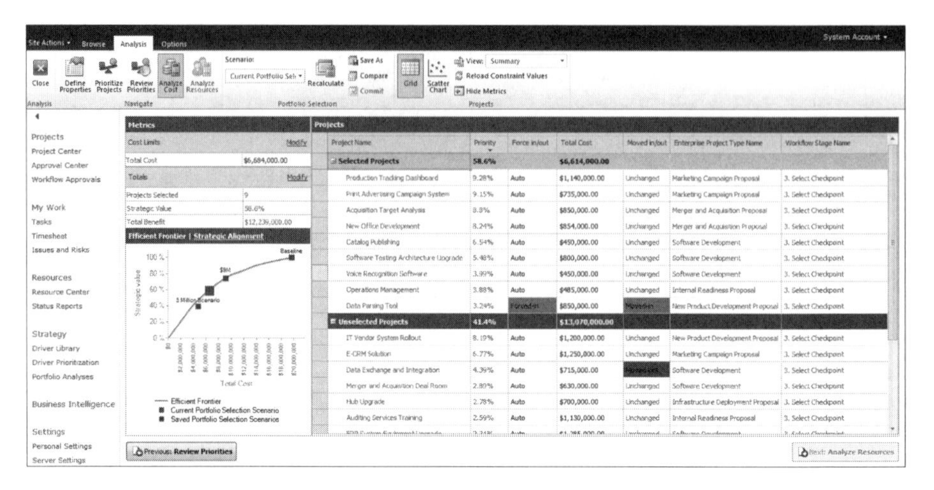

Figure 5.4 Project Server Portfolio view Source: Advisicon

This approach lends itself to the dynamic nature of projects and works much like a car GPS: It provides a constantly updated picture of where the project is headed while allowing manual course correction for unforeseen circumstances.

Enterprise Project Portfolio Management with Project Server on SharePoint

Microsoft Project Server 2010 brings robust project, program, and portfolio management together with extensive collaboration capabilities. Project Server 2010 is fundamentally a business application that is built on top of SharePoint 2010, and SharePoint is hosted centrally so that the information is available throughout the enterprise.

Portfolio Management

Project Server 2010 also unifies project, program, and portfolio management to help organizations align resources and investments with business priorities, gain control across all types of work, and visualize performance using powerful dashboards. Figure 5.4 provides a view of the portfolio analysis capabilities of Project Server. Cost and resource constraint-analysis tools provide the ability for organizations to perform what-if scenarios to determine prioritization and optimal selection of projects that will go forward for detailed planning and execution.

A common challenge today, which often is considered a resource management issue, is the work or activities that are being managed outside of a centralized work or resource planning system. This can result in the overallocation of key resources since their actual availability is not accurately managed or measured. Portfolio PM is a critical first step toward solving this issue. If properly

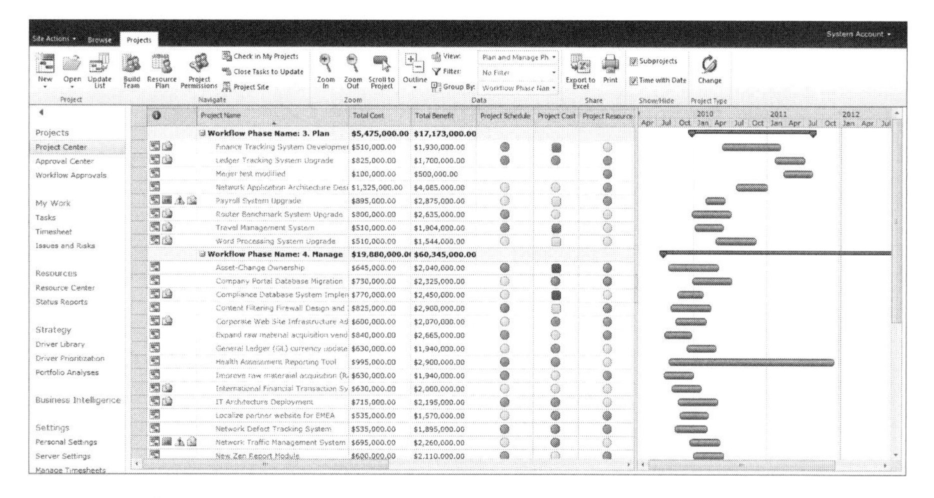

Figure 5.5 Project Server, Project Center view Source: Advisicon

implemented, it can become a control point for all work demand and resource allocation for the enterprise.

Project Management

Microsoft Project 2010 Pro desktop and the Microsoft Project Server 2010 environment both offer a fully integrated and collaborative PM information system.

Figures 5.5 and 5.6 illustrate the powerful Project and Schedule views that are easily accessible through the Web. This is great because Web-based tools accessible via intranet, extranet, or Internet provide companies with availability, accessability, and redundancy advantages over desktop-only solutions. Collaboration and communication are easier and offer a flexible, transparent PM solution.

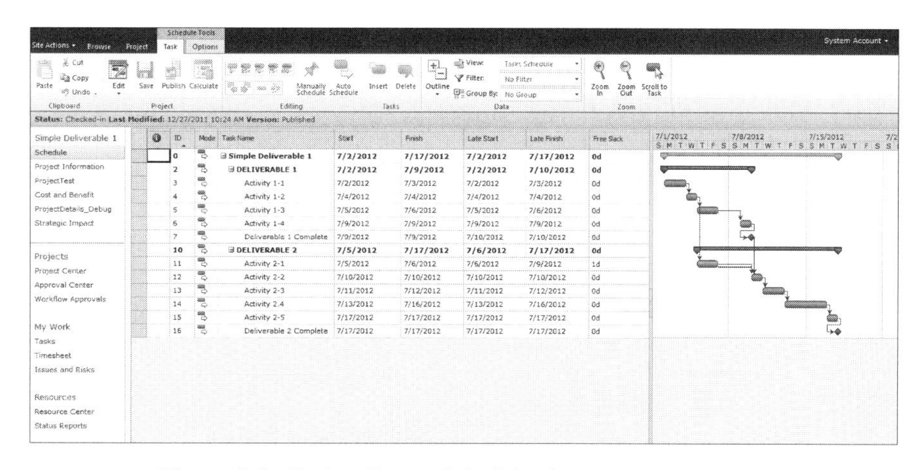

Figure 5.6 Project Server, Schedule view Source: Advisicon

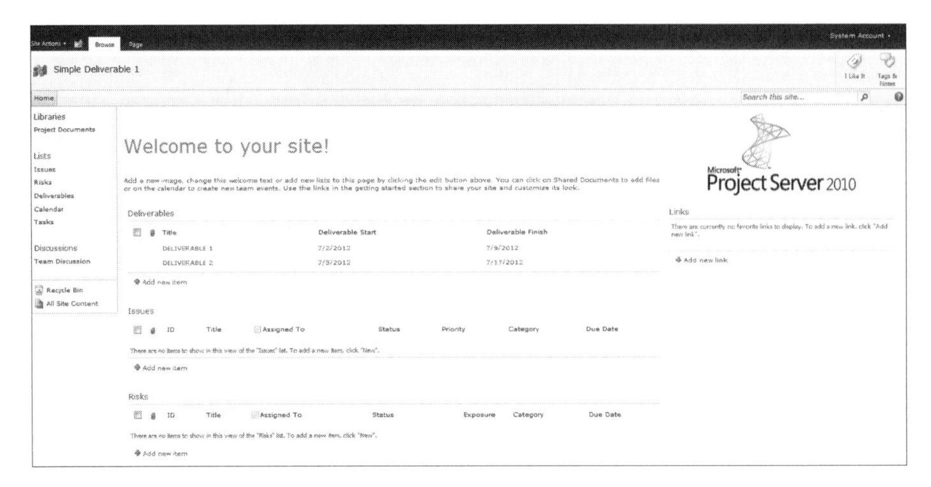

Figure 5.7 Project Server, Team site Source: Advisicon

Project Server provides a number of key SharePoint functions. These include:

- Project Sites and Workspaces
- Version control
- Task synchronization between Project Server 2010 and SharePoint Server 2010
- Governance workflows
- Tagging
- Wikis
- Discussion boards
- Connectivity with remote team members via the Web and mobile devices
- Business intelligence for dynamic reporting

PM capacities, such as risk and issue management, status reports, deliverables, team discussions, tasks, and calendars, are built into this enterprise PM solution.

Figure 5.7 illustrates the collaboration functionality of Project Server, through its integration with SharePoint Team Sites, to provide a fully centralized PM portal for managing projects.

Project managers and team members can now manage their work and deliverables using an integrated scheduling and collaboration solution.

Critical Success Factors

Here are the CSFs that we have learned about so far for PM in small business and the enterprise:

1. To be an effective organization, we need to understand and employ a PM lifecycle.
2. To attain the objectives that a project was undertaken to achieve, detailed planning is required to establish the scope of the project and define the specific course of action required to produce the required deliverables or the project product.
3. To regulate the progress and performance of the project, we must track, review, and identify areas in which changes to the plan are required and then initiate the corresponding changes.
4. To attain higher levels of successful PM, project managers need tools that are easy to learn and use.
5. Project Server 2010 unifies project, program, and portfolio management to align resources and investments with business priorities, gain control across all types of work, and visualize performance.

INITIATING AND MANAGING PROJECTS USING THE MICROSOFT PROJECT DESKTOP CLIENT

More and more organizations are outgrowing their basic tools, methods, and processes for managing work and resources. Typical tools in widespread use today include Microsoft Excel, Outlook, or Word. Spreadsheets are an excellent way to manage information at the task level; however, when you need to see the impact of one activity on another the need to move toward PM becomes increasingly apparent.

As the number of tasks and interactions increases, so does the complexity level of the scope, budget (or resources), time, and quality of the end product. When organizations move from a basic way of managing work to a more sophisticated level of rigor, they typically utilize PM methods, tools and technologies, and best practices.

Effective Work Scheduling and Resource Planning with Microsoft Project Desktop

Let's examine the key steps required to do effective scheduling and resource planning using Microsoft Project 2010 desktop.

Work Planning and Scheduling

Figure 5.8 illustrates the new Microsoft Project user-controlled "manually scheduled tasks" feature, which allows freeform entry of task data. Manually scheduled tasks provide users the ability to set the duration and the start and finish dates for a task, without any adjustment by the scheduling engine. Team members can

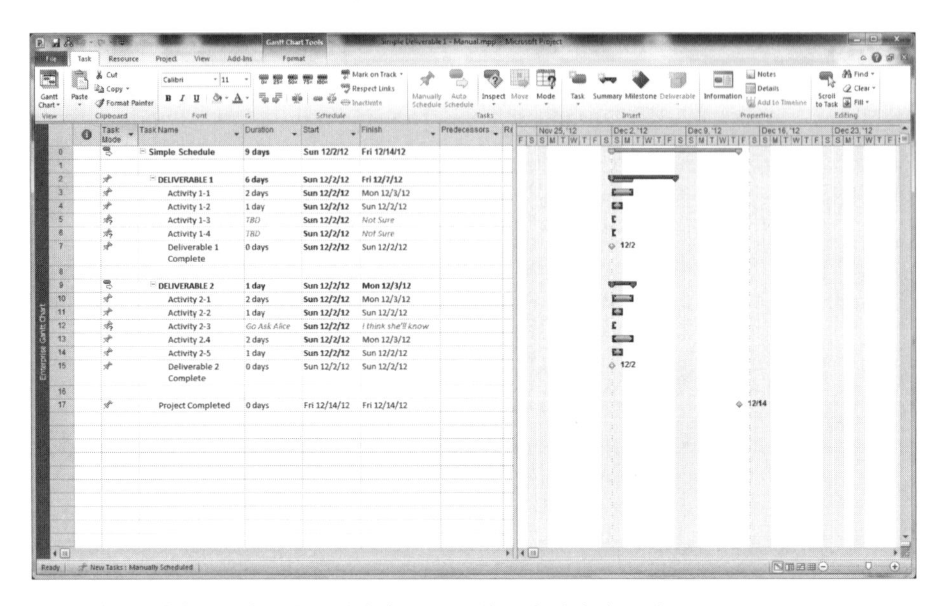

Figure 5.8 Project Schedule in Manually scheduled mode Source: Advisicon

place a manually scheduled task anywhere in their schedules, and Project 2010 will not move it.

Project managers who were accustomed to automatic scheduling with past versions of Project can turn the manual scheduling feature off for specific tasks or the entire schedule. Notice that some of the date fields in Figure 5.8 have text entered in them. This was not permitted in releases prior to Microsoft Project 2010.

This new user-controlled scheduling capability allows project managers to work through planning before allowing the scheduling engine to take over.

Project 2010 also allows for a mix of automatic and manual scheduled tasks. A project option determines which mode for newly inserted tasks.

As Figure 5.9 illustrates, we actually can improve on our estimated completion date for the project by creating an activity network and switching to Auto Scheduled task mode (a three-day improvement on the manual schedule Finish date).

Dynamic Schedules

A project manager is responsible for ensuring that the products of the project (i.e., the deliverables) are delivered on schedule at a proposed cost. Given that the schedule is the planned use of resources agreed on by all the stakeholders of the project to complete the work, it is extremely important that it can be used to dynamically determine (i.e., forecast) whether we can complete the work according to the planned dates (i.e., milestones).

In order to use a schedule to dynamically forecast the future, four factors must be taken into consideration:

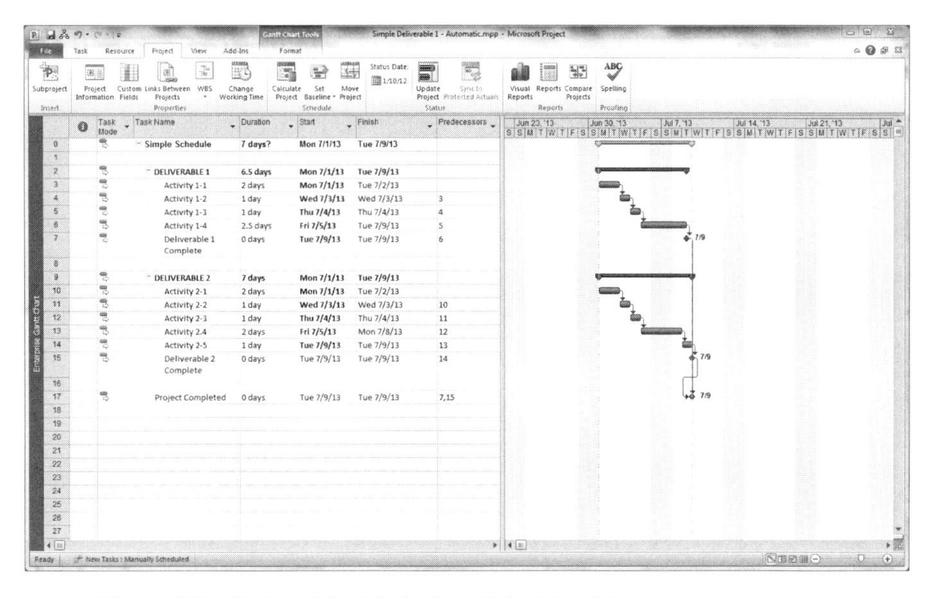

Figure 5.9 Project Schedule in Auto Scheduled Mode Source: Advisicon

1. All tasks must be connected into a "network" of activities. This allows Project to dynamically calculate end dates for activities and key milestones.

   Figure 5.10 illustrates the relationships between tasks and milestones. It is the best view for understanding the interconnected relationship and dynamic impacts of predecessor tasks on successor (dependant) activities.

   Only tasks that start at the beginning of the project or finish at the end can have open ends. This provides the scheduling software in Project with the ability to calculate start and finish dates for each of the tasks and milestones within the schedule.

2. There should be very few (if any) constraints in the schedule. Be careful to not set start and finish dates for auto-scheduled tasks (see Figure 5.11), as this will set "Start No Earlier" and "Finish No Earlier" constraints on those tasks.

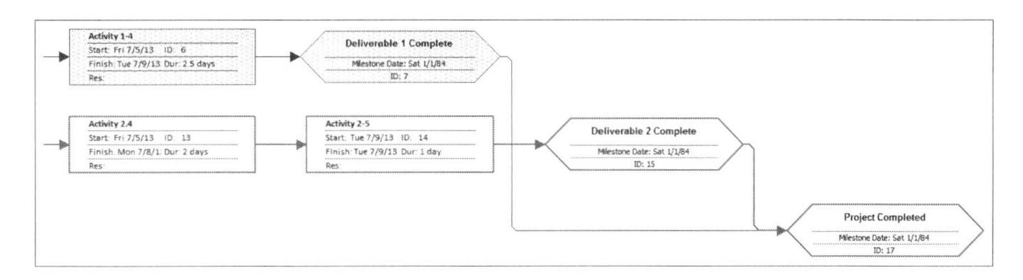

Figure 5.10 Project Network Diagram Source: Advisicon

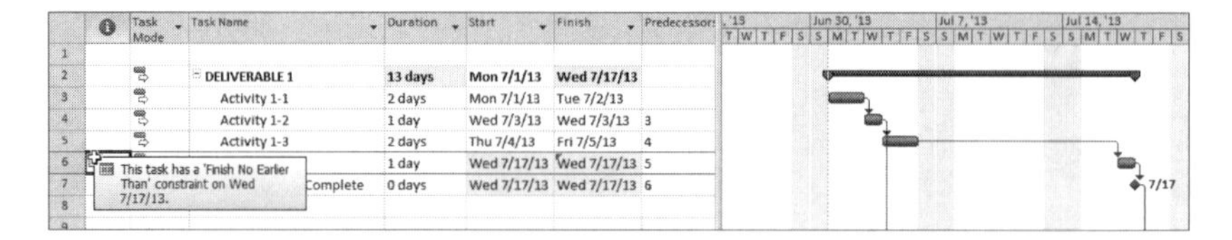

	Task Mode	Task Name	Duration	Start	Finish	Predecessors
1						
2		DELIVERABLE 1	13 days	Mon 7/1/13	Wed 7/17/13	
3		Activity 1-1	2 days	Mon 7/1/13	Tue 7/2/13	
4		Activity 1-2	1 day	Wed 7/3/13	Wed 7/3/13	3
5		Activity 1-3	2 days	Thu 7/4/13	Fri 7/5/13	4
6			1 day	Wed 7/17/13	Wed 7/17/13	5
7		Complete	0 days	Wed 7/17/13	Wed 7/17/13	6
8						

Figure 5.11 Project Finish-No-Earlier-Than Constraint Source: Advisicon

	Task Name	Duration	Start	Finish	Predecessors
2	DELIVERABLE 1	6 days	Mon 7/1/13	Mon 7/8/13	
3	Activity 1-1	2 days	Mon 7/1/13	Tue 7/2/13	
4	Activity 1-2	1 day	Wed 7/3/13	Wed 7/3/13	3
5	Activity 1-3	2 days	Thu 7/4/13	Fri 7/5/13	4
6	Activity 1-4	1 day	Mon 7/8/13	Mon 7/8/13	5
7	Deliverable 1 Complete	0 days	Mon 7/8/13	Mon 7/8/13	6

Figure 5.12 How to Delete a Constraint in Project Source: Advisicon

	Task Name	Duration	Start	Finish	Predecessors
5	Activity 1-3	2 days	Thu 7/4/13	Fri 7/5/13	4
6	Activity 1-4	1 day	Mon 7/8/13	Mon 7/8/13	5
7	Deliverable 1 Complete	0 days	Mon 7/8/13	Mon 7/8/13	6
8					
9					
10	DELIVERABLE 2	9 days?	Thu 7/4/13	Tue 7/16/13	
11	Activity 2-1	2 days	Thu 7/4/13	Fri 7/5/13	4
12	Activity 2-2	1 day	Tue 7/9/13	Tue 7/9/13	11,7

Figure 5.13 Linking Tasks in Project Source: Advisicon

To clear a constraint, simply select the Start or Finish date field and press the Delete key, as shown in Figure 5.12.

Constraints make it difficult for the Project scheduling engine to perform its job because they "constrain" the dates and the schedule.

3. Utilize a deliverable-based structure for your schedule, with five to ten key activities and a milestone. Link only the activities and the milestone and avoid connecting to the Summary bars.

   Figure 5.13 illustrates the relationship where linking to a summary activity the sub tasks can start earlier based upon their links. If the summary task had been linked Activity 2-1 would have had to start whenever the predecessor task to the summary task would have ended.

Microsoft Project Add-ins

There is a very useful add-in for Microsoft Project called WBS Chart Pro, available from Critical Tools. This utility displays a project using a tree-style diagram known as a work breakdown structure (WBS) chart. WBS charts display the structure of a project showing how the project is broken down into its summary and detail levels (see Figure 5.14).

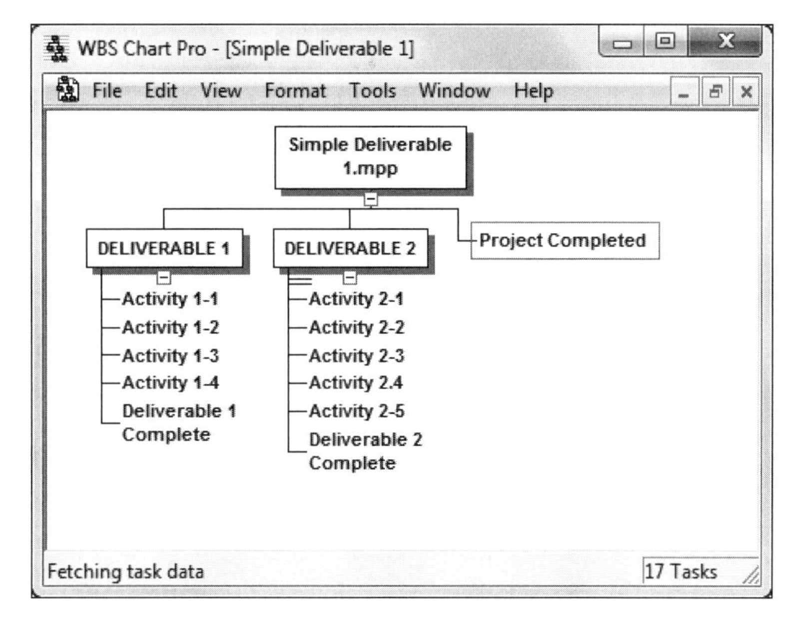

Figure 5.14 Critical Tools WBS Chart Pro Add-in for Project Desktop Source: Advisicon

Using a WBS chart is an alternate approach to planning and displaying a project than the more traditional Gantt view shown earlier.

Critical Tools also offers an add-in called PERT Chart EXPERT for Project, which is used to create PERT charts (also known as network charts, precedence diagrams, and logic diagrams). A PERT chart displays the tasks in a project along with the dependencies between these tasks (see Figure 5.15).

Both of these tools are easy to learn and use, and provide significant productivity during this critical working, planning, and scheduling phase of the PM lifecycle.

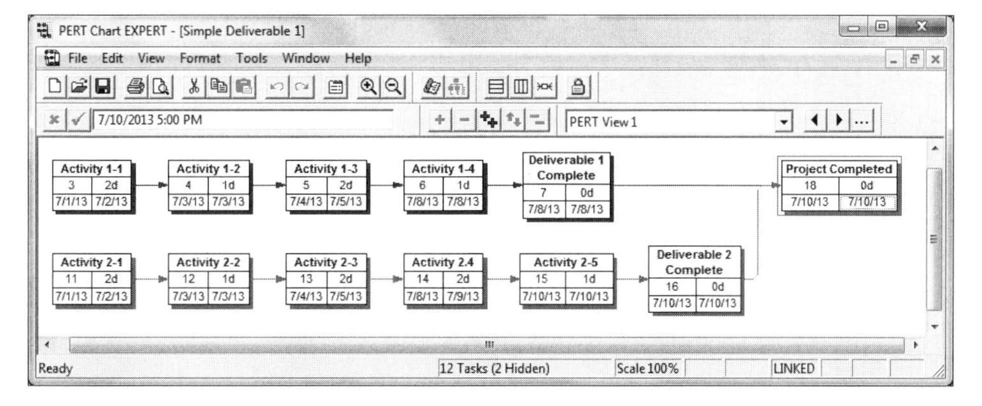

Figure 5.15 Critical Tools PERT Chart Expert Add-in for Project Desktop Source: Advisicon

Resource Planning

We now have a fairly good picture of the work that needs to be completed (and what order to complete it in) as defined by the activity network. The critical path method schedule that we have just completed is near perfect—that is, until we perform the next critical step.

The next step is to estimate the resource requirements (such as labor and costs) for the work schedule. This often-overlooked step in schedule development is critical to perform. It allows us to identify and commit the critical resources that are required to complete the planned work and ensure that they are going to be available when we need them. Critical resources are those resources that are needed to complete a project. Resources can be physical objects, such as equipment and materials, or intangible concepts, such as labor or costs. It is important for project managers to identify the critical resources as their availability can have a significant impact on a project.

Project 2010 desktop makes it easy to associate all types of resources (named, generic, or team) to work activities. The challenge is that Project 2010 can easily overload resources, as outlined in Figure 5.16.

This split view makes it easy to identify and correct overallocations within a single project. The difficulty comes when we try to manage shared resources across several projects. Later in this chapter we explain how the Microsoft Project Server Resource Pool helps us manage this challenge.

Project 2010 desktop also has a Task Inspector that helps drill into issues and jump directly to views to solve overallocation, task errors, or slippage due to

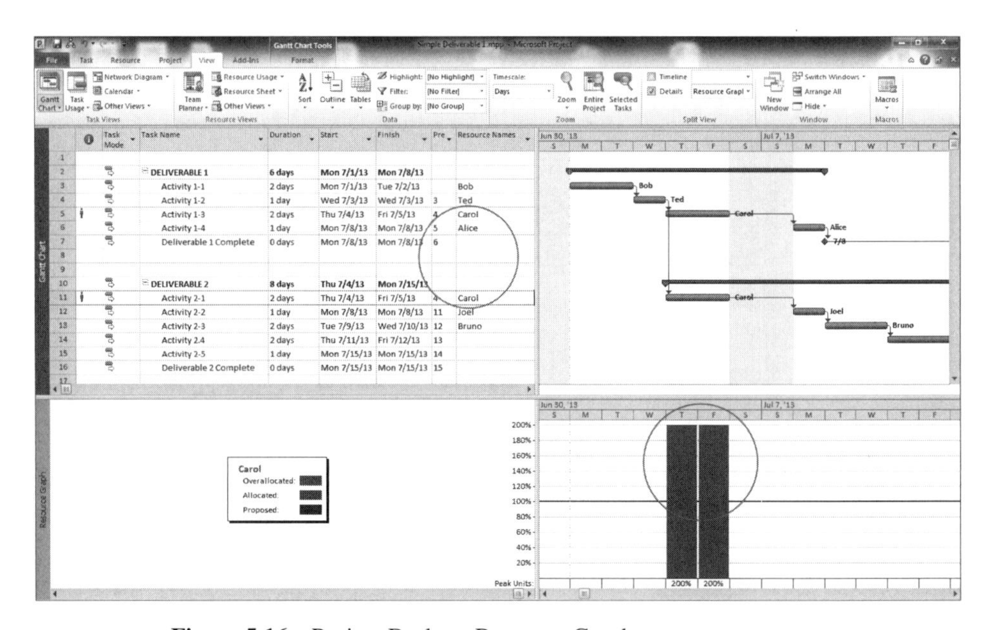

Figure 5.16 Project Desktop Resource Graph Source: Advisicon

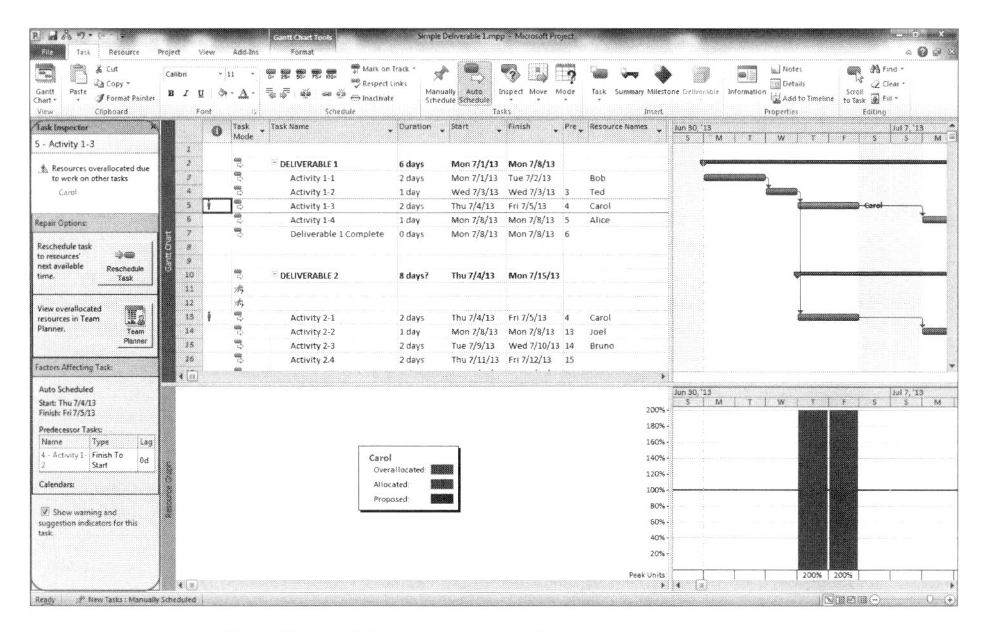

Figure 5.17 Microsoft Project Desktop Task Inspector Source: Advisicon

calendars, predecessor tasks, or resource overallocations. The Task Inspector in Figure 5.17 highlights that the overallocation is due to work on other tasks. Note the "Factors Affecting Task" and "Repair Options" sections of the Inspector.

The Team Planner is a new feature for Project Pro 2010 that gives project managers greater visibility into and control over their team's work. (See Figure 5.18.)

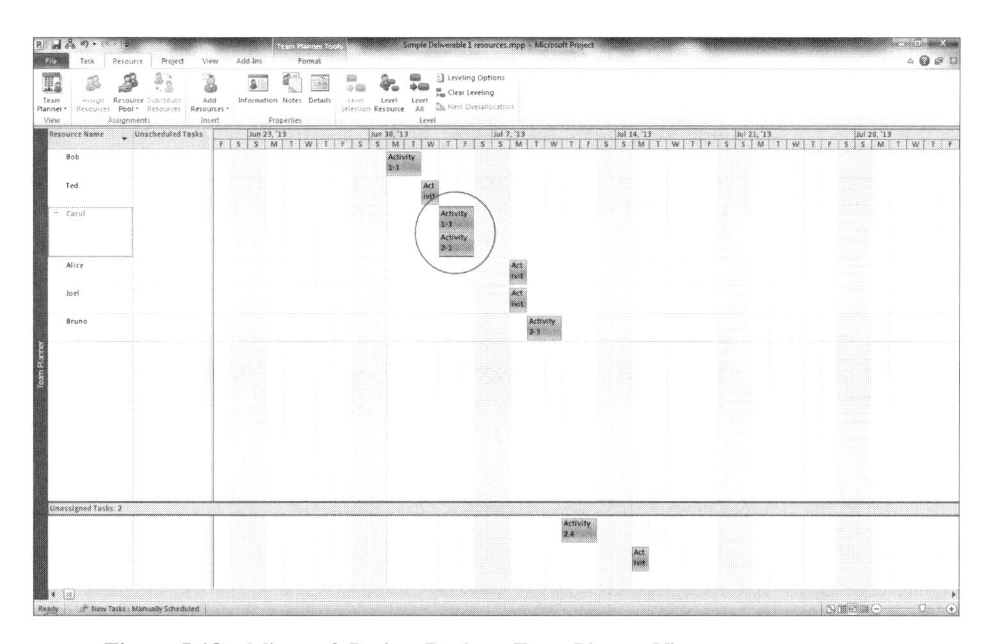

Figure 5.18 Microsoft Project Desktop Team Planner View Source: Advisicon

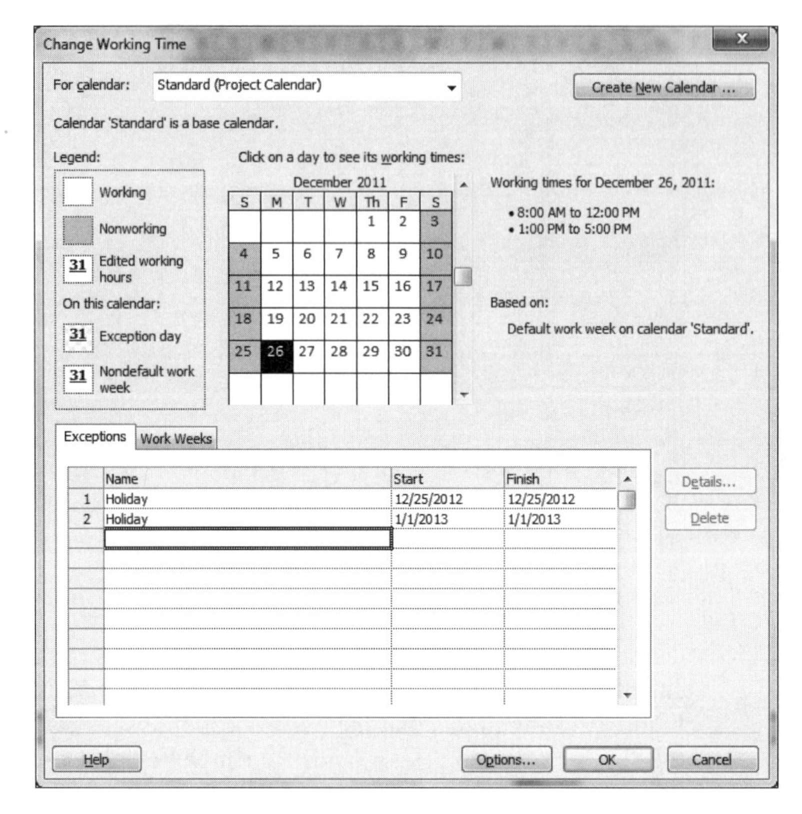

Figure 5.19 Standard (Project Calendar) Source: Advisicon

Project Calendars

To accurately reflect resource availability, you need to update the Standard Project Calendar and record any exceptions to the work week. Project calendars are located on the Project tab by selecting "Change Working Time." (See Figure 5.19.)

Schedules will not accurately reflect the true finish date of the project until all calendar exceptions have been entered (e.g., holidays, office closures, offsites, etc.). Figures 5.20 and 5.21 illustrate the impact that calendar exceptions have on the finish date of the project.

Project supports calendars at the project, task, and resource level.

Collaborative Work Management: Integrating to SharePoint

Microsoft SharePoint makes it easier for teams to collaborate on projects. Using SharePoint you can set up sites to share information with others, manage

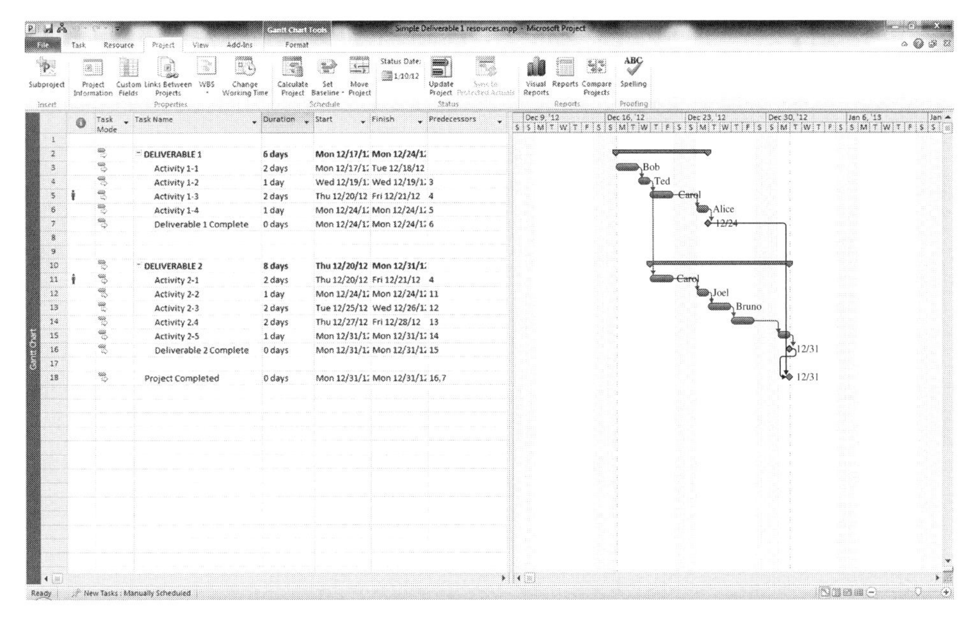

Figure 5.20 Schedule with No Calendar Exceptions Source: Advisicon

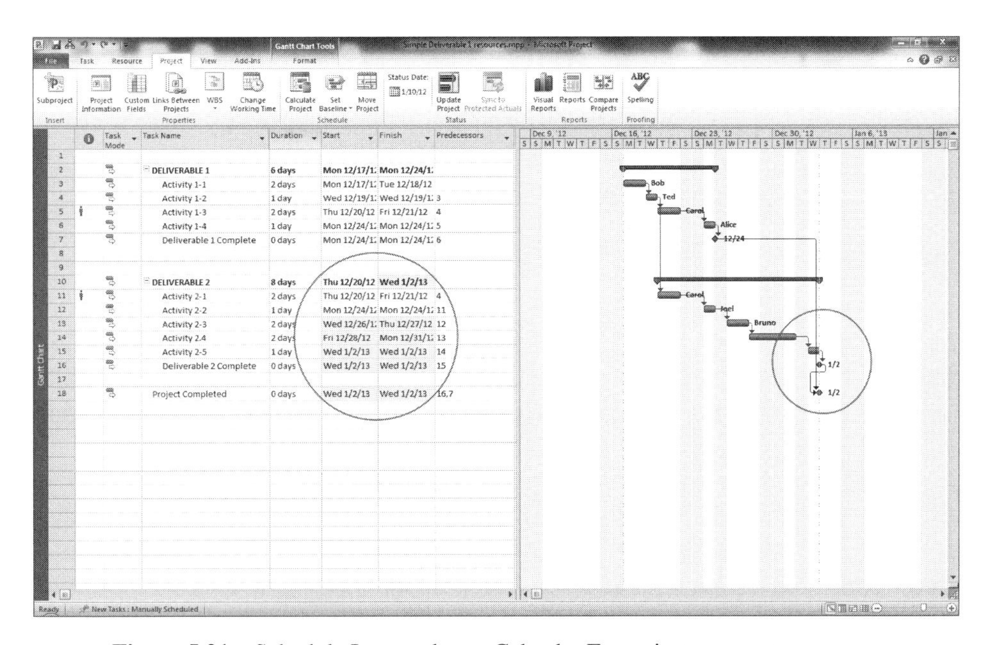

Figure 5.21 Schedule Impact due to Calendar Exceptions Source: Advisicon

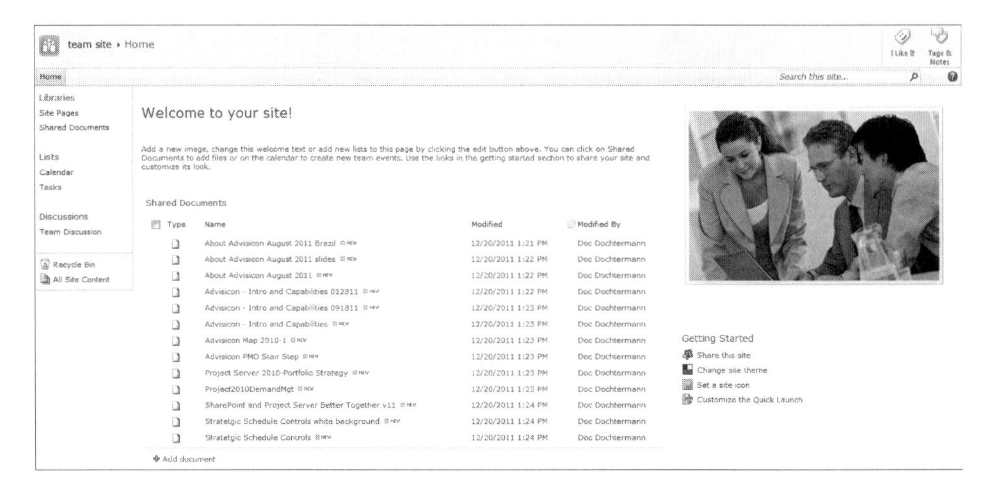

Figure 5.22 SharePoint Team Site Source: Advisicon

documents, and publish reports to help everyone on the team work better together. Figure 5.22 illustrates a SharePoint team site.

SharePoint supports task lists to offer those unfamiliar with formal PM with simple yet sufficient capability to provide a powerful PM solution. Figure 5.23 illustrates the use of the SharePoint project task list, which, once set up, can be easily synchronized with the Microsoft Project Desktop schedule shown previously.

The power of SharePoint task lists lie in their simplicity as well as their ability to share and edit them via SharePoint or Outlook. This means that team members and can view and contribute project status information.

A project manager can now use all the scheduling capabilities of Project Professional in conjunction with the collaborative capabilities of SharePoint. Project plans can be synchronized from Project to SharePoint and vice versa. Any changes made in Project or SharePoint can be easily updated into the other with the click of a button. To use this new capability, simply:

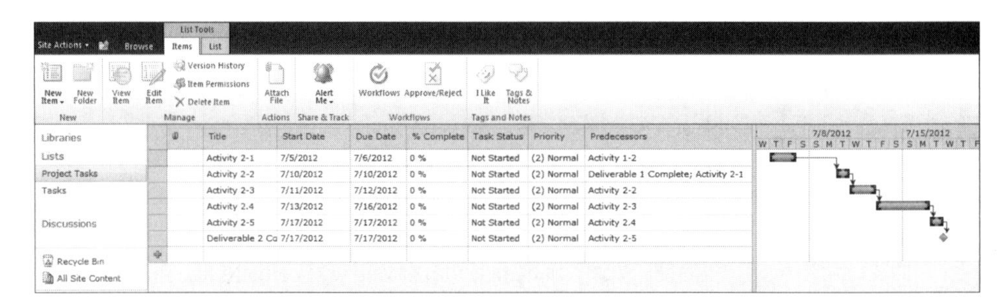

Figure 5.23 SharePoint Project Task list Source: Advisicon

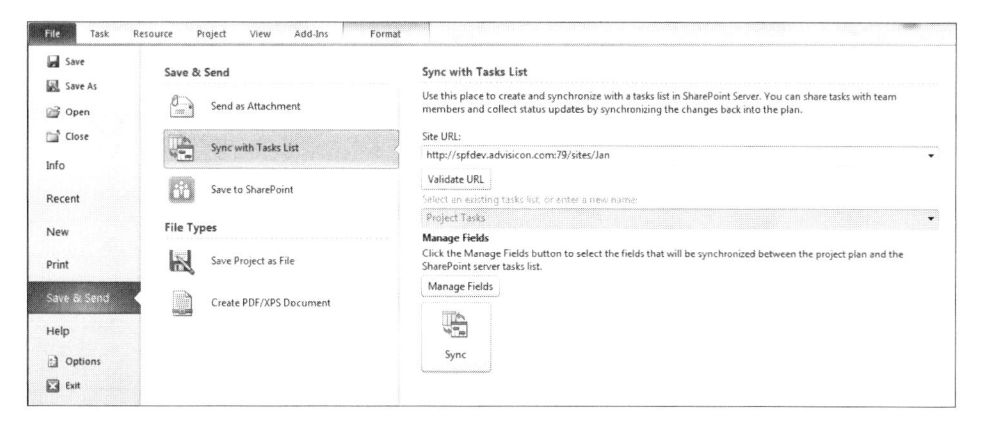

Figure 5.24 MS Project Sync with SharePoint Task Lists Source: Advisicon

1. Create a Project task list in the SharePoint site that you have authority to access (or ask your IT support team to assist you). Copy this URL.
2. Enter your schedule into MS Project desktop and then select the File menu, click on Save and Send, and select "Sync with Task Lists."
3. Paste the URL saved in step 1 into the Site URL field and click the Validate URL button.
4. Select an existing task list or enter a new name. You can also click on the Manage Fields button to select fields to be synchronized between the project plan and the SharePoint server task list.
5. Click the Sync button, and the project plan will be published to SharePoint.

SharePoint tasks can now be viewed and updated by the project team in SharePoint, and the project manager can synchronize the updates to the Project Plan. Figure 5.24 illustrates the MS Project File tab (backstage) where these steps are performed.

In the next section, we see how Project Server—in concert with SharePoint—can extend the capabilities of the Project desktop tool and provide a fully integrated enterprise-level PM solution.

Visual Reporting from Project to Excel for Charting, Graphing, and Pivot Analysis

Visual Reports is a new feature in Project Standard and Professional 2010 (O'Cull 2006) that lets you report on your project's data in Excel using PivotTables and PivotCharts and in Visio using a new feature called PivotDiagrams.

The out-of-the-box Project desktop installation comes complete with a number of prebuilt Excel and Visio templates. You can also create your own templates,

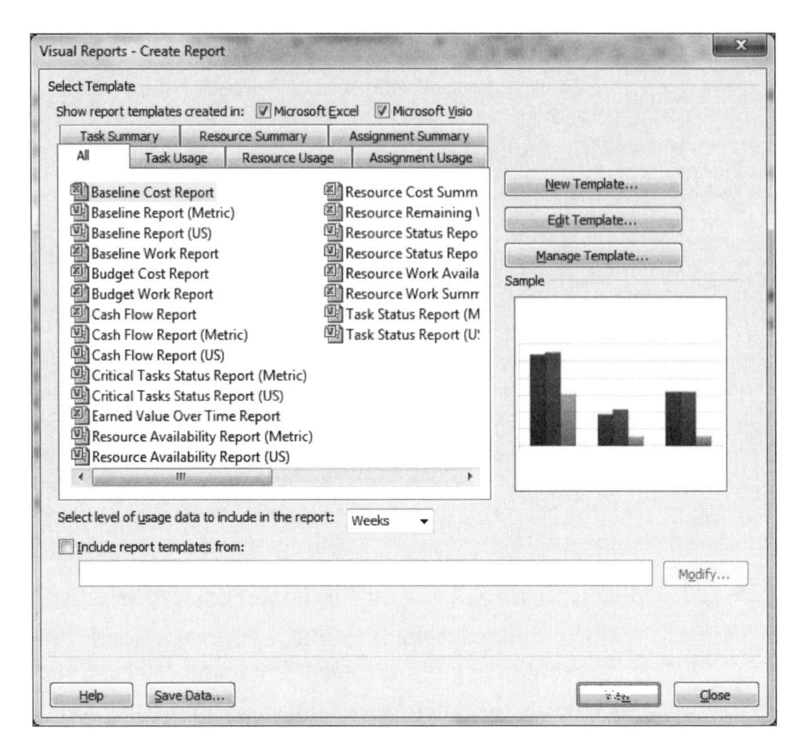

Figure 5.25 Project Visual Reports Source: Advisicon

which you can share with others in your organization. Figure 5.25 shows those Visual Reports that are included with Project.

Using Visual Reports, you can easily create reports based on data from your projects using familiar Excel and Visio formats that are already in common use by your team. You can also include templates from other locations, such as a public share. When you create or edit a template, you can specify which project fields and custom fields to include in the template.

Visual Reports works by first creating a database on the computer that contains data for a specified project. A local cube is built, then Project connects the cube to a PivotChart in Excel or a PivotDiagram in Visio.

There are six different cubes to create reports from: resource, task, and assignment in both summary and usage (time-phased) versions. These cubes are completely separate from the Project server cubes.

Figure 5.26 presents an example of an Excel report that is essentially a simple pivot table representing budget cost for a specific project. Figure 5.27 provides an example of a Visio chart depicting the critical tasks for a specific project.

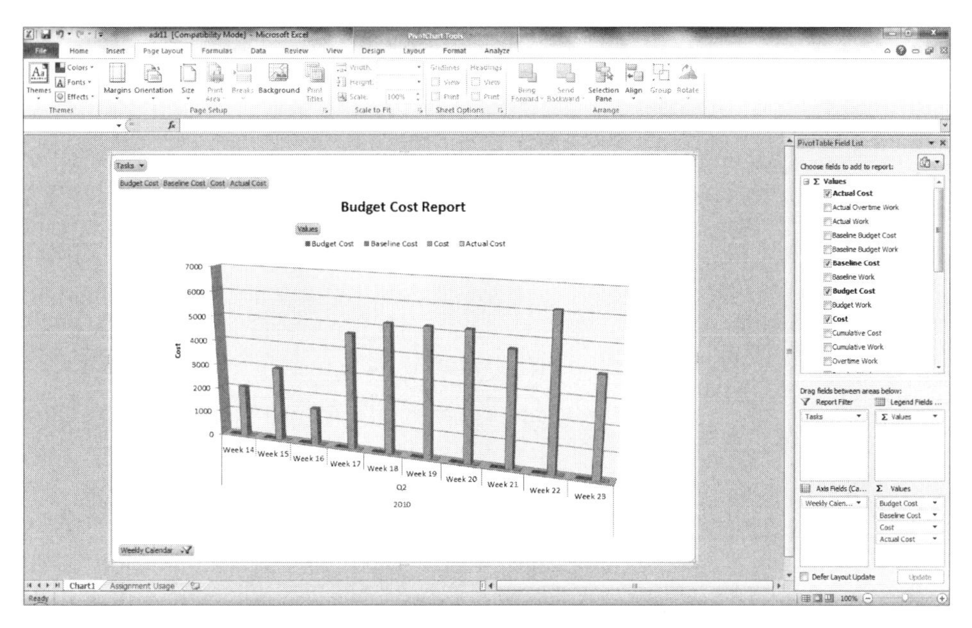

Figure 5.26 Project/Excel Visual Report Source: Advisicon

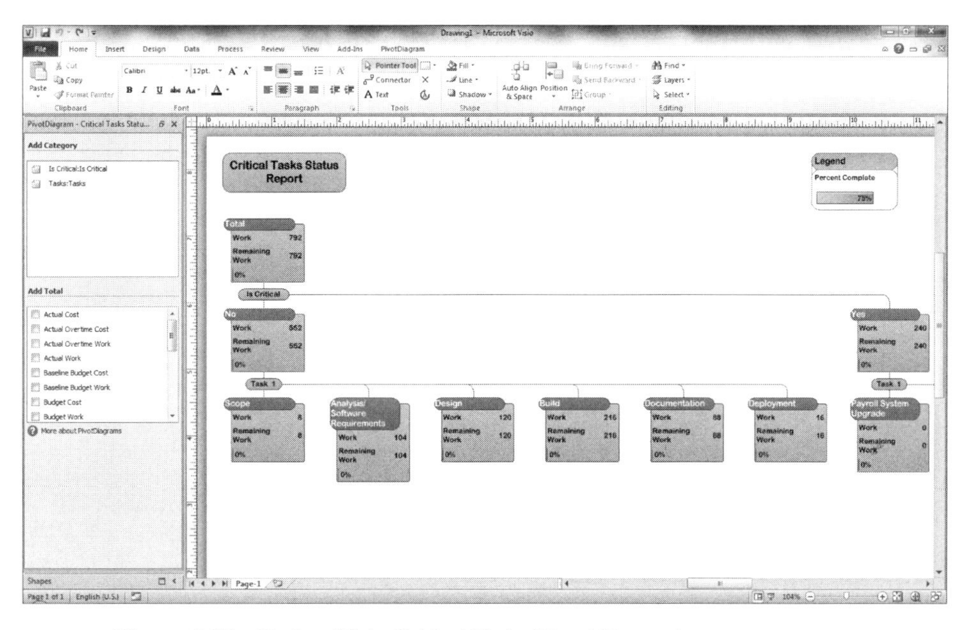

Figure 5.27 Project/Visio Critical Tasks Visual Report Source: Advisicon

Project 2010 Moving Closer to Agile Planning

We would love to clarify a common myth about agile that centers on planning (Bjork 2011). Agile is a methodology that utilizes an incremental and iterative approach to software development. Requirements and solutions evolve through collaboration between self-organizing and cross-functional teams. A software release results from multiple iterations (or sprints).

Agile, however, is not a method that avoids planning; an agile team does just as much planning as a team that subscribes to a more traditional software development methodology (i.e., waterfall). Nevertheless, there are some key differences between planning on an agile team and traditional planning:

- A traditional approach to planning requires you to gather requirements, review architecture and design options, and develop a project lifecycle plan that guides the development team to a successful outcome.
- An agile planning approach also involves studying requirements, architecture, and design. However, an agile approach puts an emphasis on getting started on the well-known and well-understood requirements versus performing a full lifecycle plan. The idea is that the team will derive greater value from starting on the known requirements than it will from developing an exhaustive end-to-end plan.

As shown in Figure 5.28, the "Manifesto for Agile Software Development" (Manifesto 2001) places a higher priority on responding to change than on following a plan.

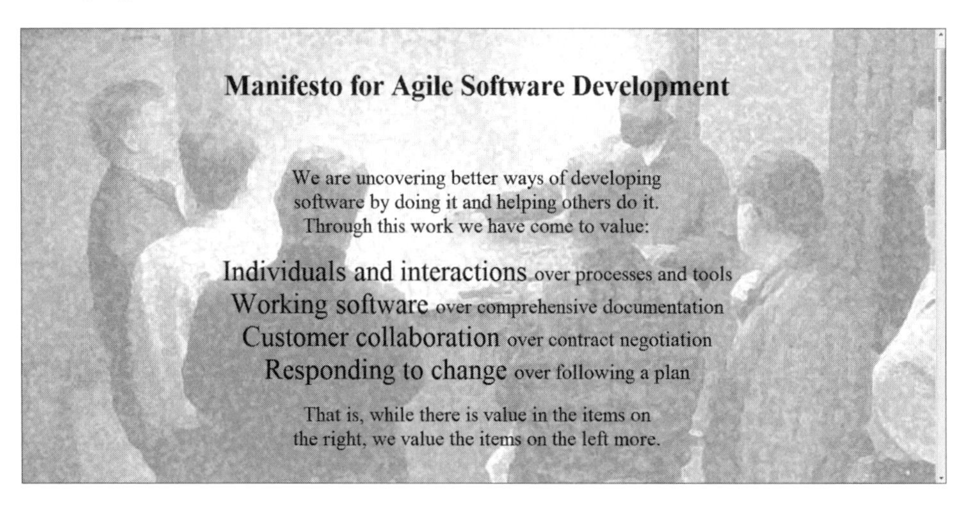

Figure 5.28 Manifesto for Agile Software Development Source: agilemanifesto.org

The key point is that an agile team expects that, once it gets started, executing the plan is more than likely to change based on what the team learns along the way. Based on its iterative and incremental approach, the team wants to have flexibility to adjust and react to changes as they occur.

This is a common and understandable stance by practitioners who have not yet had the opportunity to work with both methods. Those who have worked with both will tell you the first steps of scrum—particularly for large, complex projects—are to define high-level requirements for the entire project and prioritize them (Perera 2011).

Agile and Project Desktop

The next most common myth about agile that we would like to dispel is that you cannot use Microsoft Project to develop plans or schedules.

Agile project planning typically is referred to as release planning. An agile release plan (or schedule of work activities) plans multiple sprints that form a release of a particular solution or a product. The plan is not necessarily project oriented; however, the concept for projects is very similar.

Microsoft has developed a Scrum Solution Starter designed to provide guidance on using Microsoft Project 2010 to manage scrum projects, aiming to help individual scrum teams to start using Project to:

1. Manage product backlog
2. Manage scrum backlog
3. Track progress and generate burndown charts

The Microsoft Solution Starter focuses on the Project 2010 desktop client and on the individual scrum team experience.

Scrum is an iterative, incremental methodology for PM often seen in agile software development. Although scrum was intended for management of software development projects, it can be used to run software maintenance teams or as a general project/program management approach.

Figure 5.29 illustrates the New Scrum Project button that is installed in the Project Backstage tab when the Scrum Solution Starter is installed.

The Project 2010 Scrum Solution Starter comes complete with installation files and documentation; we will not repeat that discussion here. We will, however, highlight some of the key functions of the Scrum Solution Starter.

When a new scrum project is created, the Scrum ribbon tab is presented (see Figure 5.30) from which you can select from the various views (e.g., Product Backlog, Key Dates and Milestones, and Add New Sprint (see Figure 5.31)).

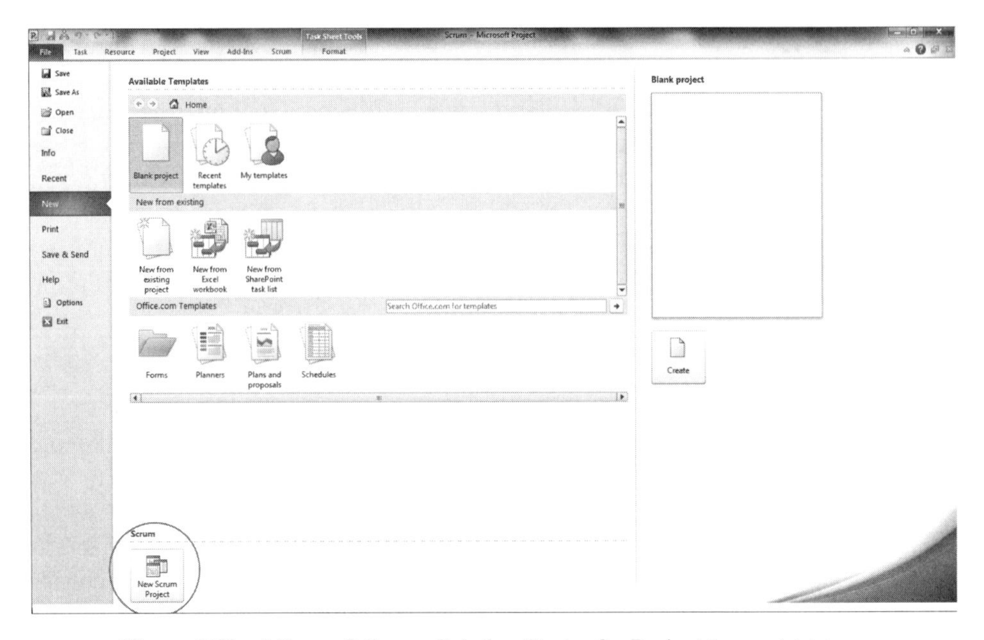

Figure 5.29 Microsoft Scrum Solution Starter for Project Source: Advisicon

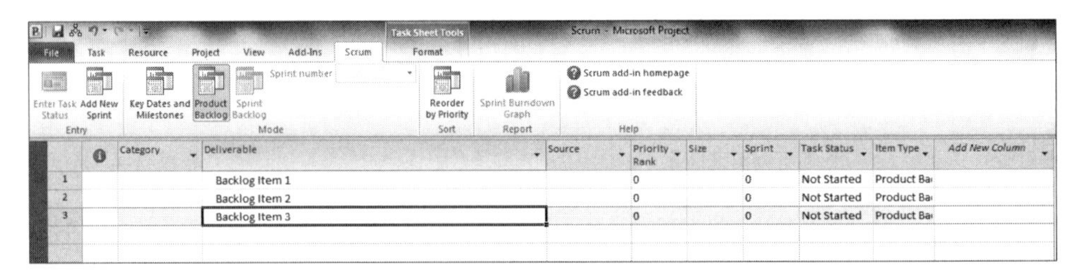

Figure 5.30 Project Scrum Ribbon Tab Source: Advisicon

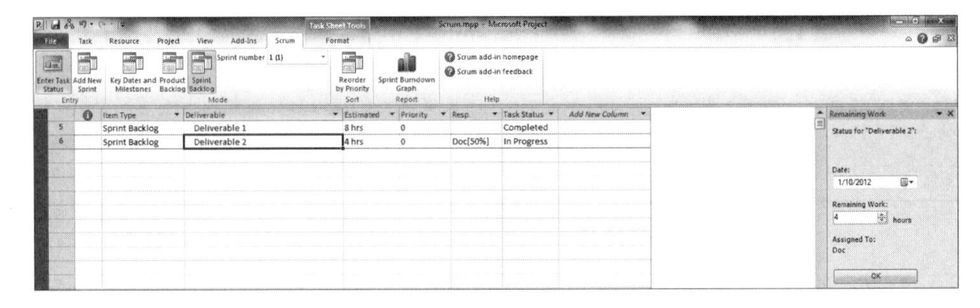

Figure 5.31 Project Sprint View Source: Advisicon

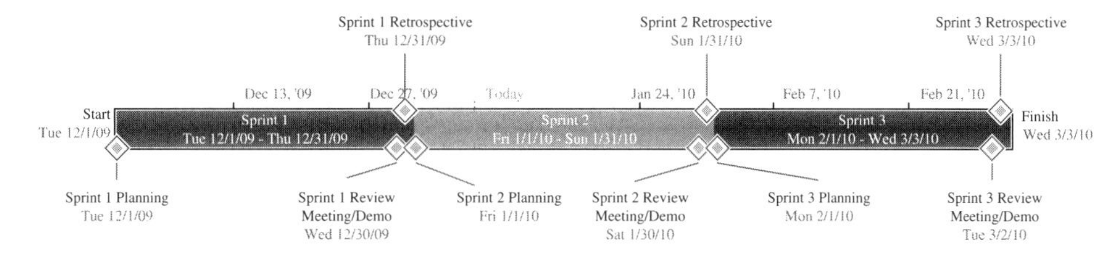

Figure 5.32 Sprint Timeline View Source: Advisicon

The Key Dates view allows users to track nonwork item dates, and visualize them using the timeline view, as shown in Figure 5.32.

Microsoft Project can also be used to assign scope to iterations (Aguanno 2011), providing a flexible yet dynamic way of managing the work of a project. Figure 5.33 illustrates this approach.

The Sprint Burndown Chart button on the Scrum ribbon tab displays the current sprint data from start to end of a sprint. Figure 5.34 illustrates a burndown chart that shows progress/trends in the current sprint.

As there is a tremendous amount of reference material on the Web describing how to use Microsoft Project to plan an agile project (Absolut Agile 2009), we will not go into specifics here.

Critical Success Factors

Here are the CSFs that we have learned so far about initiating and managing projects using the Microsoft Project Desktop client:

	Task Name	Priority	Predecessors	Complexity (in Points)	Duration
1	**Iteration 1**	**25**		**10**	**10 days**
2	Book Marketing Plan	25		7	1.6 wks
3	Back Cover Marketing Blurb	25		1	2 wks
4	Introduction	100		2	2 wks
5	**Iteration 2**	**100**		**11**	**10 days**
6	Chapter on Agile Requirements Gathering	100	2	5	2 wks
7	Chapter on Agile Estimating	100	6SS	6	2 wks
8	**Iteration 3**	**100**		**9**	**10 days**
9	Chapter on Agile Planning	100	7	7	2 wks
10	Chapter on Backlog Maintenance/Replanning	100	7,9FF	2	2 wks
11	**Iteration 4**	**100**		**9**	**10 days**
12	Chapter on Daily Meetings	100		3	2 wks
13	Chapter on Velocity	100		6	2 wks
14	**Iteration 5**	**100**		**10**	**10 days**
15	Chapter on Reporting	100	13	4	2 wks
16	List of Possible Early Endorsers	200		1	2 wks
17	Solicitation Package for Early Endorsers	200	16FF	3	2 wks
18	List of Final Book Reviewers	200		2	2 wks
19	**Iteration 6**	**200**		**4**	**10 days**
20	Book Review Package	200	18	3	2 wks
21	About the Authors	300		1	2 wks
22	**Iteration 7**	**400**		**16**	**10 days**
23	Chapter on Agile Change Management	400		6	2 wks
24	Chapter on Agile Testing	400		5	2 wks
25	Acknowledgements	500		1	2 wks
26	Chapter on Dealing with Stakeholders	600		4	2 wks
27	**Iteration 8**	**500**		**12**	**20 days**
28	Conclusion	500		2	2 wks
29	Index	500	4,6,7,9,10,23,1	10	2 wks
30	**Iteration 9**	**600**		**3**	**10 days**
31	Additional Resources	600		2	2 wks

Figure 5.33 Assign Scope to Iterations in Project Source: Advisicon

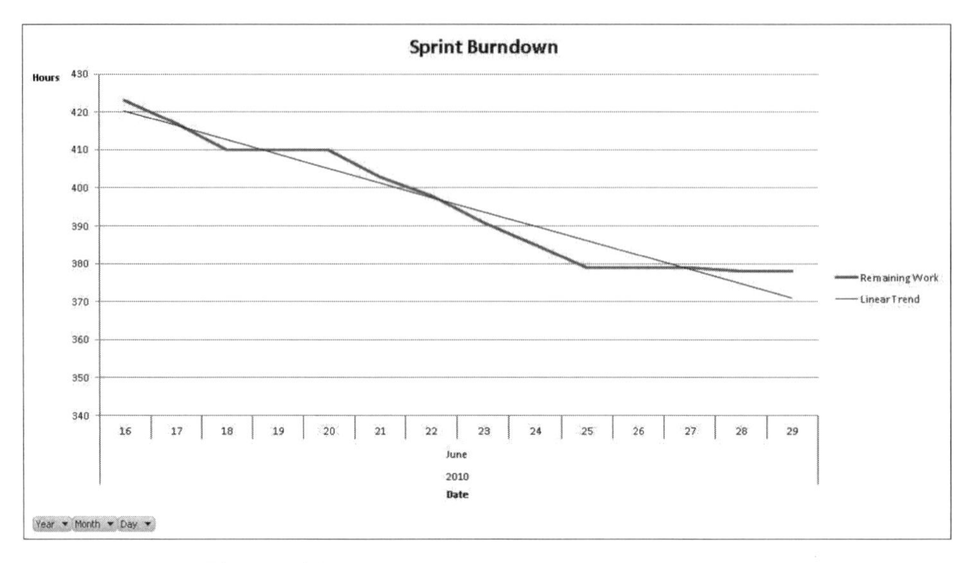

Figure 5.34 Sprint Burndown Chart Source: Advisicon

1. Add-ins available for Project through third-party providers to accurately reflect the work breakdown of the schedule and make data entry and reporting easier.
2. To assist the project manager and other project stakeholders, a properly defined and maintained schedule can be used as an effective "forecast" tool.
3. To accurately reflect the true finish date of a project, not only does the critical path need to be represented using a task activity network, but the tasks also need to have resources assigned to them to ensure that the critical resources will be available at the scheduled times.
4. To accurately reflect working times and resource availability, project calendars need to reflect any exceptions to the workweek.
5. To make it easier for teams to collaborate on projects, you can use SharePoint to share information with others, manage documents, and publish reports. This helps everyone on the team work better together.
6. To provide eye-catching reports that are also informative, using formats that are familiar to your target audience, Visual Reports is a feature in Project Standard and Professional 2010.
7. To provide guidance on using the Project 2010 desktop to manage scrum projects, Microsoft has developed a Scrum Solution Starter, which helps individual scrum teams to start using Project to:

- Manage product backlog
- Manage scrum backlog
- Track progress and generate burndown charts

BEING AN EFFECTIVE ENTERPRISE PROJECT MANAGER USING MICROSOFT PROJECT SERVER

In this section we discuss the key steps required to do effective enterprise PM using Microsoft Project Server 2010.

Initiating and Managing Projects

The PM lifecycle follows a project from the initial concept, through business case analysis and project initiation, then to using workflow to track the project through various stage gates of the project as it progresses. The PM lifecycle uses these phases as a measurement of success. PPM technologies provide electronic tools to produce accurate and timely reporting of project and portfolio results.

Figure 5.35 is an example portfolio lifecycle that illustrates four key phases (i.e., create, select, plan, and manage). A phase represents a collection of stages grouped to identify a common set of activities in the project lifecycle. A stage represents one step within a project life cycle (e.g., propose idea, request review, full business case, or deliver project).

Phases and stages are managed in Project Server 2010 through each stage by the use of enterprise project types and workflows.

Enterprise Project Scheduling

This section focuses on the Plan and Manage phase of the PM lifecycle illustrated in Figure 5.35. In this phase, detailed planning and scheduling are performed.

Saving and publishing the schedule are two key steps with Project Server to manage and share the schedule with other stakeholders.

Once the schedule has been developed in the Project desktop tool Figure 5.36 illustrates how the plan is saved to the Project Server. The key difference between

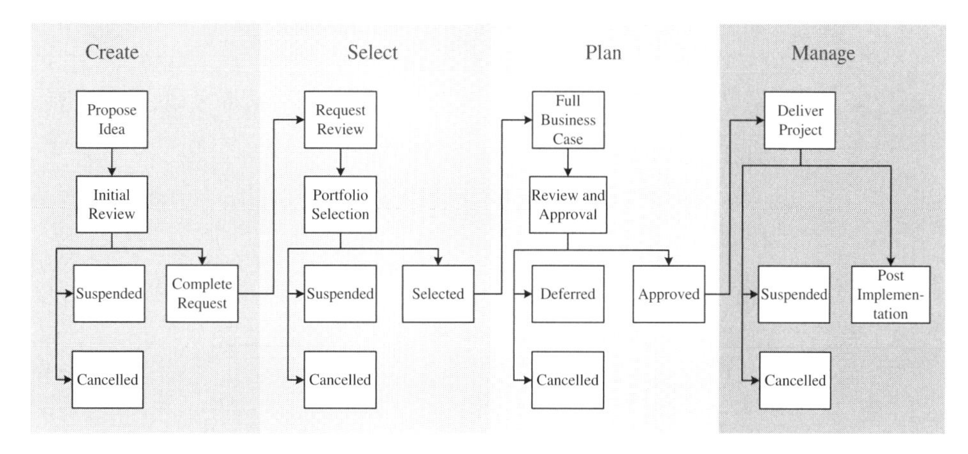

Figure 5.35 Project Management Lifecycle Source: O'Cull 2009

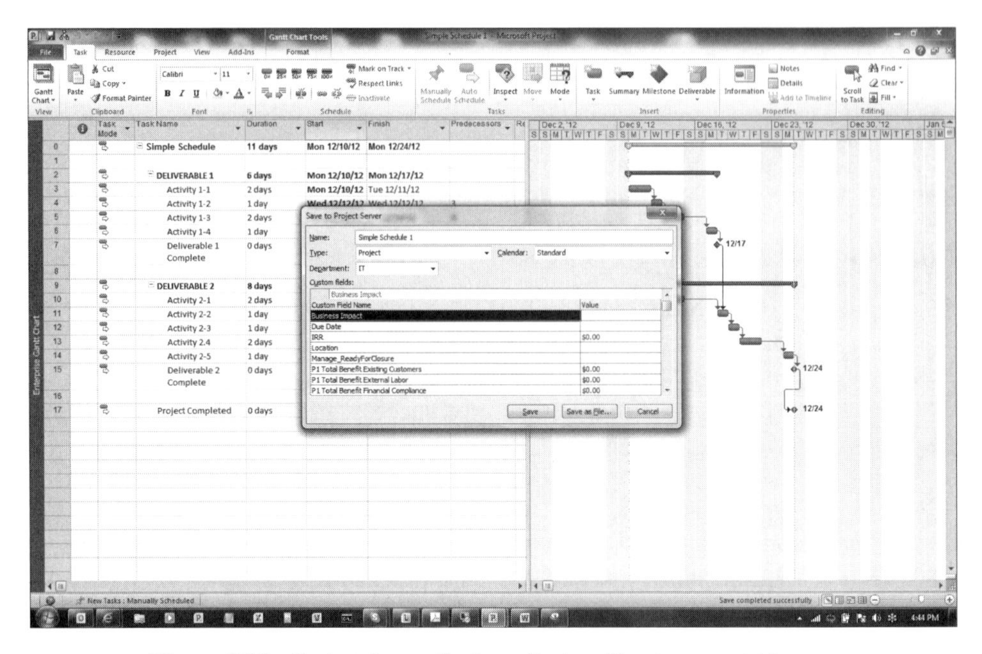

Figure 5.36 Project Server Saving a Project Plan Source: Advisicon

the stand-alone and enterprise versions is that the schedule is not saved to an.mpp file. When Project is connected to Project Server, the Save Project command causes the Project Server dialog box to open, where a project name and other attributes can be entered. The Save button then signals Project Server to save the project schedule to a database.

One of the most significant reasons that organizations move to enterprise PM is to better manage their valuable resources. As we mentioned earlier, managing resources is very challenging using the desktop tool, as there is no visibility into the other projects' utilization of resources.

Project Sever manages all resources through a centralized enterprise resource pool that contains all of the enterprise resources needed to perform project work in the organization (people, costs, and materials required to execute a project). Resource pool attribution drives the ability to track activity-based project costs as well as material consumption. Project cost drives meaningful performance measures, providing cost data that can help make better business decisions.

Project managers use the Project desktop Build Team from Enterprise tool to allocate resources to project plans (see Figure 5.37). Note: This capability is also available using the PWA in Project Server.

Figure 5.38 illustrates how the Build Team from Enterprise menu is used to match and replace the generic resource project manager with a named resource that has the required skills for a given role.

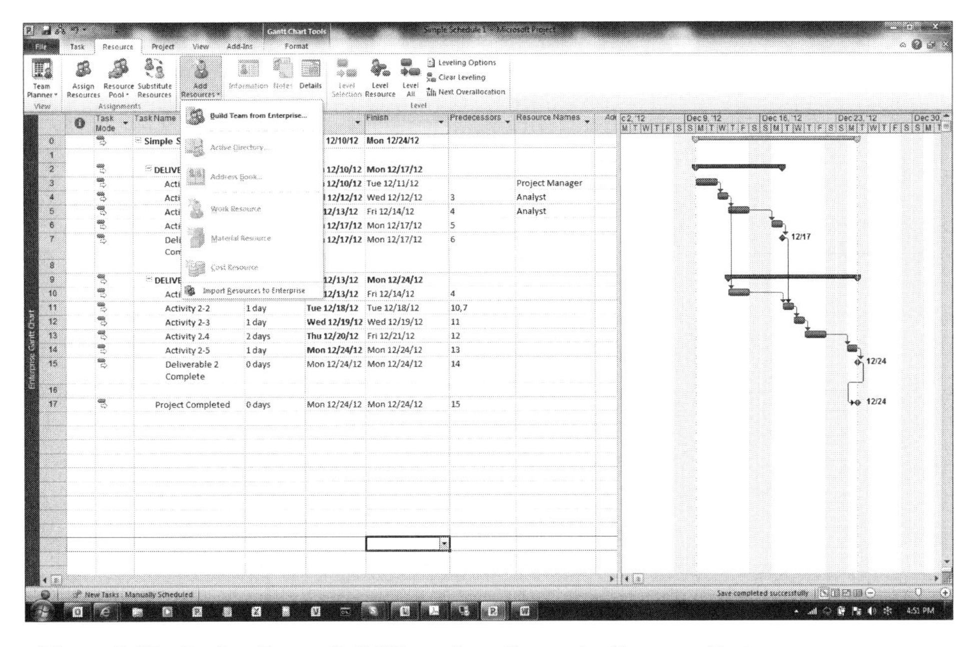

Figure 5.37 Project Server Build Team from Enterprise Resource Pool Source: Advisicon

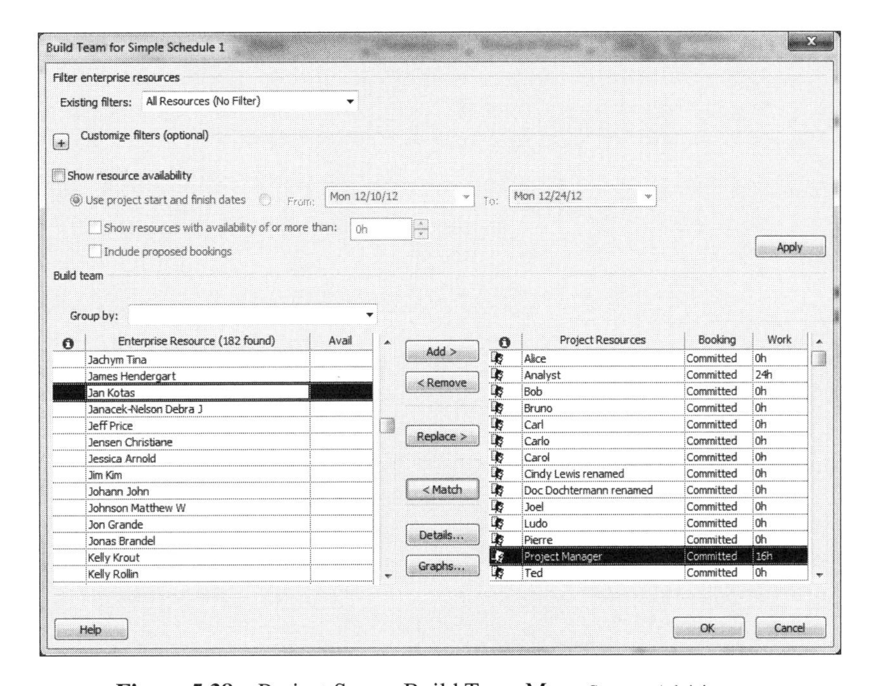

Figure 5.38 Project Server Build Team Menu Source: Advisicon

Figure 5.39 Project Server Assignments by Resource View Source: Advisicon

Enterprise Resource Planning

Project Server 2010 provides the ability to manage three basic resource types: work, material, and cost. Work resources can be used to model people and equipment. Material resources is used to represent the supplies consumed during a project's lifecycle. Cost resources can be used to track budget costs and budget expenses separate from the work resources that are assigned to tasks. Work resources can affect both the schedule and the cost of the project, while material and cost resources affect only the project cost.

The Resource Center provides graphical and textual views to assess the level of work by resources (demand) against the actual availability of enterprise resources (availability) (see figure 5.39).

Resources can also be pivoted to view assignments by Project, as shown in Figure 5.40.

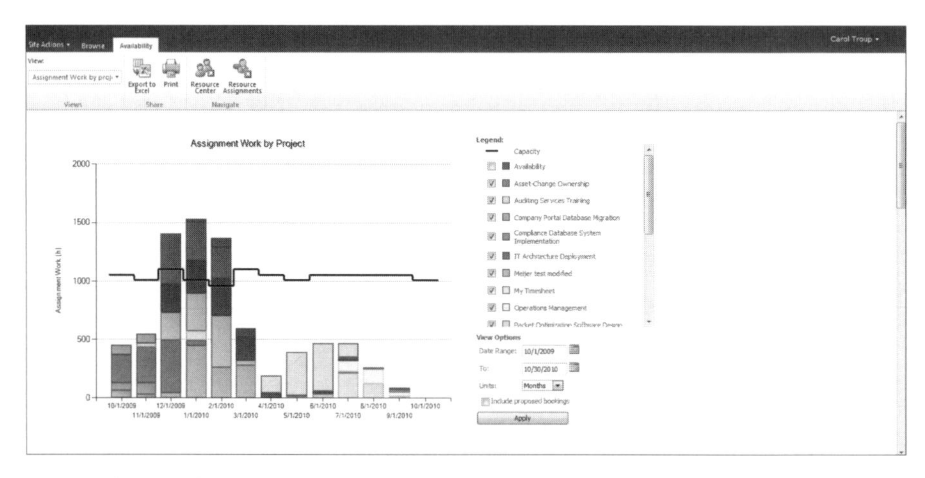

Figure 5.40 Project Server Assignments by Project View Source: Advisicon

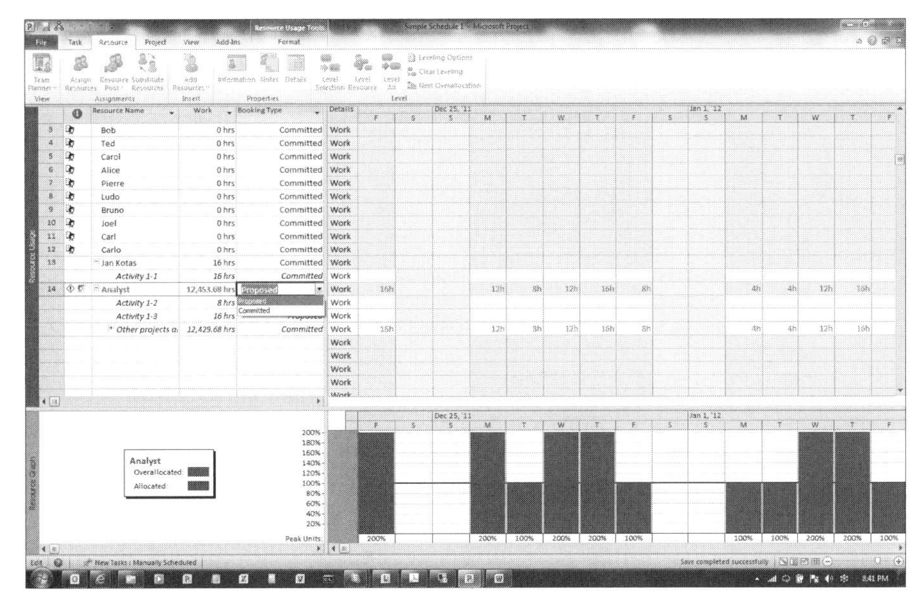

Figure 5.41 Project Resource Booking Type Source: Advisicon

Project Server does not restrict you from allocating work to resources and instead provides a capability to propose and then commit resources to a project. Figure 5.41 illustrates how the Booking Type field works in context with the Project Resource Usage view.

Proposed resources can also be viewed from the now-familiar Gantt chart/ Resource Graph split view as illustrated in Figure 5.42.

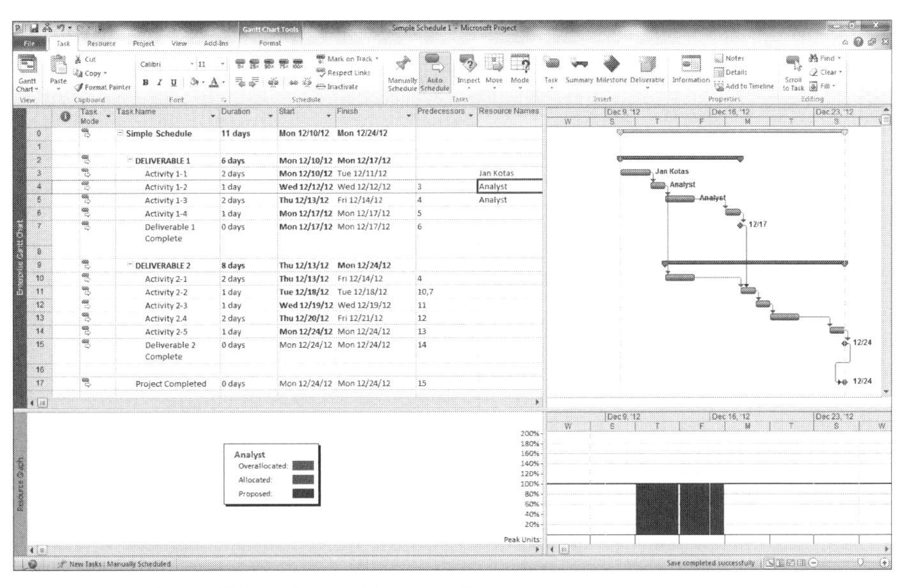

Figure 5.42 Project Proposed Resources Source: Advisicon

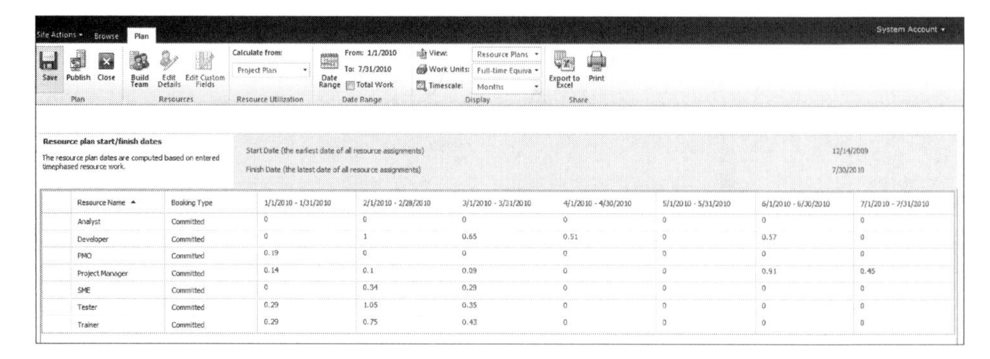

Figure 5.43 Project Server Resource Plan Source: Advisicon

In addition, Project Server supports a resource plan that can be used for capacity planning. (See Figure 5.43.) Resource plans are great ways to perform rolling wave planning prior to detailed schedule development.

It is important to understand that resource plans do not integrate with the Project desktop client. They are Project Server side only. The hours booked to the resource plan will, however, be deducted from the availability of the resource.

Here are some key factors that should be considerated if you are going to use resource plans:

1. The resource plan greatly simplifies the assignment of resources as there are no task-level assignments.
2. Resource plans are a great way to estimate resource usage (i.e., as a placeholder only).
3. Tasks should be used to commit actual resources, with assignments eventually being planned and tracked within the schedule.
4. Assignment data can be pulled from project task assignments up through a specified date. Thereafter, the resource plan assignment data can be used.

Resource plans were developed to provide a way to estimate corporate resource capacity while a number of projects are in full execution and others are still in the planning phase (Ducolon 2007). They are therefore ideal for the early phases of a project lifecycle, where the project is still just a concept or opportunity and not yet a committed and fully detailed project.

Managing Enterprise Projects

The Project Center provides a centralized view of all projects in Project Server. Filters and views provide the ability to filter, group, and display key fields. Figure 5.44 lists all projects currently in the Manage phase.

Team members update their work assignments through the "My Work" view illustrated in Figure 5.45. In this example, actual hours working on a specific

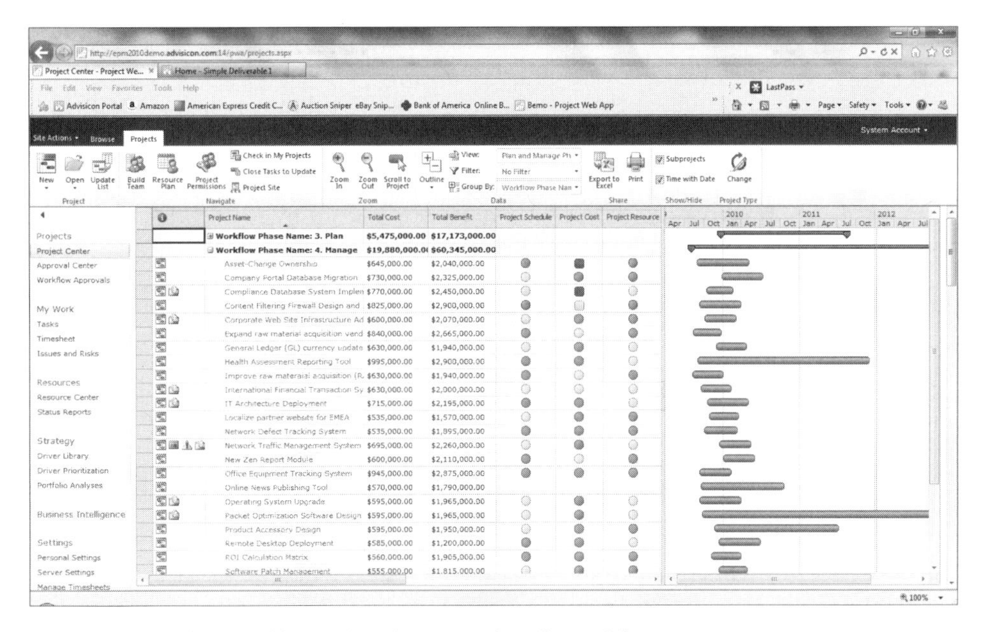

Figure 5.44 Project Server Project Center View Source: Advisicon

activity are captured, including any adjustments to remaining work or finish date, when the activity will be completed. These updates are then forwarded to the project manager for inclusion into the schedule.

Remember the dynamic Project Schedule that we introduced earlier? The view shown in Figure 5.46 illustrates the actual work and cost information against the

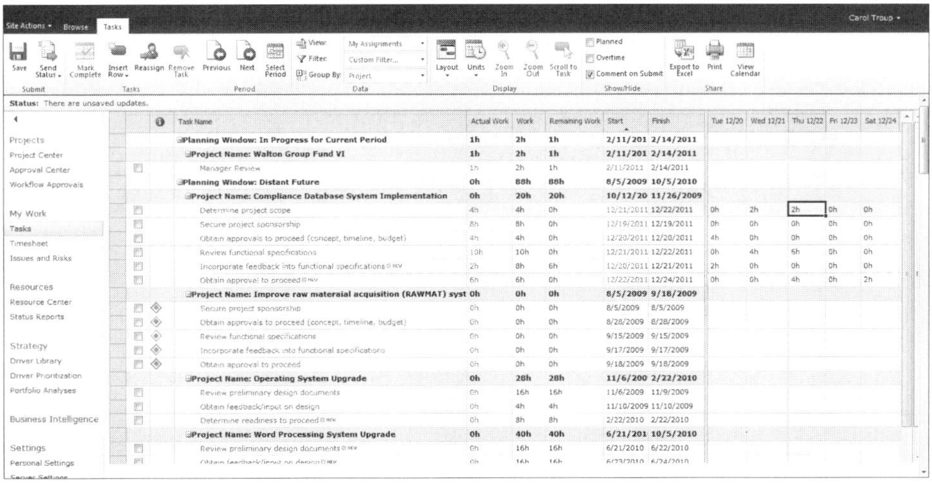

Figure 5.45 Project Server My Tasks View Source: Advisicon

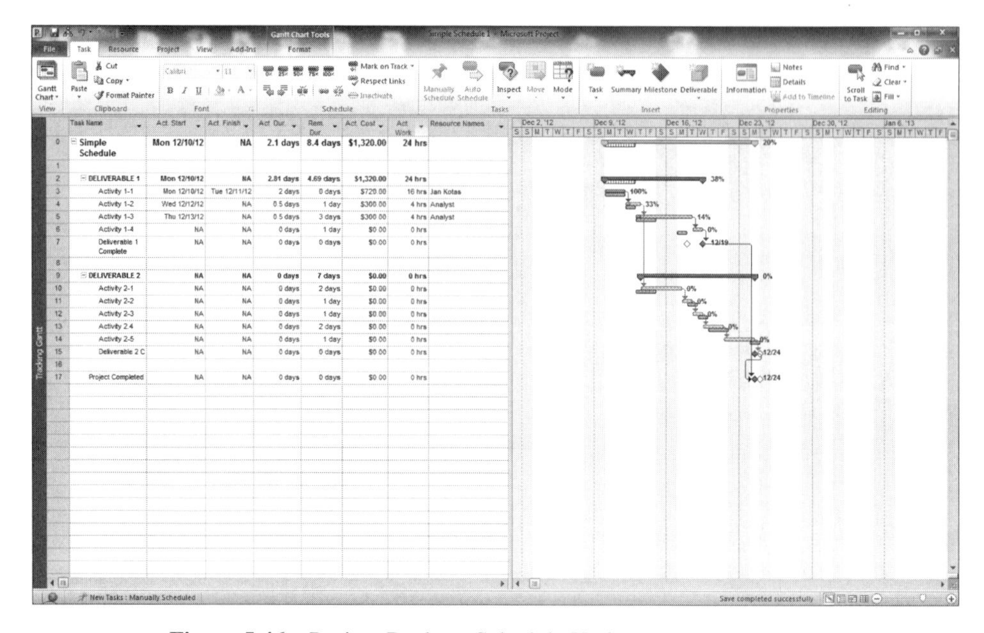

Figure 5.46 Project Desktop Schedule Updates Source: Advisicon

original plan. This baseline approach is used by project managers, as it aids them during schedule-analysis to take corrective action by pulling the project end date back into compliance.

Due to the Web-enabled connectivity of Project desktop and Project Server, PM has never been easier or more functional than in the 2010 release.

Enterprise project management (EPM) solutions are now being hosted in the cloud, making onboarding fast and economical for all sizes of organizations.

Business leaders are now seeking work and resource management, business intelligence, and analytic capabilities that are far beyond the current capabilities of their IT infrastructure. Success or failure will depend on the ability of an organization to select and implement the proper PM and collaboration tools while on its journey upward toward high levels of EPM maturity.

Enabling Cultural Adoption and Ease of Use for End Users

Implementing an EPM solution is not going to happen by accident or luck. Success depends on end user adoption of the new solution, which must be based on careful design and forward-thinking enterprise processes.

Successful and sustainable collaborative strategies must be designed around three major factors:

1. **Usability,** which directly relates to the ease of use and learnability of a solution. A compelling experience is an important part of engaging the end user toward adoption of a new enterprise solution.
2. **Impact.** An attractive, easy-to-use interface will help draw end users to the new solution. However, sustained use will come from access to valuable content, the ability to connect and communicate with other stakeholders more efficiently, and incentives that align with knowledge sharing and the new enterprise approach to managing work and resources.
3. **Organizational readiness,** which highlights the discrepancy between current and target environments. Organizational readiness showcases any dissatisfaction with the status quo of the existing project, program or portfolio process and creates a compelling vision of the future state. This increases the degree to which organizational members perceive the change as needed, important, and worthwhile.

By paying close attention to these factors, organizations can position themselves to harness the power of a collaborative solution to support the overall business objectives of the enterprise, not just the priorities of a specific business unit. By doing so, ultimately organizations will drive both top- and bottom-line growth for the business as a whole.

A well-designed and sustainable collaboration platform needs to fulfill two key objectives:

1. It has to be so intuitive to use that there is no appreciable learning curve.
2. It has to be configurable and extensible so that more advanced users can adapt it to their specific needs.

Project and Project Server 2010 fulfill these two key objectives offering best-in-class capabilities enabled by the extensibility of the platform architecture.

Role Based

There are multiple audiences in organizations: from IT professionals, to project managers, to end users. Each audience cares about different aspects and has specific concerns.

End users, for example, don't need to know all the details of Project desktop, Project Server, and SharePoint. They might be:

- Executives who want to see project or resource status and reporting (i.e., dashboards incorporating KPIs, graphical views, and other Web Parts).
- A team member who only needs to see his or her My Tasks information and the details online (through PWA).
- Team leads who need to edit, update, review, and approve tasks right in PWA versus needing Project client installed.

Project managers, however, require the power and flexibility of a desktop tool like Project to design and manage projects of any size or complexity. PMs tend to be more mobile and therefore need access to a checked-out version of the schedule and the ability to connect with Project Server to share schedule changes and updates with the team.

Agile and Project Server

For software development, Team Foundation Server (TFS) is Microsoft's software Application Lifecycle Management (ALM) tool. TFS provides a range of ALM functionality including work item tracking, and planning tools and reports, along with other functionality in the areas of Configuration Management and Team Collaboration.

Integration between Project Server 2010 and Team Foundation Server 2010 is significant for organizations that want to bridge the gap between PM and software development (Feissinger 2010). This capability further strengthens Microsoft's ALM solution by enabling PMs and development teams to work together more effectively (Channel 19, 2010) while not getting distracted or overwhelmed by each other's processes or data. It enables teams to work together more effectively by:

- Providing executives with insight into project portfolio execution, alignment with enterprise strategic objectives, and resource utilizationby leveraging the data stored in different systems.
- Bridging the collaboration between the PM office and application development by utilizing common information and agreed-on metrics to facilitate better coordination between teams using disparate methodologies, such as waterfall and agile.
- Enabling development and PM teams to use popular, easy-to-use tools such as Microsoft Project, Project Server, SharePoint, Office, and Visual Studio, to work better together to communicate project schedules and product backlogs.

Integrating ALM and PPM improves visibility across the entire application development lifecycle, enabling project managers and developers to manage their work according to their own methodologies yet have seamless connection with each other.

Critical Success Factors

The CSFs for being an effective enterprise project manager using Microsoft Project Server are listed next.

1. To provide accurate and timely reporting of project and portfolio results, a PM lifecycle is used to track projects through various stage gates of progress and final measurement of success.

2. To better manage valuable resources, organizations need to move to enterprise PM, which supports management of all resources through a centralized resource pool.
3. To properly track labor, materials, and costs, three basic resource types (work, material, and cost) need to be defined and updated for every project.
4. To implement successful EPM, sustainable collaborative strategies must be focused around three key factors: usability, impact, and organizational readiness.
5. To bridge the collaboration gap between the PM office and application development teams, common information and agreed upon metrics are required.

FLUENT PROJECT MANAGEMENT USING THE FLUENT UI: INTRODUCING THE RIBBON

In this section, we examine the new Fluent User Interface (UI) that Microsoft introduced with the release of Office 2010. In this new version of Office, all applications share a common Ribbon interface, including all Office applications, Visio, Project, Project Server, and SharePoint.

Microsoft has completely revamped the Office Fluent UI, or ribbon. The new user interface design represents a dramatic departure from the overloaded menu and toolbar design model of previous releases. Project's extensive capabilities are now organized into logical, easy-to-find groups that help you accomplish actions efficiently rather than searching for specific functions.

These new capabilities are being driven by end users' need for a simple yet powerful PM solution. User feedback clearly indicated that people had great difficulty finding, using, and understanding the vast feature set in Office.

Frontstage and Backstage to Create the Optimal Work Management Tool

End users were introduced to a whole new interface with the release of Project Desktop 2010, including the ribbon, the Quick Access Toolbar, and the built-in context menus. Microsoft originally introduced the ribbon extensibility model in the 2007 Microsoft Office system as part of the Office Fluent UI. This was a new way to customize the user interface and create custom tabs and groups that were specific to users' needs.

Office 2010 extends the span of the UI extensibility platform by providing support for customization of the new Backstage view, along with the ribbon, the Quick Access Toolbar, and context menus—referred to herein as the Frontstage.

The Microsoft design team identified that there were two distinct types of functions within the Office applications—IN and OUT functions:

1. The IN functions are the ones that most people are more familiar with. These are the functions that act on the content of the document and show up on the page. Examples include commands like bold, margins, spelling, and styles. These functions make up the heart of the application.
2. The OUT functions help people do something with the content they create. Examples include Saving, Printing, Permissions, Versioning, Collaboration, Document Inspector, Workflows, and the like. The primary characteristic is that the OUT functions don't act on a specific point in the document and their effects don't appear on the page.

Project Frontstage (the IN Functions)

Let's take a high-level look at the Project ribbon (i.e., Task, Resource, Project, View, and the special contextual tab Format).

The Task tab includes functions associated with tasks in addition to commands that are also on the first tab of other Office applications (e.g., cut, copy, paste). You can think of the Task tab as the Project desktop home tab (see Figure 5.47a).

The Project tab includes functions that affect the entire project. Notice that in addition to the standard project functions, subprojects and linkage to other projects are included here. The ability to compare projects also is included (see Figure 5.47b).

The Resource tab is where you access functions associated with resource management. (See Figure 5.47c.) The new Team Planner view is accessible from this page (see Figure 5.18 earlier in this chapter).

The View tab is where you select the view, filter what data you wish to see and how it is arranged, set up combination views, and run Project macros. (See Figure 5.47d.)

Additionally, each view has its own contextual tab, labeled Format (see Figure 5.51). This tab contains functions that are used to format the content of a particular view. The Format tab provides incredible control over the presentation of tabular and graphical information. You can adjust styles here, select text and chart styles, and even invoke Drawing tools.

Project Backstage (the OUT functions)

The Backstage view is the new end user interface experience seen when you click on the File tab in any of the Office 2010 applications (i.e., Word, Excel, PowerPoint, Outlook). While the other ribbon tabs focus on things you do when you are working with your project (add tasks, edit resources, change formatting, etc.), the Backstage view focuses on things you do to your project as overall—for example, open, save, publish print, and share (ReedShaff 2009).

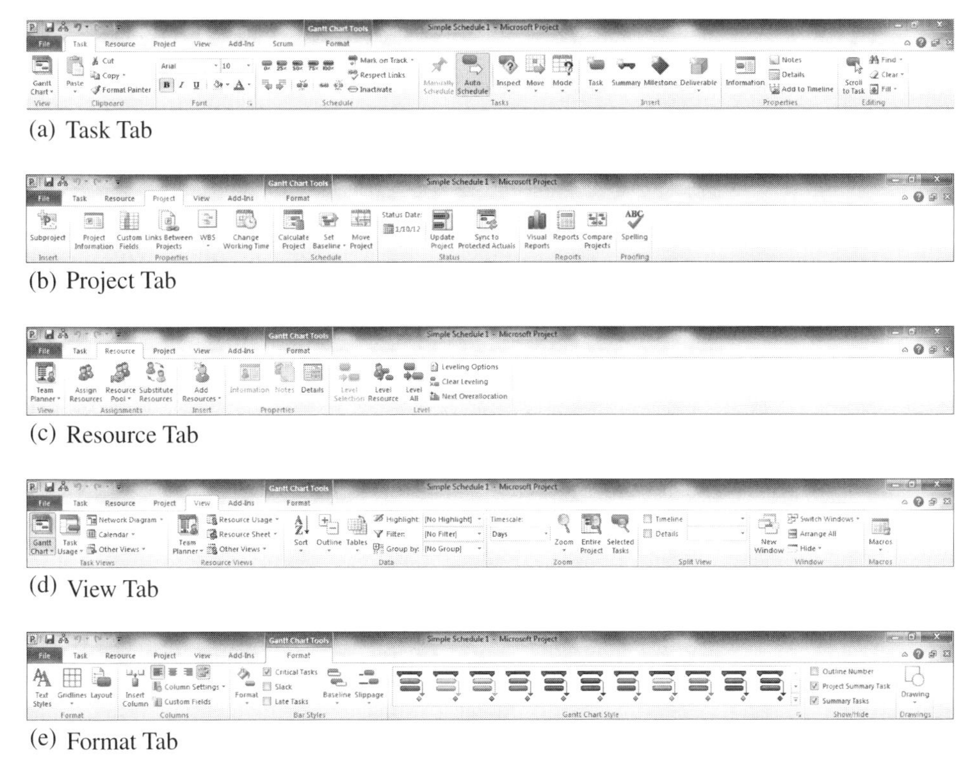

(a) Task Tab

(b) Project Tab

(c) Resource Tab

(d) View Tab

(e) Format Tab

Figure 5.47a-e Project Desktop Tabs Source: Advisicon

Figure 5.48 illustrates the Project desktop Backstage. The key functions of Backstage (Kaufthal 2010) include the:

- **Info tab,** where you can get high-level status about the project and make related changes.
- **Recent tab,** which provides quick access to recently opened projects and also allows you pin the projects you want to always keep on the recent list.
- **New tab** centralizes a number of ways to start a project (e.g., blank project, templates, from existing projects, etc.).
- **Print tab,** which combines print preview with common print settings, providing an all-in-one interface for printing.
- **Help tab,** which is similar to that of the other Office apps.
- **Options,** which includes the redesigned options interface for Project 2010.

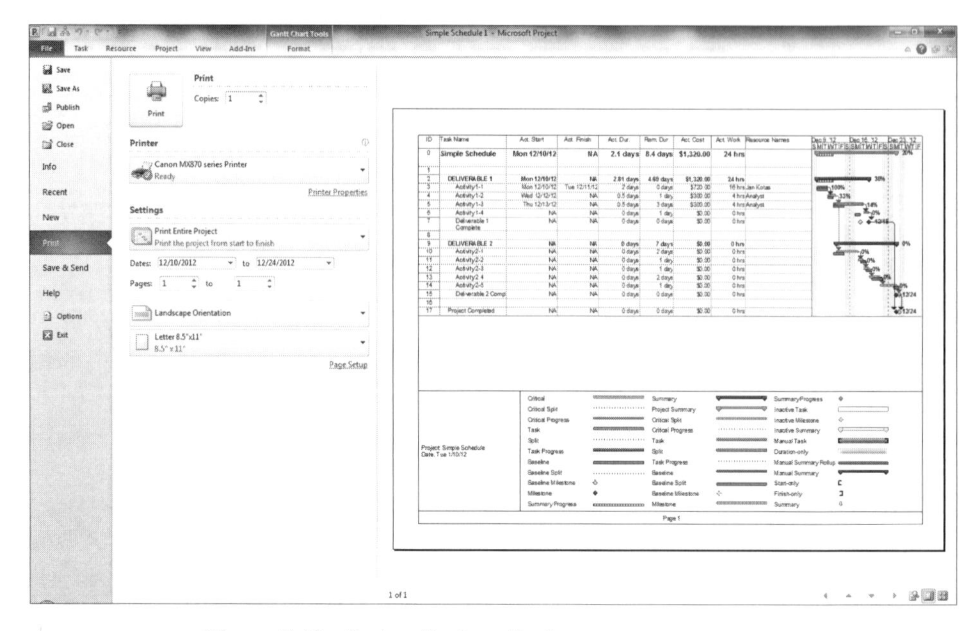

Figure 5.48 Project Desktop Backstage Source: Advisicon

Making the Project Backstage even more useful is its extensibility capability. Organizations can build Backstage add-ins for their own employees, customers, or others. For example, an enterprise could build buttons into its version of Project that integrate with the company's business processes (sending a file to a manager for review, exporting data into a database, etc.).

When the Project desktop client is connected to Project Server, a number of new Info tab options light up on the Backstage, as shown in Figure 5.49. A number of Project Server–dependent functions are now included there, such as:

- The link to the Project Web App home page
- Status of last publish to PWA and a button to publish again
- Buttons to check for updates, manage permissions, and work with the enterprise global template and enterprise resource pool.

The right-side pane now also lets you control the tracking method, edit custom-field values, and link to related information, such as documents, issues, risks, and the project site.

Project Web Application

SharePoint Foundation 2010, SharePoint Server 2010, and Project Server 2010 PWA are all adopting the ribbon user interface component. The PWA experience will be more consistent with the Project Professional 2010 desktop user experience, so project managers can work in similar ways within both desktop and server applications.

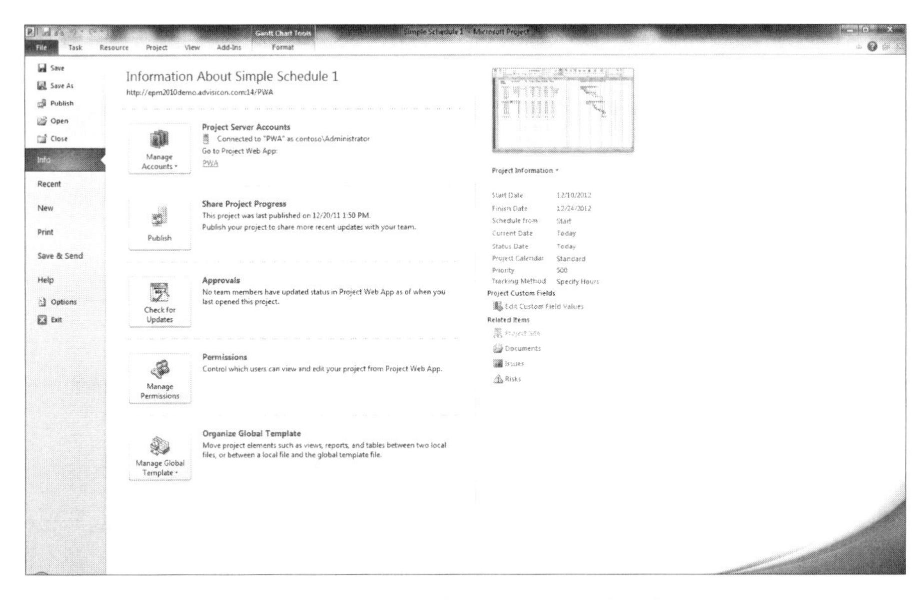

Figure 5.49 Project Backstage when Connected to Project Server Source: Advisicon

The ribbon interface also makes it easier for users who are familiar with other SharePoint Server applications to move to Project Server PWA.

Pages in Project Server that are frequently used by the project management office (PMO), project managers, resource managers, and team members use the Server Ribbon interface. Figures 5.50a, 5.50b, and 5.50c show the ribbons that project managers and team members use to access Project Resource and task information when working with PWA.

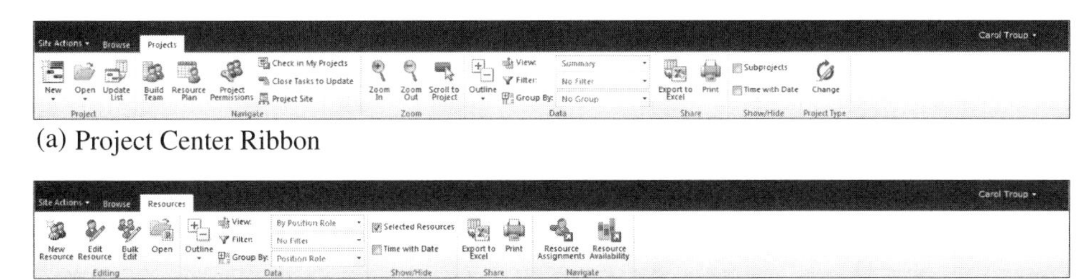

(a) Project Center Ribbon

(b) Resource Center Ribbon

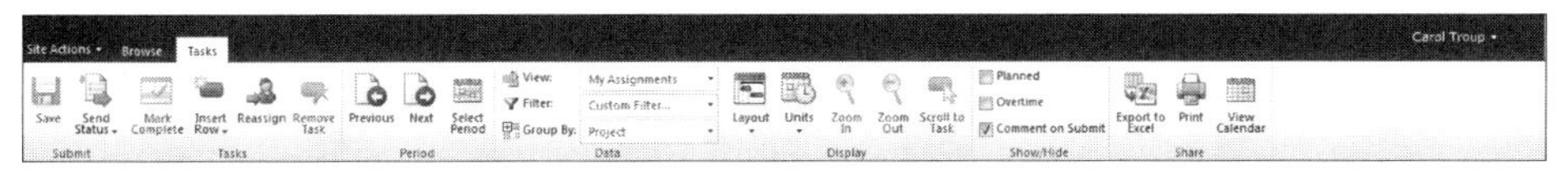

(c) My Tasks Ribbon

Figure 5.50a-c Project Server Ribbons Source: Advisicon

Customizing Is as Easy as Right Click/Left Click

The Project Fluent User Interface is fully customizable. This includes the ribbon (O'Cull 2009), the Quick Access Toolbar, and the built-in context menus. Customizations allow you to personalize the ribbon the way as you want it (e.g., create custom tabs and custom groups to contain frequently used commands).

Project Desktop Ribbon

To customize the Project desktop ribbon, you simply right-click on any menu and select the Customize the Ribbon . . . item. The menu illustrated in Figure 5.51 is displayed. Here you can create or alter tabs, add or remove commands and groups, or rename stuff.

Keep in mind that these changes are specific to your workstation and that each Office application has its own ribbon. Also note that there are main tabs as well as tool tabs. (Recall the Format tab illustrated in Figure 5.47e.)

Project Server Ribbon

Although the toolbars are implemented differently for Project desktop and Project Server, the functionality is the same. The ribbon is (and will continue to be) more consistent from Project Desktop to Project Server.

Clicking any Web Part in a SharePoint site, for example, also enables the ribbon keeping the same look and feel for every environment. This is true for

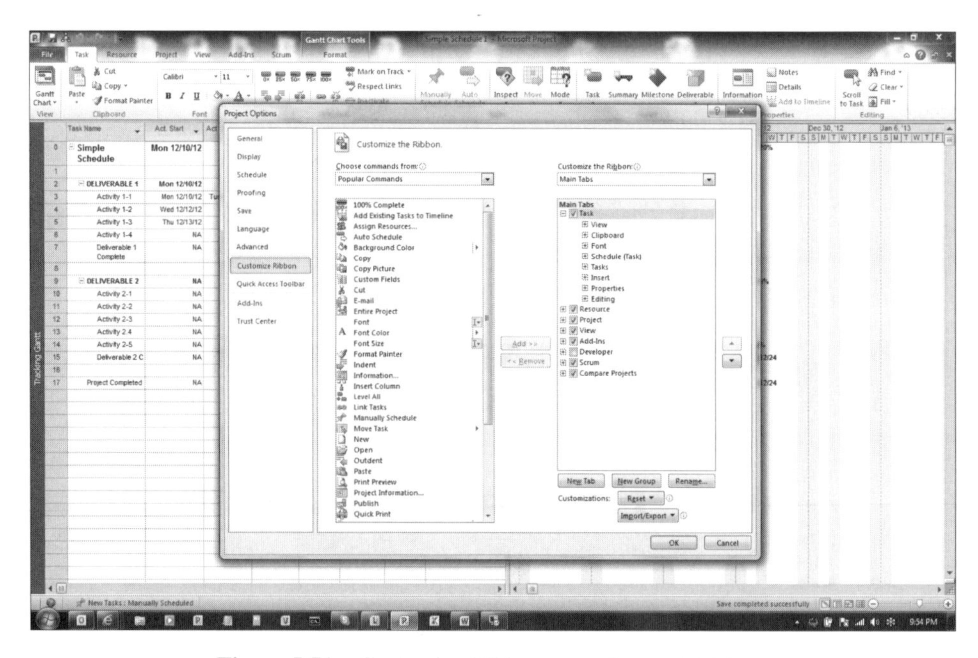

Figure 5.51 Customize Ribbon menu Source: Advisicon

Project Server, SharePoint, and Project Professional 2010 and will also be the standard for future releases of Project and Project Server.

Project Server PWA uses the core ribbon API of SharePoint Foundation 2010. Because most PWA pages in Project Server 2010 use the ribbon and Web Parts, and many of the pages use the customizable JavaScript Grid control, PWA is much easier to customize and extend than previous versions.

Some potential customization and development scenarios include:

- Add the ribbon to your own Web pages in PWA or to any other page or Web Part in SharePoint Foundation 2010 or SharePoint Server 2010.
- Design a new ribbon tab by using preexisting controls.
- Replace a command on an existing ribbon.
- Add a command to the ribbon on a specific Web page.
- Product Detail Pages (PDPs) provide a highly customizable project-creation experience. They can integrate with the ribbon user interface in Project Web app, provide Quick Launch navigation elements specific to individual pieces of project data, and dynamically filter custom fields by departmental association.

You can customize PWA Project Detail Pages by using Web Parts and a ribbon interface. Project Server 2010 includes these new Web Parts for PDPs:

- **Buttons Web Part.** Enables users to edit, save, publish, or close a project detail page or to move to the next stage in a workflow. A long page can include multiple Buttons Web Parts.
- **Workflow Status Web Part.** Enables users to check the status of Project Server workflows.
- **Project Fields Web Part.** Enables users to select or edit project custom fields for the PDP. Project summary task fields such as cost and actual work are read-only. Custom fields such as the project name, department, workflow management, start date, and owner are read/write.
- **Strategic Impact Web Part.** Includes all business driver definitions filtered by one or more departments. This Web Part enables users to rate the project impact on each driver.
- **Dependencies Web Part.** Enables users to define dependencies between projects.

Tabs that Empower the Business User

A good way to help the business user quickly adapt and maximize the use of the ribbon is to customize the ribbon by creating personalized modifications and tabs for the end user. You can place key actions, commands, macros and other commonly used features on the ribbon or on the quick launch bar (see Figure 5.52).

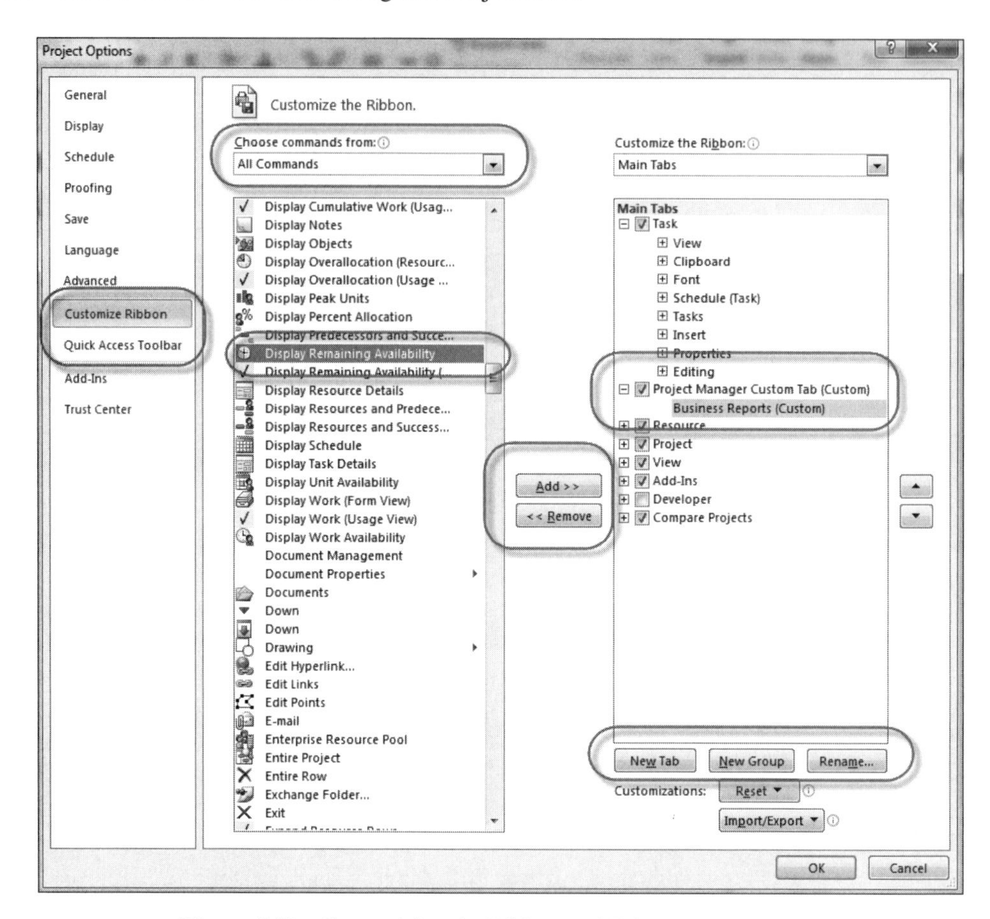

Figure 5.52 Customizing the Ribbon and Tabs Source: Advisicon

This customization helps to personalize the use of the ribbon and simplifies and streamlines key functions or tasks needed, better than all of the toolbars, buttons, and menus that were in older versions of Project.. The fact that the ribbon is easily changed or customized helps to promote personalization and use of key features.

The authors of this book all have their own personal preferences. We customize existing tabs, create new ones, and embed key features and automated functions (created with macros) to help us maximize our use of the product.

Personalization of the ribbon, commands, and buttons helps expedite end user utilization of new functions, macros, and features. Placing the features they want most or standardizing key approaches and functions designed for a PMO within a single tab will save users time.

Quickly Find and Present Information

Within Project Professional, all the columns are available, including the enterprise fields. In many cases, end users have to insert columns, resize, and then remove them from a view when done. In Project Professional 2010, users can easily add, change, or remove columns. In fact, at the end of any table (the last column) is the placeholder to add a column. Not only can you adjust columns, but you also do not have to request a change from an administrator in Project Server; as the end user, you can personalize the view by moving, hiding, and changing the columns in the PWA view to your liking.

In fact, all users can personalize a screen to their liking. This new feature allows individuals to quickly find, arrange, and present information that is pertinent to them. The best part of this new feature is that the field/column arrangements—hiding or rearranging—persists, meaning that when any user personalizes a view in Project Server, the program remembers the settings and they will be the same when that user returns.

In Project Professional, instead of adding and removing columns, we can to double click to drill down to task, graphical views and get to the data, or we can modify a table (right click in the upper right-hand corner or choose the table dropdown on the ribbon); no more inserting columns, then deleting them (see Figure 5.53).

This is a quicker, more effective way for any project manager, scheduler, or person using Project Professional to quickly get to the views, information, change data, reset the view to something more pertinent, and get back to work.

This little tip always impresses students. It saves them time and countless mouse clicks. In Project Professional and Project Server 2010, the addition of tables, views, and grouping to the ribbon helps to expedite the Microsoft Project workflow and simplify end users' experiences in working with the entire project-related data.

More Effective Options to Update and Share and Connect Information

Project provides many different options for enabling the end user to quickly access information. Some of these are: links, hyperlinks, deliverables (items flagged and posted to a SharePoint site for anyone to link, review, and consume in their schedule). All of these rapidly connect key data, files and updates into a single web page.

Updating can be done in Project Client, PWA Time, and Task Sheet views; directly in the PWA project details view; and in SharePoint if Project Professional isn't linked to Server but to a SharePoint page.

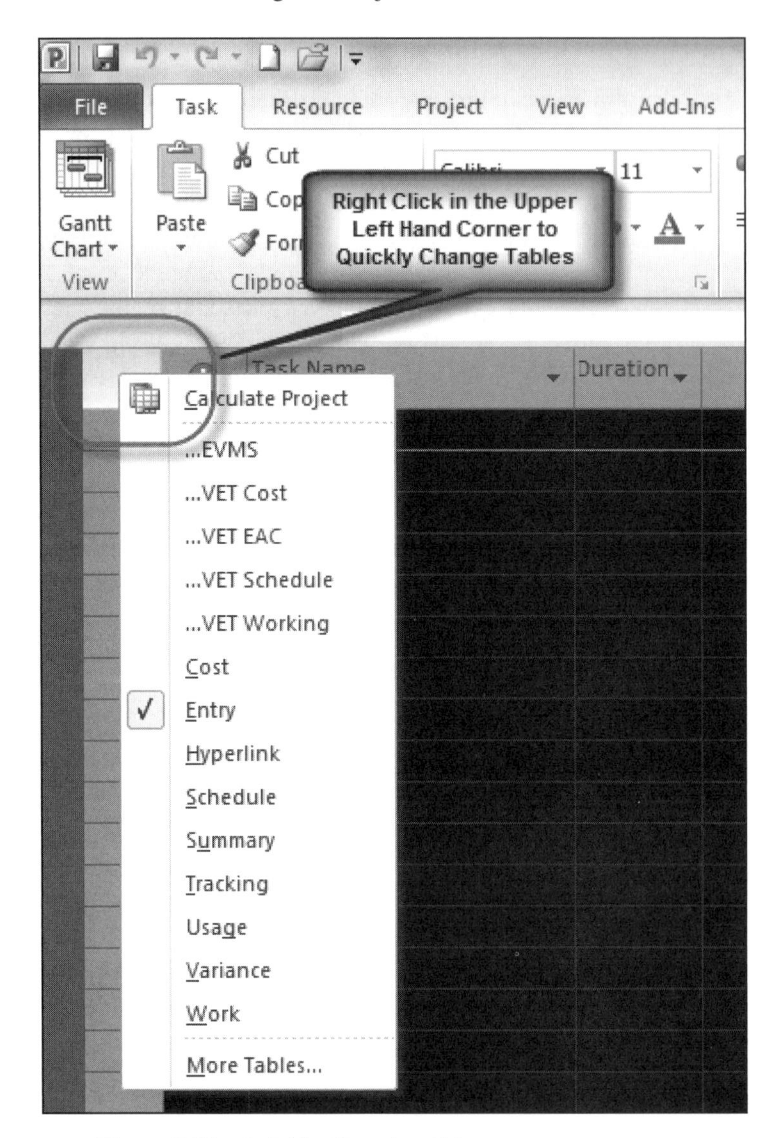

Figure 5.53 Quickly Changing Columns Source: Advisicon

Critical Success Factors

Here are the CSFs that we have learned regarding Fluent PM Using the Fluent UI.

1. To accomplish actions efficiently rather than searching for specific functions, Project's extensive capabilities are now organized into logical, easy-to-find groups on a ribbon.
2. To make the vast set of utility functions in Project easier to find, use, and understand, Microsoft introduced the Backstage with the release of Office 2010.

3. To make the new Office 2010 User Interface more valuable to end users, Microsoft provided the ability to customize the new Backstage view, the ribbon, the Quick Access Toolbar, and context menus.

IMPORTANT CONCEPTS COVERED IN THIS CHAPTER

In this chapter, we saw how the blending of specific processes, technological capabilities, and end user competencies will allow business users to address critical success factors, using Project 2010 for:

- Project management in small business and the enterprise.
- Initiating and managing project with the Project desktop.
- Effective EPM using Project Server.
- Using fluent PM with the new Office Fluent User Interface.

There is a critical demand for a simple yet powerful PM solution. The fully integrated PM solution comprised of Microsoft Project 2010, Project Server 2010, and SharePoint 2010 is clearly capable of meeting that need.

Key Summary Points

Key summary points are highlighted here to remind the reader of some of the vital points covered in this chapter.

- Organizations of all sizes utilize some form of work, resource, or PM methodology and tool set.
- Organizations are outgrowing their basic tools, methods, and processes for managing work and resources. There will be a need for a more sophisticated PM method and set of tools as organizations move toward a more mature PM approach.
- There are effective ways to use PM tools in both small businesses and large enterprises, from the individual desktop client to the departmental or enterprise server.
- The five key factors that can positively impact organizational effectiveness and end user satisfaction are:
  1. Scalability
  2. Configurability
  3. Ease of use
  4. Integration
  5. Collaboration

- As organizations gain in their understanding of how to get things done, an effective set of processes, tools, and technologies is typically implemented, usually resulting in improved organizational efficiency.
- A properly defined and maintained schedule (i.e., all tasks are connected to a network, a minimal use of constraints, deliverable milestones have been defined, etc.) can be used as an effective forecast to assist the project manager and other project stakeholders.
- End users are looking for a powerful yet easy-to-use solution for managing their work, resources, and critical timelines.
- Microsoft has completely revamped the Office Fluent User Interface based on feedback clearly indicating that users had a great deal of difficulty finding, using, and understanding the vast feature set of Office.
- The right technology can shorten the distances between different teams and make collaboration possible. The end users adoption of the technology is critical, however—choose wisely as you may not get a second chance.

REFERENCES

Absolut Agile. 2009. "Microsoft Project Tutorial Part 22—Planning an Agile Project." Accessed December 28, 2011. http://annaforss.wordpress.com/2009/05/11/microsoft-project-tutorial-part-22-planning-an-agile-project/.

Aguanno, Kevin. 2011. "Building an Agile Release Using Microsoft Project." Accessed January 11, 2012. www.agilepm.com/.

Bjork, Aaron. 2011. "Agile Planning Benefits." *Visual Studio Magazine*, September 20, 2011. Accessed December 28, 2011. http://visualstudiomagazine.com/articles/2011/09/20/agile-planning-benefits.aspx.

Channel 19. 2010. "Application Lifecycle Management: Microsoft Project 2010 and Team Foundation Server 2010, Better Together." Accessed December 29, 2011, from http://channel9.msdn.com/Events/TechEd/Europe/2010/OFS201.

Cooper, Robert G., Scott Edgett, and Elko J. Kleinschmidt. 1998. *Portfolio Management for New Products* (2nd ed.) New York: Basic Books.

Critical Tools. 2011. "Project Planning and Graphing Software." Accessed December 23 2011. www.criticaltools.com/.

Ducolon, David. 2007. "Resource Plans Explained." Official blog of the Microsoft Project product team. Accessed December 27, 2011. http://blogs.msdn.com/b/project/archive/2007/02/21/resource-plans-explained-i-hope.aspx.

Executive Office of the President Office of Management and Budget. 2011. Circular No. A-11 Preparation, Submission, and Execution of the Budget. Accessed December 23, 2011. www.whitehouse.gov/sites/default/files/omb/assets/a11_current_year/a_11_2011.pdf.

Federal Register. 2011. "Major System Acquisition; Earned Value Management. A Proposed Rule by the National Aeronautics and Space Administration on 02/10/2011." Accessed December 23, 2011. www.federalregister.gov/articles/2011/02/10/2011-2756/major-system-acquisition-earned-value-management.

Fiessinger, Christophe. 2010. "Announcing Visual Studio Team Foundation Server 2010 and Project Server Integration Feature Pack Beta." (A collection of excellent Project Server posts). Accessed December 28, 2011. http://projectserverblogs.com/index.php?s=agileandpaged=4.

Kaufthal, Jon. 2010. "Introducing the Backstage View." Official blog of the Microsoft Project product team. Accessed December 28, 2011. http://blogs.msdn.com/b/project/archive/2010/01/20/project-2010-introducing-the-backstage-view.aspx.

"Manifesto for Agile Software Development." 2001. Accessed December 28, 2011. http://agilemanifesto.org/.

National Defense Industrial Association. 2005. "National Defense Industrial Association (NDIA) Program Management Systems Committee (PMSC) ANSI/EIA-748-A Standard for Earned Value Management Systems Intent Guide." Accessed January 10, 2012. www.srs.gov/general/EFCOG/02GovtReferences/03NDIAANSI/NDIAIntentGuide.pdf.

O'Cull, Heather. 2006. "Visual Reports." *Micosoft Project 2010. Official blog of the Microsoft Project Product Team* (May 8, 2006). Accessed December 28, 2011. http://blogs.msdn.com/b/project/archive/2006/05/08/visual-reports.aspx.

———. 2009. "Project 2010: Introducing the Ribbon." Official blog of the Microsoft Project Product Team. Accessed December 28, 2011. http://blogs.msdn.com/b/project/archive/2009/09/24/project-2010-introducing-the-ribbon.aspx.

Perera, David. 2011. "Agile Doesn't Mean No Long-Term Planning, Auditors Tell PTO." FierceGovernmentIT Government IT News Briefing, October 22, 2011. Accessed January 17, 2012. www.fiercegovernmentit.com/story/agile-doesnt-mean-no-long-term-planning-auditors-tell-pto/2011-10-22.

Project Management Institute. 2008. *A Guide to the Project Management Body of Knowledge (PMBOK® Guide)* (4th ed.). Newtown Square, PA: Project Management Institute.

ReedShaff [Clay Satterfield]. 2009. "Microsoft Office Backstage (Part 1—Backstory)." *Microsoft Office 2010 Engineering: Official Blog of the Microsoft Office Product Development Group* (July 15, 2009). Accessed December 29, 2011. http://blogs.technet.com/b/office2010/archive/2009/07/15/microsoft-office-backstage-part-1-backstory.aspx.

Uyttewaal, Eric. 2010. *Forecast Scheduling with Microsoft Project 2010.* Ottawa Canada ProjectPro.

CHAPTER 6

THINKING LOCAL, GOING SOCIAL: PROJECT TEAMS CAN THRIVE USING MICROSOFT PROJECT SERVER 2010

IN THIS CHAPTER

We review implications and trends that can help predict a successful implementation and the opportunity to maximize the effectiveness of a project portfolio management (PPM) implementation.

Organizations that wrestle with project, program, and portfolio management can realize significant gains through leveraging Microsoft's 2010 PPM product and integrating it with the flow of their business and corporate culture for managing projects.

What You Will Learn

- The importance and impact of PPM technology on project, program and portfolio management
- The impact of collaboration on maximizing project speed, metrics and momentum
- How to leverage the PPM lifecycle to maximize the value of project, program and portfolio management leadership
- The key factors to avoid the kiss of death in PPM implementations
- How to integrate the business lifecycle with the PPM lifecycle for best results

PROJECT MANAGEMENT LOOKING AHEAD

Today, as the workforce expands globally, the natural evolution of project management (PM) is advancing and leveraging updated technologies. The changes and capabilities brought on by the enterprise environment have enabled PPM stakeholders access to services and applications that quickly allow individual users and groups to take full advantage of the same workflows of managing, tracking, and reporting projects but now in an enterprise and consolidated work management environment.

Microsoft Project 2010 is an example of one type of technology that includes integration of project-specific data along with the evolving enterprise workforce to include and embed applications such as project blogs, project wikis, issues, risks, and document libraries with their project schedules, collaborating with the project manager, project teams, and executives. Companies are embracing the leveraging of critical resources around the world. These virtual project teams now can work together much more efficiently and often are grouped by skill set, location, or other project-required attributes. With Microsoft Project Server 2010 residing within the SharePoint platform, Microsoft Enterprise Project/Program Management (EPM) 2010 is changing the traditional practices of PM processes. Leveraging the advanced features of Microsoft EPM 2010 parallels the recent dramatic shift toward collaborative PM practices. The transformation appears in the role of the project manager and in the interaction with stakeholder classes.

Traditionally, the project manager controlled the project lifecycle as if he or she was the center of all work defined and completed. The project progress was contingent on the project manager's actions of collecting, updating, changing, and reporting the project status. Therefore, enterprise stakeholder groups were at the mercy of the project manager's time, effort, and competencies to gain insight into the health and progress of the project. Organizations are now utilizing PMOs to address this potential source of problems in their PM processes.

Organizations are looking to the PMO as a more critical business component that derives project-specific data as part of the organization's progress ability. With the technology changes like those found in Microsoft EPM 2010, companies are rapidly adopting a PM culture that encourages collaborative spaces; everyone involved in the project is able to contribute work in the communal space. A project is led and developed by the whole team, and each team member has complete access to all information on the project. Project progress is visible to everyone on the team. In this environment, the project manager is transformed from a taskmaster to a project visionary, choosing the right direction for the project development. Project stakeholder groups now have better insight and can leverage collaboration to drive a project. People and business processes are more efficient. They can build and complete work around a PM system rather than having a PM system built around the work. The Microsoft EPM 2010 technologies use social, collaborative, and team-centric project data to produce collective project intelligence.

PPM as a Critical Part of the Business

The PPM as a critical part of the business has created an unprecedented interest in PPM methods. Historically companies find that after they have determined the strategic plans and defined types of projects that need to be initiated, they have more projects than resources available to implement them. PPM can help to address this issue. However, PPM can place a huge burden on an organization as PPM is a lifecycle process that is embedded in the functions and practices of an organization as part of doing business. Moreover, as companies embrace collaborative approaches to delivering more return on investment (ROI) and are acutely tuned to projects that contribute to the company's bottom line, there are newly discovered components that project and non-PMO stakeholders should take into consideration.

Collaborative Project Management: Avoid Negative Project Momentum

Companies strive to create a buzz, or excitement about the projects, and influence the workforce culture to drive strategic results. Leveraging EPM technologies and building cultural acceptance to work on projects together offers a powerful means for project managers to create momentum. A risk of this management technique is that it allows momentum that creates negative forces, activities, or direction. Historically stakeholders gathered to evaluate a variety of project details that ultimately lead to seas of spreadsheets and many hours of analysis. The result was many wasted hours gathering information or tracking details that never really helped drive the project and, more often than not, never got used for anything. Here are some ways to avoid negative project momentum using Microsoft EPM 2010 and newer PPM cultural practices:

- **Collect only relevant project data.** The Microsoft EPM 2010 system is a role-based PPM technology that allows a PMO to slice project metadata residing across the entire enterprise and deliver information to stakeholders that is relevant to their functional roles.
- **Solid project analysis.** Spreadsheets and other desktop business data tools have been used to model, analyze, and determine scenarios almost to the point of analysis paralysis. The Microsoft EPM 2010 technology provides custom configurations for user and stakeholder groups. Users can control what they need to analyze when they need it—all from real-time project data.
- **The right amount of automated project processes.** PMOs continue to take advantage of workflows, communication options, and centralized project data that may contain updates from around the world and around the clock. Some project stakeholders only want specific updates at a specific time, while others require near real time streaming of updates. As the PMO demands the right amount of project data, it also needs to be able to define the level of automation, from simple steps that are user driven, to complex channels of data linking together multiple stakeholders at various points of the PPM process.

PPM LIFECYCLE

Creating Business Value through Project Management Processes

PM exists because practitioners agree to and follow steps that are interrelated. The concept is simple and able to adapt to a variety of business environments and corporate cultures worldwide. PPM lifecycles are complementary to practices such as software development, agile methodologies, and business processes and standards, such as Sarbanes-Oxley. Although business lifecycles can be detailed and complex, the concept of a PPM lifecycle is straightforward and designed to create an orderly flow of information and activities. Ideas to consider when adjusting your business processes and PPM processes into a more cohesive ecosystem are listed next.

- **Adopt PPM processes that follow steps or phases.** Create a system that defines lifecycle states, such as create, selection, planning, managing, and closure. Other attributes may include various statuses within each state of the lifecycle, such as budget request, work remaining for a critical project, compliances, and the like.
- **Project relevance to business requirements.** Show how each project is aligned to objectives at the strategic level, and define elements that justify considering the project delivery as successful.
- **Separate high-performing projects from failing initiatives.** Ensure visibility into projects that are progressing efficiently, and change or kill projects that are trending toward ineffective delivery or low business value.
- **Perform project and program analysis using key performance indicators.** Determine and implement scoring of projects for the affected business stakeholders.
- **Capacity forecasting and management.** Resources are one of the critical constraints in a PPM lifecycle. Ensure that resources are taken into account during project updating and evaluation by leveraging a central, single repository of resource capacity that all project work draws from.
- **Predicting the project path to completion.** A project is truly successful if it delivers to the objectives defined in the project charter and maintains a narrow path of change over the course of execution. Projects that have required massive changes over the course of the project's life may indicate weak project initiation practices or lack of scope control.
- **Project completion delivers business success.** Projects should be evaluated as if they are part of the business lifecycle. They should be measured against fiscal efficiencies, investments into strategic objectives. These projects should be rated or ranked, just like any other corporate asset, on how they enable the company to compete as a viable entity.

Successful Project Management Leads to Portfolio Leadership

Throughout the history of PM as a defined practice, the importance of the project manager role, and its relevancy to the company, has been questioned. In days past, project managers had control over project resources, budgets, and progress. Currently, project managers find that they are embroiled in a web of corporate hierarchies, where various resource managers, self-managed groups, and other business stakeholders all have some level of ownership over the project. As all project activity rolls up to the portfolio level, there are some important points for project managers to consider while leveraging technology like Microsoft EPM 2010:

- **Focus on adding project value at the portfolio.** Project managers need to consider acting like portfolio leaders, focusing on adding value through project execution instead of just executing projects. They must start to act like business owners, constantly striving to make improvements.
- **Portfolio strength though communications.** Portfolio leaders don't wait for systems, people, and processes to tell them the answers. They leverage tools and initiate processes while encouraging behaviors that facilitate discussion. They also openly encourage communications (of many types).
- **It is about the relationships.** Portfolio leaders are using PPM collaboration systems to reach across the enterprise to connect with key stakeholder groups and influential departments within the company.

Keeping these in mind will go a long way to maximizing the value you bring to the organization as well as to helping you understand what it is you are working on.

The business environment and decision makers across most companies and in most industries are committing to a PPM approach to planning their business. Technology will still play a critical factor, and for the foreseeable future, people will continue to leverage the inefficient and weakly linked spreadsheets while working toward implementing and taking advantage of the many newer and effective systems that have been specifically designed to facilitate and enhance PPM. As organizational maturity of companies begins to increase and they start leveraging newer technologies, they must keep a link between simple and more complex environments for organizational adoption and assimilation of these newer technologies to work. Project Server was designed to help bridge the gap between the simplicity of Excel-based functionality and Web interface with the more robust enterprise portfolio and program relational database environments.

Information about the pros and cons and successes and failures of PPM has spread rapidly thanks to social media and global forums. People continue to be voracious in their consumption of information from many directions and in many different formats. Social media and mobile applications are prime examples of the different mediums in which consumers require information in more diverse ways. More and more information is added to the field of PPM daily, yet business problems continue to plague companies and decision makers still struggle to facilitate change. What is preventing companies from adopting the use of metrics, reporting and project information at the same rate as has been seen with social media? In most cases, it is the lack of standard metrics, collected and presented in an environment where the key project data are tied not only to the actual project itself but also to the portfolio planning and expected results of that project. This is compounded by the slow rate at which most companies update project management technologies and fold the newer tools and reporting technologies back to senior management.

As the global business climate continues to change, problems of the past continue to plague companies today. Briefly, too many organizations are attempting to adopt new technologies while retaining older ways of doing things. In the case of PPM, we see organizations attempting to expand their project lifecycle to embrace the idea-to-benefit mantra of PPM. They are adopting structured methods to remove politics from project selection and to build portfolios that contain projects that are fully aligned with strategies and maximize benefits and ROI. This review of projects, both pre- and postselection, needs to include the risks, issues, schedule information, and project costs (actuals and estimated) and demonstrate the impact of limited resources. As companies are improving methods, efficiencies, and communications for the execution of projects within the portfolios, they fail to consider or connect the metrics to the global market, people, processes, tools and culture. Essentially, companies are implementing PPM practices without looking at their organizations in their entirety. As a result, they aren't seeing the improvements they expected.

What do Microsoft EPM 2010 and PPM practices bring to the enterprise? Primarily, PPM vastly expands the universe of PM. Microsoft EPM 2010 provides an extensible technical platform that ties the human connection, such as social media, blogs, instant messaging, and wikis, with the processes and business data streams. The following list cites capabilities and benefits provided by PPM and Microsoft to address traditional PM weaknesses:

- **Naturally supports project lifecycles,** covering the entire project life span, from the identification of an idea, need, or opportunity, through the development and execution of the project, to the realization of benefits.
- **Leverage PMO socialization and connectivity to the community** to create a governance partnership between the PMO and the executive/operations side of the business.
- **Mitigates risks of project failure** by placing greater discipline on project selection. Politics are removed from the process and are replaced by reason and order, involving the partnership of the PMO and a governance (or investment) board.

- **PM is integrated** with strategic initiatives, ROI goals, and resource/capacity planning.
- **Performance of active projects is tracked and communicated** to all stakeholders in ways that clearly display status and identify critical performance areas.

Expanded PPM Lifecycle

The success of a project is influenced strongly by the initiation phase or pre-initiation and planning phases as it prepares for execution or the delivery of the project. That influence goes back well before the project planning phase, well before the approval stage, and as far back as the original identification of the idea, the need, or the opportunity. Technologies like Microsoft EPM 2010 allow for ideation, ad hoc task captures, and agile planning while remaining fluid in capturing critical elements for project success. PPM provides the structure for this expanded lifecycle. It adds entire sets of processes, taking these proposed projects through a rational workflow by promoting improved and consistent business cases, by evaluating alignment with strategies, and by quantifying benefits to the business.

Effective PPM operations that allow scalability to manage complex business environments function best within a PPM lifecycle that allows for multiple phases and stages within each. Additionally, each of these facets draws on a common set of resources, which can include people, funding, and facilities.

Common complaints among executives are the lack of effective oversight of prospective projects, impact against constraints, and in-flight projects. Contributing to these deficiencies is a lack of consistency in project measurements, standards, governance and reporting. An important element of PPM includes processes and reporting practices to address these deficiencies. Early adopters of PPM have reported outstanding improvements in project performance and increased efficiencies in resource allocations. Poorly performing projects are discovered earlier, while there is time to take corrective action or to terminate them sooner in the investment cycle, thus releasing scarce resources for more beneficial assignments. The increased efficiencies easily justify the investment in PPM, and executives are pleased with the vast improvement of both qualitative and quantitative information needed to make important and timely decisions about project investments.

Why is "portfolio" so important in PPM? If a high-functioning organization leveraging PM is primarily focused on projects, the responsibility for projects lies within the PMO. The PMO will maintain a robust automated system to process data relative to schedule, resources, and costs. These systems and reports are entirely lacking in the very important information related to strategic alignment and project benefits. Furthermore, PMOs rarely consider the impacts of proposed projects to an organization or, at the end of the project's lifecycle, evaluate whether the ROI was realized. An even greater shortcoming is the inability of executives to leverage the systems that are in place to get key metrics for insight across the organization. Essentially, there is a disconnect between the projects and their goals and measured results and the strategy as it ties to the business drivers and metrics associated with the organizational choice to do the projects to begin with.

It may seem intuitive enough to assume that simply implementing PPM bridges the gap, bringing executives into the process, via a governance or investment board. But in reality, the critical elements of technology implementation will be maximized by business organizations defining and tying the key success definitions and leveraging the visibility of these success factors in regular reporting. Doing this will create not only organizational adoption but also commitment to use the systems and processes, thus reinforcing the integrity of the data being tracked and reported. Microsoft EPM 2010 for PM addresses the issues of alignment, value, ROI, prioritization of proposed projects, and allocation of scarce resources. Furthermore, specialized PPM systems provide vastly improved communication for executives as well as the PMO.

Expanding from PM to PPM requires the implementation, and business user adoption, of several new processes as well as a culture of change. Some insights into options for expanding from PM to PPM are listed next.

- **Top-down planning.** The ability to build high-level plans and resource demand pictures
- **Alignment of strategic plans** with enterprise technology and process architecture
- **Prioritization and ranking** of candidate projects
- **Computation** of benefits, risks, and ROI
- **Ability to address income** (or cost savings) as well as costs
- **What-if analysis** (effect on resources of adding or removing projects)
- Executive-level **display of data** used to support the selection of new projects
- Executive-level **display of project health data** and by-exception reporting of poorly performing projects that are in progress

All these processes usually are ignored within the normal PM operation. Traditional PM software does not support these functions. If you are still using spreadsheets for PM, you will need to develop new spreadsheet models.

Change accompanies PPM for real results; cultural change and PPM adoption must have executive champions. Executive buy-in is essential. Senior executives must clearly state that PPM is a way of life in the firm and build a business case for the benefits from PPM.

The business case should be comprehensive, accurate, and relevant, and it should state expected ROI.

Implementation of cultural change and PPM systems cannot occur all at once across the enterprise. Start small, perhaps with a pilot implementation, directed by some managers who are less change resistant. The success of the pilot will help sell PPM to the rest of the organization. The leaders of that success will serve as additional champions and mentors as the methodology spreads.

Obviously, there are several areas of change as we move from basic PM to the wider scope and enlarged involvement of the stakeholder community. We have new processes on top of modified practices. There are changes in roles and responsibilities, and there must be changes in the management culture as these roles change. As an organization moves from a nonexistent or traditional PM company to a PPM company, small but structured steps will lead to a growing level of maturity in PPM. Many of the changes lead to improvements in communication—especially in the communication of information that will assist management decision making as it applies to projects.

Whether this potential is realized will depend, to a large degree, on how the organization uses its PM systems. There is an opportunity to leverage the ROI gained from early steps to fund and secure support for the future actions, as those future actions may be larger in scope, cost more, and demand more from the company's resources.

Early in this change process, organizations must to investigate and acquire the best systems to support their PPM initiative. A common mistake is to delay the move to PPM-specific software until the firm has reached some specified level of PPM maturity. Maturing in PPM can be significantly aided by the implementation of a well-integrated, robust, PPM-specific tool set. A well-designed system incorporates the best practices in the industry.

A good PPM tool set should be presented to the respective business user groups within a role-based environment that is predefined to serve their needs and assist with their existing or future workflow around project, program, or portfolio management. Once validated to maximize the ease of use and adoption by the organizational resources using the system, the tool set should be optimized to a set of user-configurable templates that stakeholders can quickly select and utilize within different work roles in the PPM environment.

PPM Kiss of Death: Making Decisions with a Lack of Interrelated Data

Spreadsheets are subject to significant flaws because the user designs the data structure, flow, and computational regimens. There is no audit trail or guarantee of consistency between worksheets. Much time is wasted when not working in a centralized data system that has synchronization across all initiatives. There needs to be a consistent story across all reporting documents. Using spreadsheets for PPM is inappropriate and considerably less effective than using specialized PPM software. Spreadsheets get in the way of integration and standardization. In other words, developing internal spreadsheets is not free.

In moving to PPM, we have much to gain from the improved processes and governance. PPM benefits from highly graphical presentations of project, portfolio, and resource information, featuring alarms and highlighting, that directs stakeholder attention to out-of-tolerance conditions.

Why not stick with what is working? Basic PM software (such as desktop PM software) tends not to support the full lifecycle with full integration. These systems are optimized for current projects and are not designed to deal with proposed projects. They generally lack the ability to consider ROI and benefits, and they can handle costs but not cost savings or income.

Typical desktop PM tools are optimized to support the classic project approach triple constraint (which primarily addresses time, cost, and scope). To support the scalable PPM demands, organizations need systems that promote validation of alignment with strategies, evaluation of benefits and risk, optimization of limited resources on proposed as well as approved projects, and prioritization of pending and active work.

Another shortcoming of the traditional PM approach using desktop scheduling PM software is that it is suited primarily for users who are directly involved in projects. Senior managers and relevant business stakeholders have limited access to viewing the metrics coming from projects in both static and real-time reporting. Their lack of access prevents these managers and stakeholders from realizing the benefits of dynamic online reporting over the rigid static reporting software of the past. Scalable PPM software, such as Microsoft's EPM 2010, helps business users work with project data to reach decisions in a manner they find comfortable and easy to use, with a common and simple interface or collaboration portal.

Microsoft EPM 2010 is specialized project portfolio software that adds all of the functions for diverse business stakeholder groups and classes to expand the project and inevitable business lifecycles. It provides support for structured selection of projects for the portfolio and allocation to resources based on knowledge and prioritization rather than politics. Structure and integration are earmarks of a robust PPM system, wherein the PPM process serves as a hub for all of the project-oriented business activities.

Knowledgeable decisions regarding acceptance and prioritization of project-based and independent work items is possible only with a robust, interactive system that contains an up-to-date inventory of work requests, active work, available resources, and unutilized resource time. The demand and capacity data must be integrated with any other resource and with project and service request modules, so that the information is seamless, timely, and consistent. Microsoft's inclusion of the PPM Lifecycle in Project Server enables taking an idea-to-launch approach. This is central to new product development applications, and is the basis for a establishing a structured process for ranking, rating, selecting, and monitoring projects through their lifecycle. A good example and popular component of this process is the Stage-Gate technique, developed by Dr. Robert Cooper, which reinforces reviews at key checkpoints of a project or proposal lifecycle.

This Stage-Gate enables a strong and reinforced process and checklist for evaluating a project as it goes from idea to launch. Microsoft EPM 2010 product Project Server has incorporated this capability for supporting a more robust PPM solution and enables end users to incorporate support, automation, and custom fields for stages and gates reinforcing industry best practices.

Initiating change, including the move to a more comprehensive PPM technology and process adoption, is a huge commitment. But what if the change was so intuitive that it streamlined the learning curve? Specialized PPM software will make this migration much easier. By incorporating all of the needed capabilities in a seamless package, and by incorporating established best practices into the various modules, the proper tool set will help to guide you to a mature PPM capability. Sunk costs are just that: in the past. You can't justify staying with obsolete systems just because they are paid for.

Building a PPM capability on top of the appropriate technology will have a big payoff for your business. Updating your system of tools, with a robust PPM engine at the core of a full-featured, integrated system, in support of enhanced PPM practices, enthusiastically embraced and promoted by an enlightened executive, is the ticket to success. Figure 6.1 showcases the new features of Project 2010 to help prioritize and select projects based on a rating and ranking system.

In leveraging Project Server as a PPM tool, it is important to get insight into the characteristics of the processes in an organization or department's project

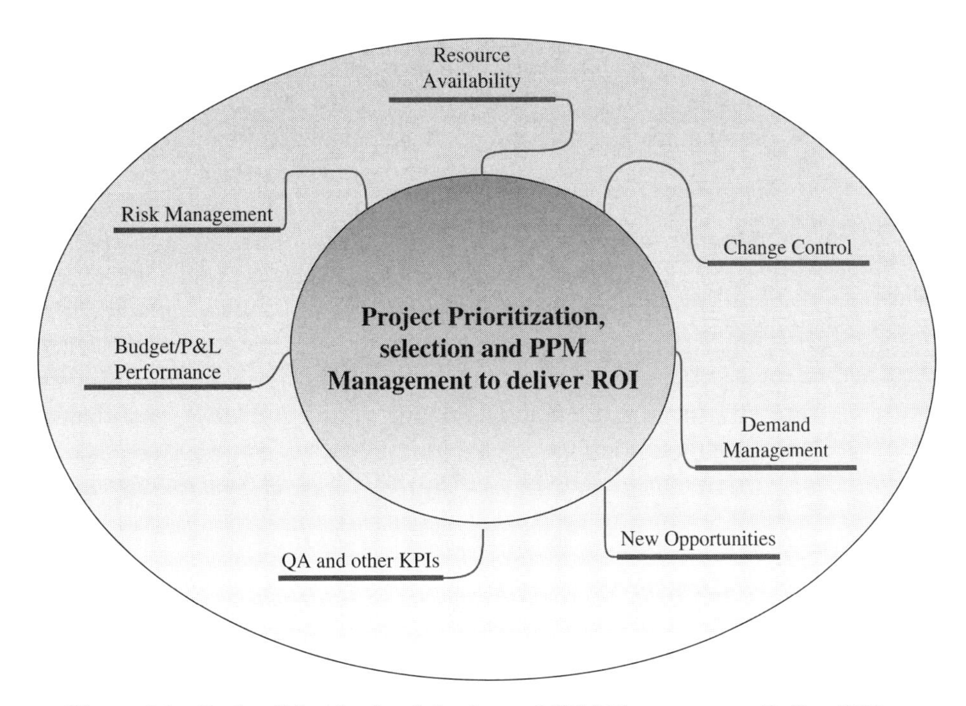

Figure 6.1 Project Prioritization Selection and PPM Management to Deliver ROI

workflows and the value they provide as well as their impact and use throughout the project's lifecycle.

How do companies get visibility to project information pathways? Process maps are created to meet a variety of needs covering stakeholder class orientation, navigation, user situation analysis, and the like. A process map's utility is derived primarily from the relationships between business requirements and user capabilities. A good process map will accurately depict and predict the behaviors of information and how people use and make decisions based on that workflow. These process maps can be integrated into the PPM system for reinforcing and evaluation criteria through a project or program's lifecycle.

What relevance does process mapping have to PPM and in particular to creating project budgets?

Typically business stakeholders cannot detail up front how much something is this going to with regard to project initiation. Often project sponsors are looking for prompt answers to the question "How are you going to do this?" But what is really being asked is "How much exactly is this going to cost me?"

When creating the business case to select which project approach to take to meet the demands of a specific business problem, or when choosing between competing projects to allocate a finite pool of funding, the formulas that are most commonly used to measure the expected financial return of a given project are time-adjusted rate of return (TARR) and net present value (NPV). These formulas explicitly acknowledge the effects of duration and timing on when costs are incurred or income is booked by the project. Similarly, earned value management techniques such as cost variance, which measures the difference between the estimated value of the budgeted work accomplished and the actual cost of the work accomplished, help project managers determine whether the project is on budget.

TARR and NPV (as examples) are constraints that can be used for portfolio analysis. But while such metrics are indispensable to managing the project throughout its lifecycle, we as stakeholders or business managers sometimes overlook the fact that the project budget also has a lifecycle of its own. What characterizes that lifecycle is the corresponding refinement in the magnitude to which cost and cash flow estimates can be made as we accumulate more detail about project scope. Thus, at whatever stage you are in of your project's lifecycle, always be explicit about the order of magnitude (bigger than a bread box, smaller than a building) you have used to derive your budget estimate. Detail the basis for the magnitude of the budget—that is, the factors that have influenced the size of the numbers presented.

Order of magnitude estimates should also include a complete list of project assumptions, risks identified to date (including an assessment of probable impacts should any risk event occur), the scope of the project as represented by the degree of specificity in the current work breakdown structure, and any other gaps or unknowns in project information that remain to be filled in. During this phase, often in the initiation phase (also commonly referred to as the create phase), provide some detailed context about how the estimate was derived. Also identify what is required to improve the precision of the estimate.

The goal at this point of project iteration is to establish the synchronicity between the PMO and the business groups, where the PMO will have the opportunity to discuss options for improving cost burn rates while delivering high-performing projects. This is the time when the PMO and the business groups will be fully engaged and ready to determine if refinements to the budget or project requirements are needed.

Using the orders of magnitude that are referred to in the budget process is an effective way of managing stakeholder expectations about what may potentially drive variance in project costs at later stages in the project lifecycle.

The control parameters for a project are believed to be budget and work or debt limits and resource limits. Product functionality determined by PPM requirements and time are control parameters, but money is a constraint. Also assume that productivity (resource work invested) and quality are the two control parameters that represent the costs aspect. These four parameters lead to what we like to call the PPM efficiency quadrant, where constraints are balanced by requirements, allowing results or outcomes to be measured and ranked.

Whenever a budget is determined, it will be applied to allocate resources to the project. A high-functioning PPM environment will enable the project manager and other stakeholders to determine resource allocation, including skill set, location, availability, and others. This collaboration leads to more effective establishment and control of the budget depreciation against work completed across the enterprise. The choices made also have an influence on project progress and an impact on quality. With the same amount of money, different choices can be made that will lead to a different type of project governance.

The requirements and the duration or elapsed time of a proposed project (or in some cases changes to in-flight projects) determine the amount of budget required. The productivity and the quality determine the way the budget will be used. The four parameters are dependent on each other. Whenever one parameter needs to be changed, the others need to be retuned.

Changes based on scope change or constraints present issues that impact the project. Examples include but are not limited to:

- New requirements identified that lead to scope review or update
- Portfolio analysis exposed opportunity at either a time or cost level
- Crash the project to meet market changes
- Quality or deliverables updated
- Resource changes

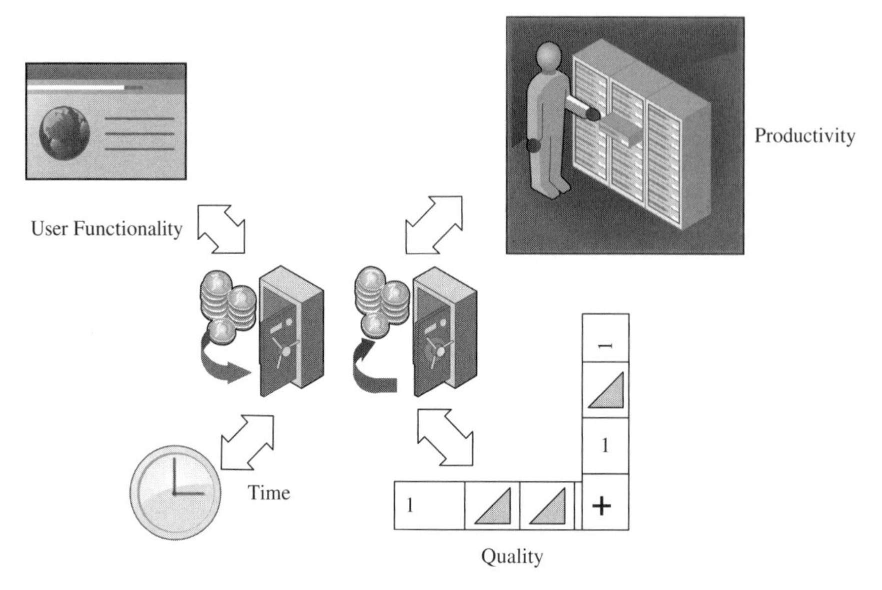

Figure 6.2 PPM Efficiency Quadrant

The changes always can be interpreted individually or collectively. The changes either enforce or weaken each other.

Figure 6.2 depicts the interdependencies between the four parameters.

The square is balancing at one corner, symbolizing the delicate balance required among the four parameters. An effective PPM environment will enable project managers and other key stakeholders to find the right balance and translate the analysis to PPM planning and the governance structure. Using the model shows that there are parameters to accommodate the change—for example, quality.

Productivity (aka resource utilization) is another way to control project requirements and changes without raising the budget. There are two ways to raise productivity:

1. **Add resources of less skill.** Increasing the team size will increase costs. In other words, the cost per resource needs to go down. This can be obtained by using less skilled resources. There are two disadvantages to increasing the team size:
   a. More members in the team will increase communication and reporting efforts. These hours are a reduction in productivity, and therefore there will still be a rise in costs.
   b. Less-skilled resources will be less productive and make more mistakes. These mistakes are then discovered in a later stage of the

project (e.g., during one of the test cycles) and need to be fixed at a higher cost.

2. **Reduce resources or swap with higher skill sets.** This method achieves the opposite effects. Overhead will be reduced, and thus productivity will increase. The team members have to be more senior; the productivity will rise since the amount of mistakes will most likely drop.

Using resource allocation and forecasting methods, time can be leveraged to accommodate PPM changes within the preset budget. However, simply increasing a project's duration will not deliver a reduction in over-allocation. The project will take longer, and with no additional measurements or use of fixed work tasks, the project and resource costs will increase.

There are several ways to address over-allocation, including creating a more detailed or resource-leveled project or proposal, changing the start time for that project or proposal. It is important to note that in the early phases of evaluating a new work proposal or project, often proposals get denied or excluded. The intent with Microsoft's PPM system is to allow end users to spend less time in the early phases and scale the level of detail, tasks, and resources as a project is moved through to its approval and selection.

Extending time on a project can follow the same pattern. Providing more time for the project allows project managers to reduce the project's team utilization percentage, thus increasing productivity for other projects that have been proposed or in the existing work portfolio. Whenever a change in a project's scope or functional deliverables can be mitigated by increasing project duration, that increase can offer the opportunity to reduce team size by adding resources of a higher quality. This mitigation is especially helpful at early stages of the project (create phase) and will provide the opportunity to gain momentum against costs or resource constraints. An example may be using a team that is smaller than planned yet more highly skilled to produce more for the same costs.

In a PPM environment, the quality of resources and work delivered can be diminished by project managers who give way first to meeting the end date or to delivering all functionality. When this happens, the quality of exercising and supporting the PPM process and approach is also compromised, especially in the create phase or manage phase, when timing and budget pressures increase. The result is an architecture that is less resilient or less robust. As they say, garbage in, garbage out.

Often the changes and results prove to work against the PMO and alter the forecasting and planning at a portfolio level. The easiest option for handling the demand for increased quality is to reduce the project scope and requirements. Another option is to scope the project into milestones or add stages in each phase of the project. You can add program phases to create program dependencies across the portfolio.

Increasing quality with no change in duration requires increasing resources if changing scope and business requirements is not an option. There are two options:

1. Reduce the team and increase the timeline and duration of the project.
2. Reduce the team and increase the skill set of the team, allowing for more senior resources.

If the PM chooses to compromise quality, project success needs to be achieved with a minimum of risk. A way of achieving this is by leveraging indicators, dashboards, and collaboration across the enterprise in relation to business stakeholders' needs. Another option is to prioritize projects and programs at the portfolio level, where business requirements and strategic objectives will be categorized, organized, analyzed, and prioritized. These metrics need to be agreed on by representatives of the business stakeholder class.

If the timelines for the project are constrained and delivery at a certain date is mandatory, one option is to agree to delay noncritical projects. Another way of meeting the deadline is to prioritize the projects based on critical strategic requirements and time to market. Overall, this process will help to remove the pressure on a limited or constrained budget. In order to do this, projects will need additional prioritization measurements so that the rating, ranking, and reprioritization of projects in a portfolio can be done quickly and efficiently. Either quality reduction needs to be achieved—in most cases, this is not desirable—or an increase in resource allocation (number or skill level) is required.

A strategy to address this issue and to meet the deadline, increase productivity, and remain within budget is to reduce the size of the overall team and ensure that higher-quality resources are allocated to the project. Studies have shown that the number of defects increases with the square of the team size. Essentially, more hands do not mean more accuracy or faster performance. On the contrary, in many cases it results in teams doing duplicate work or rework. This is a good reason to keep teams lean and mean.

In the pursuit of increasing productivity, you might ask how to increase the productivity of an existing team. There are many motivational theories, but project or portfolio managers have at their fingertips the ability to measure and track key metrics related to the project or the project schedule.

An option for increasing productivity is to leverage collaboration and create competition by measuring productivity through the schedule or related project metrics. It is exciting and tactically stimulating to resources to see the results of their efforts reflected in the regular progress reports associated with the project team's work activities.

This feedback leverages current business practices and technology to their fullest and gives visible and constant progress reporting on metrics that ultimately provide project success. Introducing results of measurements and logic improvements will boost productivity; it also reinforces the importance of the metrics an organization needs to succeed. As Peter Drucker stated, "That which gets measured gets done."

A great way to see this success and a good long-term organizational strategy for ensuring it is to introduce some healthy stress by defining milestones in the near future. Goals that are within reach increase productivity. William James of Harvard found that motivated employees work at 80 to 90 percent of their ability while unmotivated employees work at about 20 to 30 percent of their ability. A nearby milestone that is just achievable will increase the level of productivity up to 110 (or even 120) percent. The reason for this phenomenon is that people tend to relax until it is almost too late to achieve the objective. A kick is required to make things happen. Of course, we are not advocating increasing stress levels, as the negative effects of stress have effects on productivity, creativeness, and health. Rather we are encouraging the establishment of processes and technology to ensure accountability. These processes and technology help keep attention focused on critical activities and will improve performance and results that are clearly traceable and measurable.

By planning your milestones so that the cycle of stress and relaxation reaches its optimum (between four to six weeks), your productivity will increase. People tend to focus on deliverables and the near future.

Finally, offshoring to a low-wage country will increase the productivity while keeping you on budget. When offshoring is considered, you also need to calculate for additional overhead and management and a more rigid governance structure due to distributed delivery. As a rule of thumb, you need to add between 15 to 20 percent of the hours spent in the offshore location due to a loss in productivity. This number depends on the experience the project team has with offshoring.

Adopting the PPM Lifecycle as a Component of the Business Lifecycle

All projects follow more or less the same pattern. In the beginning of the project, during requirements gathering and the design phases, there is a focus on the needs of business user groups, strategic alignment, high-level costs, timing, and quality. During these phases, the milestones and specific resources are less of a concern. The project manager is often a facilitator, gathering all functional and nonfunctional requirements. During the create phase, the scope becomes clear, and an end date is agreed on. The focus changes to time and resources, which are the axis for analysis and movement of the project through stages and phases.

IMPORTANT CONCEPTS COVERED IN THIS CHAPTER

This chapter's emphasis was focused on learning how to leverage the different lifecycles surrounding project, program, and portfolio management blended with PPM and the business lifecycle. Some key concepts were:

- Blending the right amount of automation and centralization can radically improve the business and processes surrounding PPM implementation.
- By following a PM methodology and folding that into the native features of PPM, you can establish a reinforced and supported workflow that yields the best results for metrics, reporting, and visibility and communication.
- PM can be supercharged through a PPM implementation, establishing metrics for pre–work portfolio management, existing work portfolio management, and postproject completion analysis of a project against stated goals and objectives.
- There are key gotchas to avoid in a PPM implementation. We call these the kiss of death. Being aware of them is critical in ensuring a successful deployment.

CHAPTER 7

BETTER TOGETHER: MICROSOFT PROJECT 2010 WORKSITES USING SHAREPOINT SERVER 2010

IN THIS CHAPTER

This chapter highlights some of the features available with Project Server 2010 in a SharePoint collaborative environment. The ability for the project teams to communicate and centralize information has rapidly expanded the adoption of SharePoint even without project or portfolio tools.

Project Server 2010 is the only enterprise application created in SharePoint that leverages not only project, program, and portfolio management capabilities but also the full spectrum of collaboration and reporting that SharePoint brings.

What You Will Learn

- Project Server 2010 integrates the increase in social communication, simplification, and centralization of information, enabling faster viewing and getting to the right information.
- Project management (PM) practitioners have a significant advantage with the PPM 2010 product due to the availability of tailored views, ribbons, and easily accessible, modifiable content search and reporting.
- Project communications and project-related information is dynamically linked and geared for better social integration with reading, responding, and managing.
- Automation and workflows no longer require deep programming knowledge.

INTEGRATION OF COLLABORATION, SOCIAL MEDIA, AND PROJECT-RELATED INFORMATION

Up to this point, we have focused on "what's in it for me" with regard to a PPM environment's key stakeholders' relationship to supporting tool capabilities such as enterprise project types (EPTs), workflows, reports, actuals versus estimates, program/project/task interdependencies and relationships, and resource capacity planning. Now we set our sights on the one aspect that cannot be ignored but is the most challenging to manage: people. More specifically, we focus on managing people's behavior.

If we think about social media and the growth of multiple ways to accomplish the same activity, we can see that there is a significant drive to allow flexibility in managing work or work tasks. People who are responsible for doing an assignment (e.g., those who test a part and document the results) may perform the task differently from their colleagues. By allowing for flexibility and making use of multiple avenues to perform work (SharePoint Lists, My Tasks, timesheets to progress activities in Outlook or MS Project), we enable a wider audience and organizational culture to take advantage of the full Microsoft stack (the wider offering of multiple Microsoft products).

Social media is more than a craze or fad; it is a new way to leverage information for business purposes. Imagine predicting that one of your team will be out of the office next week, because you saw a travel update on Twitter. As a project manager, you could subscribe to Really Simple Syndication (RSS) feeds from project-related blogs and stream them right into your SharePoint Workspace, where you can share and exchange the information among team members. Being able to capture the behavior of resources opens the door to a whole new set of data and the ability to collect and manipulate that data for extensible information.

When SharePoint is used in a PPM environment, alerts help to tell the story, and other included media (pictures, charts, discussions, etc.) illustrate and articulate the evolving situation. Additionally, geospatial tracking, tagging, ratings, "I like it" sentiments, and others are now critical pieces of insight that fill in the blanks. Think of these forms of metadata or visual cues like irrational numbers in the math world. They are values that exist but cannot be expressed rationally. Human emotions, actions, and gestures communicate a significant amount to any environment. Project managers for the most part agree that communication is one of the more important components in a PM environment, and communication is defined as both verbal and nonverbal. Poker players read their competition by looking for the nonverbal cues, or tells, that may give them just enough insight to make an informed decision. Together, SharePoint and Project offer tells to decision makers related to the health, progress, status, and future of their PPM environment.

Consider the wealth of honest, open conversations that happen among employees in any given day at the water cooler. While a lot of it may be of a personal nature, the work- and project-related bits are of a very high quality. The second that conversation ends, that information dies on the vine, and its value ceases to exist. By leveraging the power and familiarity people have with social networking, you can collect, store, and use this high-quality, relevant information to further the project and eventually the business.

As organizations begin to realize the benefits of a fully networked environment, more and more methodologies and systems will become integrated. As the lines that used to define a nonnetworked legacy approach continue to fade, exceptional instances of these seamless integrations rise to the top of the pack and become evident.

Such is the case of pairing SharePoint and Project Server. Project Server 2010 combines the power of both technologies to provide a powerful solution that effectively manages project, operational, development, and all other types of work in an organization. Project Server 2010 leverages SharePoint Server capabilities including workspaces, forms, Excel Services, social networking, workflows, enterprise search, and more to meet customers' unique requirements. A common and familiar platform for both Project Server 2010 and SharePoint 2010 will help organizations get more value for their investment, streamline operations, save money, and have direct impact to business results.

Project Management as Practitioners Want It

Managing projects, information, and stakeholder objectives is complex, and complexity elevates risk, which threatens project delivery and the ability to meet strategic objectives. Complexity also elevates opportunity. Project 2010 democratizes PM by enabling individuals in teams of all sizes to quickly plan, manage, and deliver work on time and on budget. It brings the ease of use from Excel to the Project platform to connect to these broader PM audiences. Project Professional connects to two servers to facilitate teams to collaborate around work. The first connection is to the SharePoint server tasks list which allows project managers to publish their plans where their project team lives and then automate status collection through distributed task lists. The other server connection is to Project Server 2010, which provides for a unified PPM solution (See Figure 7.1).

The gap between these two business elements (project and portfolio management) was long considered to be insurmountable. Now Project Server 2010 offers an integrated and efficient solution (see Figure 7.2).

Dynamic Worksites and Collaborative Workspace

Now, within the SharePoint Server environment and along with creating project sites, any user can create workspaces for project team collaboration. A workspace is contained within the Project Site and allows the team to work together on a document (e.g., project charter, communication plan, etc.). There are also meeting workspaces that store meeting minutes, agendas, and calendars and can record meeting decisions. These workspaces allow for a "conference room" within Project Sites where collaboration is more focused on a particular document or meeting.

This degree of collaboration-heavy features plays into the earlier points of capturing the living information from the team that can be lost in traditional

Sites

- Ribbon UI
- SharePoint Workspace, SharePoint Mobile
- Office Client Integration
- Standard support

Communities

- Tagging
- Social bookmarking
- My Sites
- Blogs and Wikis

Content

- Enterprise content type
- Audio and Video content type
- List Enhancements
- Enterprise metadata

Search

- Phonetic search
- Navigators
- FAST Integration
- Document Preview
- Enhanced Pipeline

Insights

- Project Services
- Excel Services
- Charts
- Visio Services
- SQL Integration

Composites

- External Lists
- Workflows
- SharePoint designer
- Business connectivity services

Figure 7.1 Project Server 2010 Part of SharePoint Server 2010 Universe Source: Advisicon

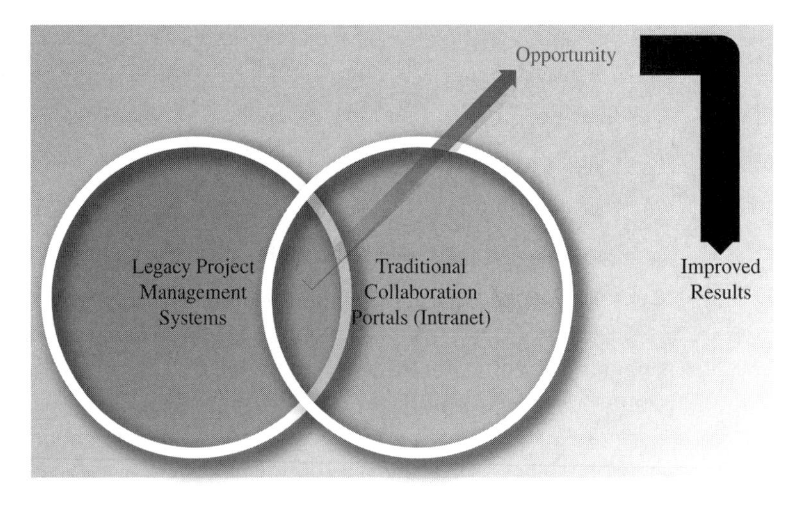

Figure 7.2 Opportunity Translates to Business Results Source: Advisicon

water-cooler discussions. Project sites aim to provide all natural discussion tools a project manager could ask for in a single, one-stop experience for the team.

These workspaces are governed by the permissions granted to team members. For example, the project manager may have formed a core team to discuss and collaborate on the project charter prior to releasing it team-wide. Through permissions, only the core team may have access to this workspace. This control is crucial to managing work and team stakeholders efficiently.

Project Team Discussion

Discussion boards provide a forum for conversing about topics that interest your team. For example, you could create a discussion board for team members to suggest activities. Team communication and discussion is the key to project success. Discussion boards allow project team members and project stakeholders to discuss open-ended issues and proposals, give feedback, and ensure a smooth flow of communication at all levels. They are a good way to manage escalation and team interaction. And since the discussion board is housed within the Project Site, the communication is captured as a historical project asset.

Discussion boards can be set up using SharePoint 2010 to share and discuss topics. The discussion board shows the most recent discussion first, as well as the number of replies for each discussion. That way, you can quickly see which discussions have the most recent activity and which ones are the most popular. Members can also customize their own views of discussion boards. Participants can even add items to discussion boards from their e-mail application, making participation as easy as sending an e-mail. Multiple participation venues (through the intranet, Web, or e-mail) also widen the audience, as people can contribute remotely. So whether team members are in the office or in the field, participants have an easy way to contribute content.

Discussion comments can be displayed in either a flat or threaded view. Flat view displays all comments in the order in which they were created. Threaded view lets users view comments by conversation. The discussion groups are organized by topic. Anyone can post a message to a discussion, and multiple people can respond in a free-form manner. Figure 7.3 showcases an example of a discussion

Subject	Created By	Replies	Last Updated
Conference Plans NEW	System Account	1	4/6/2012 10:05 AM
Potential speakers NEW	System Account	2	4/6/2012 10:04 AM
Session Paper NEW	System Account	1	4/6/2012 10:04 AM
Yearly Sales NEW	System Account	0	4/6/2012 10:04 AM
Proposals NEW	System Account	0	4/6/2012 10:03 AM

Figure 7.3 SharePoint Server 2010 Discussion Example Source: Advisicon

WebPart, very commonly used on Project sites to help showcase frequently asked questions or posed questions with answers.

SHAREPOINT SERVER 2010 OFFERS CRITICAL BUSINESS CAPABILITIES

Project Server has really opened the door for business functionality to be driven to a single portal.

Later in this book, chapters 8, 9, and 10 we review some of the dashboard and business intelligence (BI) capabilities. However, some of the immediate critical needs are the day-to-day activities of project team members, managers, and other stakeholders within the project/program environment.

Things like searching and having alerts and other key information driven to the stakeholders, without them having to manage or go and look, save countless hours and enable project teams to focus on exceptions or issues rather than having to manage and review all of the collaboration portal information surrounding a project or a portfolio of project information.

Search Capabilities Meeting Business User Needs

The search capabilities, as shown in Figure 7.4, are one of the most powerful features of Project Server and SharePoint. With proper setup, an organization can search across all projects and quickly locate key information, files and other important information.

Searching for key information can be a nightmare on a file share. For project team members, searching within Project Server is simplified and mimics how users search within a Web browser. As Project Server is based on SharePoint, it supports several types of search. Users can use the Search box on each page of a site or on the Search Center site, or create a detailed query by using the Advanced Search page. Users can look for content by searching for keywords, a specific phrase enclosed in quotation marks, or by values that are assigned to properties.

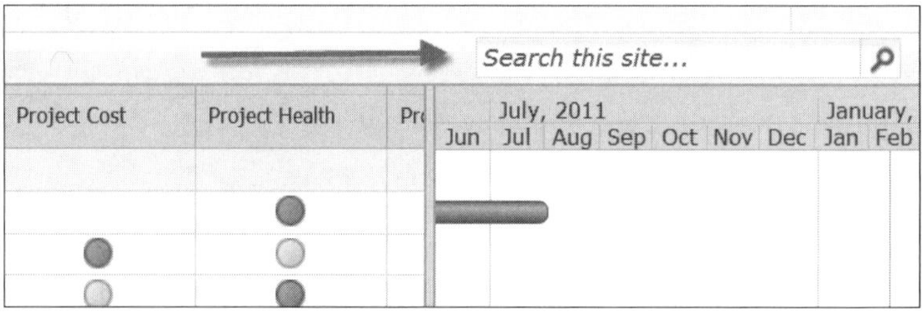

Figure 7.4 SharePoint Server 2010 Contextual Search Source: Advisicon

When users are looking for content but are not sure where it is located, they can start their search at the highest site SharePoint site (parent site) where they think the information might reside. Then refine the results to find the information. Users can look for content by entering keywords or a specific phrase enclosed in quotation marks.

If a more specific query is needed, users may want to use the advanced search page, which is available from the search results page. By using advanced search, users can choose to display or exclude results that include certain words, filter search results by language or type, and search on content properties. Thus, it is very easy to search for data on the site without wasting lots of time.

SharePoint 2001 supports these types of search:

- **Boolean search.** Users can use Boolean operators—AND, OR, NOT—to combine text and construct a meaningful query.
- **Wildcard search.** Search queries can now match a wildcard at the end of a text string. For instance, users can search for "tenn*" and get results that have the word "tennis" in them.
- **Faceted search.** When a search query returns many results, the faceted search functionality displays a refinement panel on the left side of the screen. In the panel users can refine the results based on criteria such as result type, site, author, modified date, and tags.

Advanced Workflows without Needing Deep Programming Abilities

Project Server working as an application within SharePoint offers significant automation capabilities.

Workflows can be created easily and efficiently in Project Server. Many third-party tools are available to speed this process or are waiting to be deployed.

This section does not delve deeply into setup, configuration, and programming instructions for creating workflows. Rather, its aim is to help those who are contemplating automation of manual steps understand the overall capabilities of Project Server with SharePoint and what is possible with workflow automation.

The concept behind workflows reaches directly to a growing and maturing organization that wants to automate key steps, alerts, field updates, and the progress of schedules using workflows. A key step in getting the most from workflow automation in both Project Server and SharePoint is the process map. Identifying business and PPM processes lays the foundation for rapidly identifying areas that can be automated. This identification process will become a focal point and also a test case for any technical or development team to create workflows in either Project Server or SharePoint.

As this section suggests, you can get to simple and even complex workflows without having to build them yourself or become a.NET developer. (Being a techie or developer can be extremely useful for expanding what already exists, however.)

Project Server's Solution Starter Kit comes with two workflows that have features and functionality that end users can edit or leverage for excellent automation and workflows. These are the Demand Management Dynamic and Demand Management Morphing workflows.

Project Server provides an out-of-the-box workflow named Sample Proposal Workflow that can be leveraged for the phases, stages, and other enterprise project types. This workflow is very easy to leverage and actually is a good demonstration as it includes all of the usual things you might see in such a process: validation, approval, selection, and so on.

Good interfaces for managing workflows can be created with InfoPath forms, which enables nondeveloper types straightforward and nontechnical interface creation abilities (see Figure 7.5).

Hide Details **Initial Proposal Details** Add Stage

Stage Settings:

Stage Selection:
Select a stage from the drop down.
Note: The same stage cannot be selected twice in the same workflow.
Initial Proposal Details

Stage Status Message:
Type a message to show up in the stage status. When a project enters a stage, the workflow stage status page will display a section where this custom messages will be shown.

Force Submit on Stage:
By selecting "Yes" the workflow will always wait on this stage, even if there are no required fields in the stage. Selecting "No" will mean that if there are no required fields that need to be filled out in the stage, the workflow will simply move forward without requiring an end user "Submit".
Yes
No

Approval Settings

Approval Required?:
Checking "Yes" will cause the workflow to send approval requests for this stage. The user will not be able to move beyond this stage unless an approval is received.
Yes
No

Approval Type:
Indicate the type of approval for this approval point.

First Response:	Selecting this would mean that the very first (Yes/no) response will be used, and the workflow would not wait for any of the other responses
Majority:	Selecting this would mean that a majority of the approval responses will be used as the final decision
Consensus:	Must receive 100% of the same response for the decision to be final.

First Response
Majority
Consensus

Approvers:
Select the Group/Users that will receive approval requests for this stage.
Project User Groups:
Administrators
Executives
Portfolio Managers
Project Managers
Resource Managers
Show Users

Filter Approvers by Department:
Checking this option will mean that only users that are in the same department as the project will receive approval requests.
Yes
No

Approval Stage Status Message:
Type a message to show up in the stage status when a project is waiting for approval.
Note: If this field is left blank, the default message will be: "The project is currently waiting for approval. The approvers are: <list of approvers>"

Figure 7.5 InfoPath Workflow Example Source: Advisicon

For those who are interested in going beyond the standard or canned workflows, you can edit these out-of-the-box workflows, which may be easier than trying to build them from scratch.

There are different options for creating or leveraging programmed or nonprogrammed workflows. The distinction between SharePoint (Project Workspace) workflows and the workflows within Project Server should be understood.

In general, to build new Project Server automated workflows, you would use Visual Studio. Some of the technical resources out there have already been doing SharePoint workflows, but since Project Server workflows are a bit more complex and there's no graphical user interface, you don't get to use SharePoint Designer with Project Server out of the box.

If you want to modify the workflow or create custom workflows, you are going the Visual Studio route. One option is to purchase a third-party tool. Nintex Workflow for Project Server 2010 tool is a common one that many have found easy to use.

This is a great way to leverage workflows since it handles the approvals and also provides options that cover the workflow needs of most organizations at the beginning of the move to automation. It is good to note that as more advanced, branching workflows are needed moving to a Visual Studio or a programmed solution will become necessary. However, getting started and creating workflows with InfoPath is an excellent way to provide scalability to establishing growth of a PPM system.

The steps for associating an InfoPath form with a workflow are:

1. Once it is installed, go to the location—e.g., to http://server_name/pwa_name/_layouts/WrkSetng.aspx) on your server.
2. Click Add a workflow.
3. Create a new workflow based on the DM DynamicWorkflow template.
4. Configure each of your phases/stages using all your precreated stages with your approval requirements, then submit to finish.
5. Now return to PWA and in Server Settings create or assign the newly created workflow to your ETP.

These five steps allow Project Server administrators or power users to take workflows and connect them with existing enterprise projects in a matter of minutes.

Note that automated Project Server workflows or more complex branching iterative loop reviews and the like typically are developed in Visual Studio with the.NET Framework 3.5 or higher. SharePoint workflows may also be done in SharePoint Designer, but these would be dedicated to SharePoint itself, not the programmability in Project Server's Project Server Interface (PSI) engine.

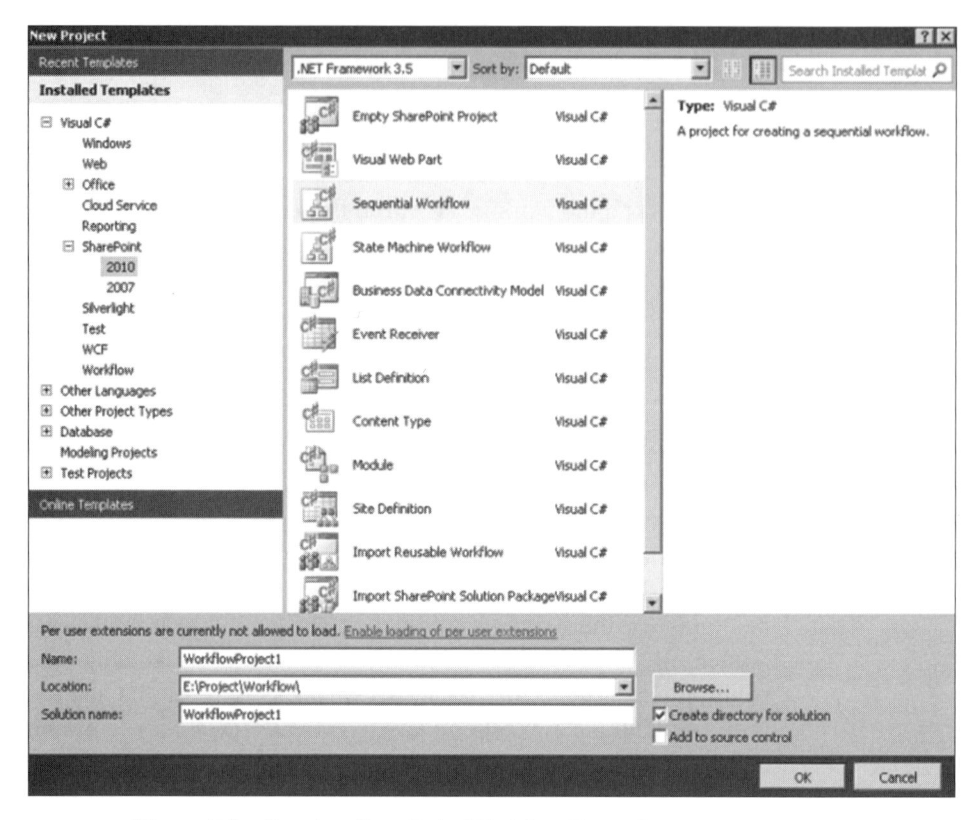

Figure 7.6 Creating SharePoint Workflow Example Source: Advisicon

When using Visual Studio, you can easily create a SharePoint solution package (.wsp file) for a Project Server workflow. Once a workflow is created, Visual Studio creates the Features and Package folders in the workflow project that contains the package definitions, files, and code used by the workflow. Figure 7.6 illustrates the creation of a workflow example.

Using Windows Powershell, you can install and run the workflow packages. Note that any changes require that you restart a workflow to test it out, especially if you are using a branching workflow. Figure 7.7 shows the project server workflow list or package that can be referenced.

Remember that using Visual Studio 2010 to develop and debug Project Server workflows and other solutions for Project Web App requires development on a Project Server computer. In fact, for best practices, we recommend that you have a development, test, and production instance of Project Server.

Enough technical details. TechNet has many great examples and samples for the technical team to review and test for creating workflows. The Solution Development Kit has code samples and examples that can be deployed readily.

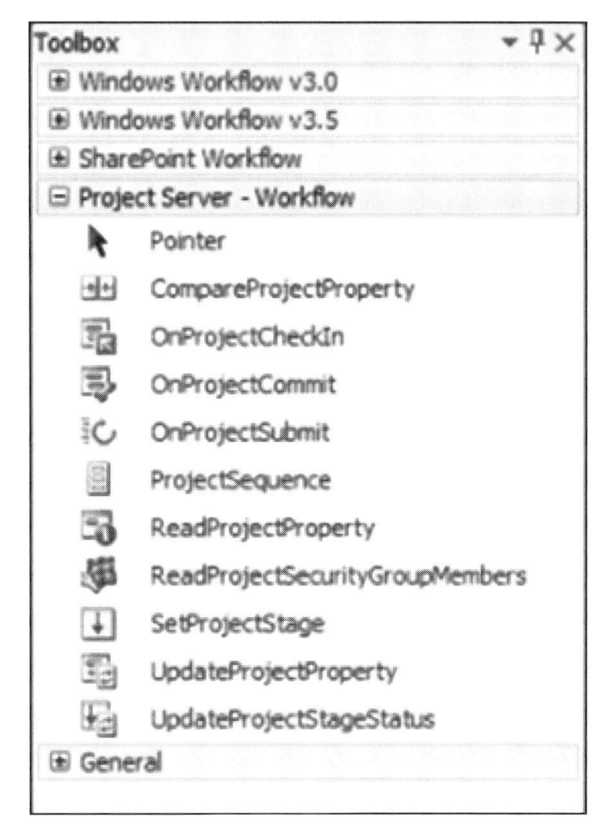

Figure 7.7 Project Server Workflow List Source: Advisicon

We hope that as you, the reader, review the options to move to automating workflows, you definitely consider these points:

1. Create a process map of the business process, inputs, and outputs.
2. Leverage the out-of-the-box existing workflows provided.
3. Review TechNet and other online references.
4. Utilize technical resources as you need them (developers), but don't be afraid to tackle the connecting and automation of your business processes. Nothing ventured, nothing gained.

Leverage Integrated Communities to Build and Manage Knowledge Assets

Communication is a major success factor for any project, regardless of size or type. To enhance communication, the project team must function more like a community. SharePoint 2010 has made great strides in bringing the social computing elements to users in an easy-to-use format. This collaboration is key to project success and inherently develops the project team not only by using business knowledge about team members but by creating personal ties to enhance internal team relationships enterprise wide.

Many organizations, regardless of size or type, have different types of communities in terms of the work each group performs or areas the company has to develop. SharePoint allows organizations to create collaborative Web spaces for these communities to interact and improve understanding of the activities they are developing. In SharePoint, you can create surveys and link them to reporting tools, such as SQL services, Crystal Reports, and other business reporting products. All of these tools (which can connect to the data behind the scenes) can be integrated into SharePoint and Project Server and easily analyze, quantify, organize, and present the key trends or data elements customized to specific audiences.

SharePoint 2010 is the ultimate Swiss Army Knife for collaboration with smart connections between people and across teams. It enhances the set of collaboration and social networking tools for both organic and managed communities across your organization with the following features:

- **Collaborative content.** SharePoint 2010 offers improved blogs and wikis, calendars, discussions, tasks, contacts, pictures, video, and much more. With Office 2010, multiple people can simultaneously author content on a SharePoint site.
- **Social feedback and organization.** SharePoint 2010 enables involves organizing, finding, and staying connected to information and people through bookmarks, tagging and ratings, searching, navigation, profiles, feeds, and more. It combines informal social tagging with formal taxonomy so you can choose the right approach for a given set of content.
- **User profiles.** SharePoint 2010 enhances user profiles to reflect colleagues, interests, expertise, either via explicit tagging or recommendations based on Outlook and Office Communicator. The model is opt-in so users can manage what information is shared publicly. Users decide when an interest is something that they want to share or be asked about by others in the organization.
- **MySites.** MySites give quick access to your content, profile, and social network while continuing to let you customize, target, and personalize pages to the needs of different roles and users in your organization. The enhanced newsfeed helps track interests and colleagues.
- **People Connections.** SharePoint 2003 contained a universal person hyperlink and presence icon, so that you could always navigate to a user's MySite, send e-mail, start an instant message, call, and the like. SharePoint

2010 enhances the user interface in conjunction with Outlook and Office Communicator. Also, colleague tracking and people search features have been greatly improved, with new algorithms and user experience leveraging expertise, social data, and more. In larger companies, organization chart browsing via the address book is one of the most popular features in Outlook.

Collecting Information from Multiple Sources

Surveys provide an efficient and cost-effective way to collect feedback on everything from how satisfied customers are with your product offerings to whether employees are satisfied with key result areas. Users can use surveys to ask team members what they think about issues, how to improve processes, and many other topics. Surveys yield the substance needed for future product, system, or process improvements and development. They enable you to present specific questions and collect the answers in an organized format, similar to a poll. Surveys can be implemented using SharePoint 2010 (see Figure 7.8).

SharePoint offers Web-based surveys that can be completed by anyone who has access to a Web browser; even mobile devices are supported. The responses can be named or anonymous, real-time results are available, and you can apply analysis tools. Results can be collected using several different types of questions, such as multiple choice, fill-in fields, and even ratings.

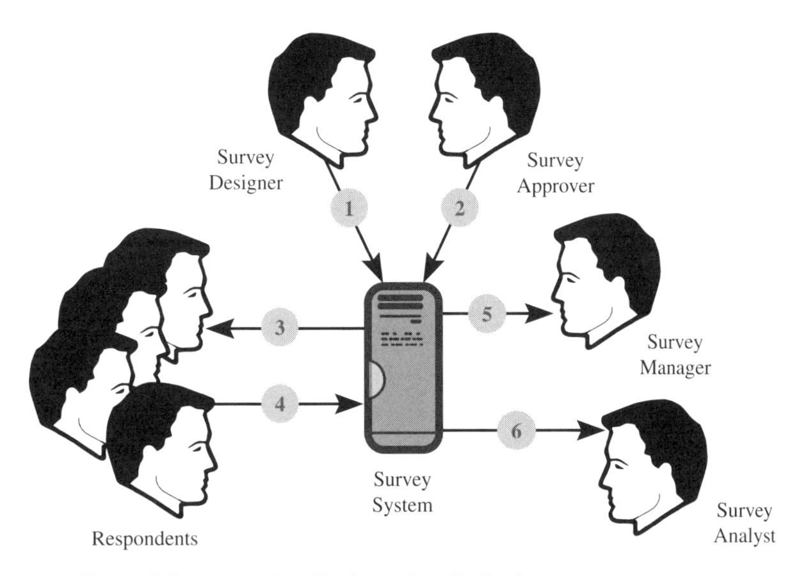

Figure 7.8 Example of Information Gathering Source: Advisicon

Figure 7.9 SharePoint Server 2010 Survey Results Example Source: Advisicon

Depending on how a survey is set up, you may be able to respond to that survey only once. If you are the person creating the survey, you can add branching logic so that the survey changes according to responses to specific questions. In a survey that branches, questions appear only if they apply to someone's situation based on previous responses. If the questions don't apply, the survey skips that set of questions or the respondent is offered a different set of questions. Figure 7.9 shows the survey results from end users answering questions on the survey.

After people respond to the survey, you will want to see the results, including any patterns visible. Surveys provide a graphical review of responses, similar to a chart, or you can export your results to another program, such as Microsoft Office Excel. Surveys also can provide individual responses or a list of responses.

BEING SOCIAL IN A PROJECT ENVIRONMENT

Today's project teams no longer work in the same building, floor, let alone the same region or continent. With the growth of cell phones and a more electronically connected workforce, the need for project teams to leverage the ability of social networking tools, concepts, and capabilities is increasing.

In 2009, a senior member of the Microsoft Project Marketing team asked a large group of Northwest stakeholders if they knew a good use for social networking tools. Many of these senior project, program, and portfolio managers really didn't have a good response. Now, this senior marketing team member was being a bit coy, but if the managers' sons or daughters were posed the same question, the response would have been very different.

Here we are, just a few years later, and the impact of socialization technology is everywhere. Different tools from blogs, networking sites, Twitter, Facebook, and LinkedIn surround the project community and those who are trying to communicate concepts, ideas, jobs, and marketing ideas.

Social Computing and Communication

SharePoint Server 2010 includes social computing tools, such as MySite Web sites, and social content technologies, such as blogs, wikis, and RSS. These features are built on a database of properties that integrates information about people from many kinds of business applications and directory services. You can adapt content to each user while enabling administrators to set policies to protect privacy. SharePoint 2010 introduces ready-for-prime-time social networking for the enterprise, with a rich feature set around social content, activity feeds, social and people search, and more.

Every team member can customize unique MySite pages. This page provides information about the team member, everything from their direct supervisor and subordinates, biography details, interests and skills, to contact information. It is a virtual address book. Metadata tags supply necessary data for searching. Now one member can look for a specific skill set or business acumen, and find someone possessing those skills internally.

The idea of communities is to bring the project team closer, enhance internal collaboration and cooperation, and streamline communication. It is also another tool for people to see what others are working on, what experience they possess, and how their skills can assist in projects.

Social computing also provides communication platforms for things such as a chat feature, which shows an employee's status (e.g., whether in a meeting or not, or available to contact for assistance, etc.). There is also a posting feature similar to the postings seen on other social networking sights. Employees can post items to their walls and others can comment.

Developing camaraderie among team members enhances communication and helps get teams through the forming and storming phases of team building and on to the more productive norming and ultimately performing phases much more quickly than before.

Business Users and Their Customized Pages

The ability for end users to move columns around in views as well as to hide columns from an existing view is one feature that end users find especially helpful.

While on the surface this ability may strike fear into the hearts of the PMO (which is trying to standardize or create a uniform look and feel for people working in Project Server), it actually isn't that bad.

End users cannot turn on columns of data that they do not already have access to. The feature simply allows end users to drag and drop columns in the order that they prefer or to hide columns from a view that they personally may not find useful. A good example of this is when team members work on laptops. They don't have a lot of real-estate space on their screens for viewing views, and they can hide columns or move the columns around to help them get to the key information they need. This doesn't affect any other user and is localized to their log-in only.

This quick formatting allows team members and other stakeholders to manage the PPM views quickly and easily, without having to bring in the Project Server administrator to build new or customized views for them. After one migration from a 2007 PPM environment to a 2010 environment, the Project Server administrators were excited at how little they had to tweak and fine-tune views for end users. A little end user training saved the administrators hours weekly adjusting and helping adjust views for specific stakeholders. The end users were happier and so were the administrators, who now could focus on some of the BI and other new features enabled by PPM 2010.

Other components also allow business users to customize pages. When business users or stakeholders go to a project workspace or essentially a SharePoint site dedicated for a project, they also can custom tailor that page to their personal view.

Again, this doesn't meant that they are able to add or change the SharePoint objects associated with that site; they just can hide or unclutter the page in order to help them see or get to the most important content they are looking for.

Many organizations choose to disable this ability, but the ones that provide a little training around this feature find that the stakeholders tend to prefer tuning the content for their viewing and/or relative to the size of their screen resolution.

What is nice about this feature is that with a click, business users can reset the page back to its original view or how it was updated by the Project site owner or SharePoint administrator.

Active Updating as Users Can Tag and Rate Content

Social tags enable users to tag and track the information they are most interested in. Users can also leave impromptu notes on profile pages of a MySite Web site or any SharePoint Server page. You can now tag any source on the Internet (or intranet) that has a URL. This tag is stored in your tags section on your MySite and also appears in your activity feed. Other users can post notes relating to your tag, which effectively creates a discussion board around the tagging activity, allowing conversations around something that has been tagged.

One of the key points of tagging is security trimming. Let's take this example: What happens if you tag a document that someone else doesn't have access to? Social tagging uses the search index to provide security trimming on content that is stored in SharePoint. It enables senior managers to tag confidential documents, but those tags are not visible to anyone who doesn't have read access to those documents. Figures 7.10 and 7.11 illustrate the tagging or the adding of meta data to any SharePoint object so that it can be found via search or the web crawler search features enabled in SharePoint.

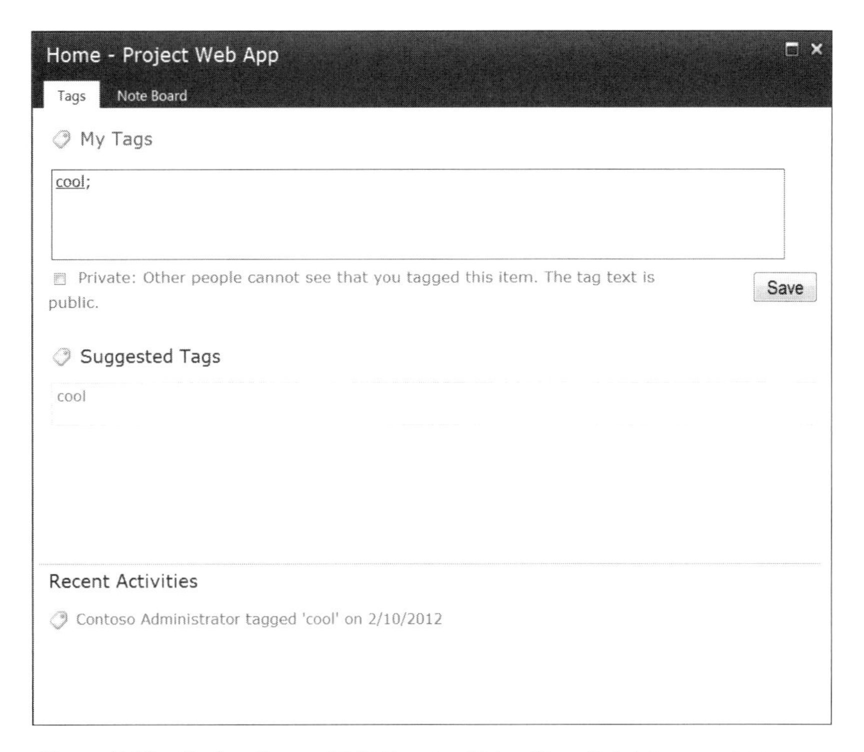

Figure 7.10 Project Server 2010 Tagging Using SharePoint Source: Advisicon

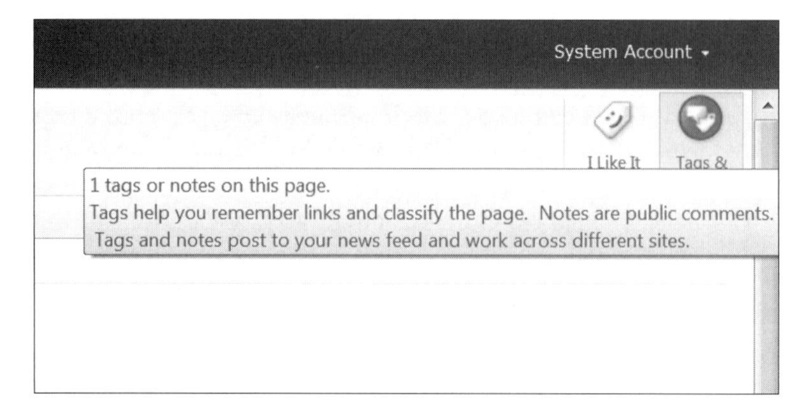

Figure 7.11 Project Server 2010 Tagging Notes Example Source: Advisicon

On top of this is a Ratings feature, where users can rate content within SharePoint lists. Thus, SharePoint 2010 now has social feedback functionality, in that you can tag and rate content and other people can interact with that tag, creating a discussion. Figures 7.12 and 7.13 illustrate the ability to rate content. This will become more important as social media and new sorting, ranking and search features are being used by organizations.

Creating Highly Connected Teams through Alerts and Notifications

A key component of both SharePoint and Project Server is the ability to notify end users of changes, updates to key documents, lists, or other important project related information. This helps to connect the user community to the information around projects, documentation, and key project collaboration. The ability to create pull process versus pushing or managing communication to different stakeholders can be easily managed from the Project Workspaces or SharePoint Project documentation. A project manager can manage and provide access to

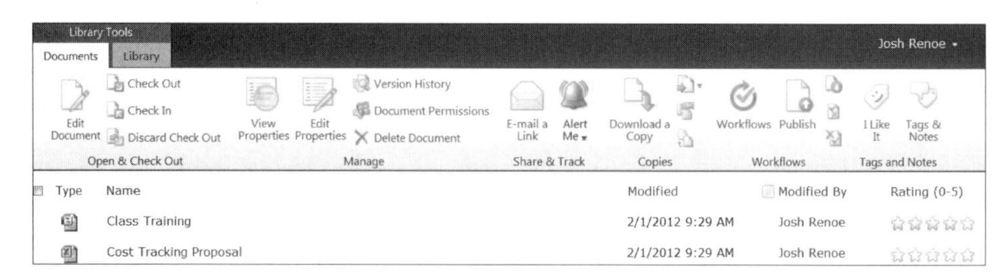

Figure 7.12 SharePoint Server 2010 Ratings Example: The Ribbon Source: Advisicon

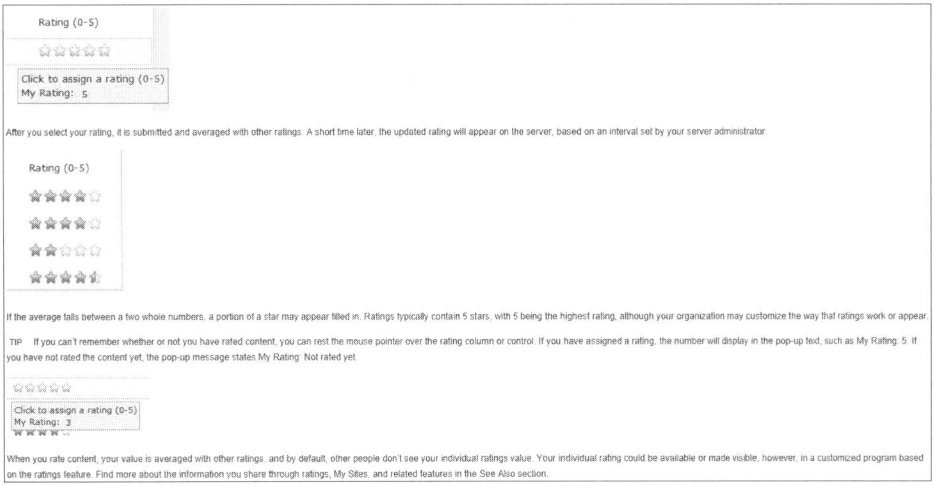

Rating (0-5)

Click to assign a rating (0-5)
My Rating: 5

After you select your rating, it is submitted and averaged with other ratings. A short time later, the updated rating will appear on the server, based on an interval set by your server administrator.

Rating (0-5)

If the average falls between a two whole numbers, a portion of a star may appear filled in. Ratings typically contain 5 stars, with 5 being the highest rating, although your organization may customize the way that ratings work or appear.

TIP If you can't remember whether or not you have rated content, you can rest the mouse pointer over the rating column or control. If you have assigned a rating, the number will display in the pop-up text, such as My Rating: 5. If you have not rated the content yet, the pop-up message states My Rating: Not rated yet

Click to assign a rating (0-5)
My Rating: 3

When you rate content, your value is averaged with other ratings, and by default, other people don't see your individual ratings value. Your individual rating could be available or made visible, however, in a customized program based on the ratings feature. Find more about the information you share through ratings, My Sites, and related features in the See Also section.

Figure 7.13 SharePoint Server 2010 Ratings Example: The Menu Source: Advisicon

documents on a Web site, and interested stakeholders can be notified by e-mail when site content changes. This type of notification is called an alert. Alerts notify you of changes to documents, list items, document libraries, lists, surveys, or search results—helping you to be more effective by staying up to date with relevant information. If you have access to a SharePoint site, you can add, modify, and delete alerts from Microsoft Outlook. You can create alerts for lists and document libraries as well as for individual items and any files in them.

Alerts and reminders can also be set for changes made to task or status reports in Project Server. Each resource is able to configure the frequency of e-mail alerts and reminders for upcoming and overdue tasks; in addition, project managers can configure their resources' e-mail reminders. In such cases, the project Manager's settings override those set by the resource.

IMPORTANT CONCEPTS COVERED IN THIS CHAPTER

This chapter's emphasis was on the social and collaborative features that Project Server has in a SharePoint environment. The better-together story enables and empowers the project team to work at a higher and more productive rate, enhancing and centralizing communications.

Here are some of the key features highlighted and discussed in this chapter:

- Dynamic and Integrated SharePoint Workspaces help integrate key communications and documentation directly with the work being performed and the projects they fall in.

- Issues, risks, and documents are linked and dynamically connected so that team members or stakeholders are never more than one click away from information.
- Workflow capabilities empower end users who do not have to have a developer on hand to automate information or optimize activities.
- Workspaces can be customized for end users, helping to minimize IT dependence and allow end users more ownership of the collaboration portals.
- Alerts and notifications, while possibly being set by a manager, can be tailored to a team member's specific needs, minimizing overcommunication.

CHAPTER 8

EFFECTIVE TRANSITION OF STRATEGY AND EXECUTION: PROGRAM MANAGEMENT USING MICROSOFT PROJECT SERVER 2010

IN THIS CHAPTER

In this chapter, we discuss the importance of integrating strategy with the practical needs of execution in a project portfolio management (PPM) implementation. The needs of an organization's solutions or products being managed can put a strain on how the project is managed, yet what does Project Server 2010 do to help ensure good visibility, accountability and quick drilldown capabilities in this world of time-to-market driven corporate goals?

What You Will Learn

- The challenges and how a project management office can address them
- How to managing information for project success
- How to leveraging collaboration for program success
- Workflows for supporting project lifecycle management (PLM)
- How to incorporate issues, Risks, and deliverables with PPM
- Graphical (static and drilldown) capabilities enabled in PPM 2010

A project management office (PMO) rarely achieves its full potential. Project managers find themselves short on resources and budget or incapable of delivering projects on time. Without visibility into business goals, the PMO is unable to align project priorities with the long-term strategic requirements of the enterprise. Some PMOs even fall short of their initial promise to define and standardize processes and practices, which makes it difficult to leverage institutional knowledge or to reduce the effects of common operational challenges. We examine why some PMOs fail to deliver much beyond modest increases in efficiency and tighter monitoring of projects while others become centralized hubs for integrating general business management processes and enabling the free flow of information across the organization.

PROJECTS ARE THE "HOW," PROGRAMS ARE THE "WHY"

Being more strategic requires PMO applications, or support tools, that provide visibility, accountability, and control, including resource requests and time and financial management.

In the mid-2000s, there was a major push for corporations to establish PMOs with the goal of instilling much-needed project management (PM) discipline in every department across the enterprise, especially within information technology (IT) groups. This trend was driven partly by the passage of Sarbanes-Oxley (the legislation spawned by the Enron, Tyco, and WorldCom scandals that cost investors billions of dollars) but more often by the desire to define and standardize PM practices and facilitate project portfolio management (PPM) as well as the desire to determine methodologies for repeatable processes.

From an organizational perspective, a project, program, and portfolio management office can be one of three types:

1. An **enterprise PMO (ePMO)** spans multiple departments to integrate processes across business units.
2. A **departmental PMO** typically is established in (IT) departments, but is also found in marketing, research and design, and other department-level organizations.
3. A **special-purpose PMO** is created for a single major project or set of projects.

There is also a wide variety of governance and organizational structures. Some enterprises have PMOs that operate as a unique entity within their organizations while other enterprises have some combination of multiple PMOs that are operating independently, are organizationally aligned, or are based on the division of PMO functional responsibilities.

What Is the Purpose of the PMO?

The basic definition of the PMO in a business or professional enterprise is *a permanent organization* tasked with one or more of these objectives:

- **Define and maintain** the guidelines, policies, processes, and standard documentation around projects.
- **Encourage and sustain repeatability** related to project management.
- **Provide central, coordinated management and oversight** into the initiation and strategic planning of projects.
- **Coordinate and develop PM training** for continuous organizational improvement.
- **Offer a broad range of services** from budgeting, to product management, to direct project leadership, to support functions such as coaching, consulting, and marketing.

- **Support the prioritization of strategic projects** to ensure that the organization is working on initiatives aligned with strategic business goals
- **Provide oversight across the resource pool** to support the assignment of resources to the highest prioritized initiatives

Enterprise PMOs can have an even wider scope of responsibilities that includes all planned work and comprehensive resource management, including operations.

What Are the Challenges of the PMO?

Each of the business management services just listed has one goal: delivering the highest-priority projects on time, on budget, and within scope.

This is the first and most important challenge of the PMO and the best measure of a PMO's effectiveness. If projects are being delivered late or over budget or do not meet objectives, the PMO has room for improvement.

Unfortunately, "room for improvement" is the case at most companies. There is a significant gap between the current ongoing practice of managing, tracking and reporting on projects and what can be done. According to the Standish Group's most recent report*, 32% of all IT projects succeed (i.e., the projects were delivered on time, on budget, with required features and functions), 44% were challenged (i.e., these projects were late, over budget, or delivered with less than the required features and functions), and 24% failed (i.e., canceled prior to completion or delivered and never used). One of the biggest determining factors in the success of a PMO is its relative level in accepted process maturity models, described next. As the PMO matures, its general effectiveness increases accordingly.

- **Level 1.** Most business processes are informal or undefined.
- **Level 2.** Most business processes are defined but not well adopted.
- **Level 3.** Most business processes are defined, repeatable, and followed.
- **Level 4.** Most business processes are aligned, and performance is measured.
- **Level 5.** Most business processes are optimized and continually improved.

Maturity comes not only with time but also with the ability to overcome a host of other challenges, including:

- Insight into ongoing operations
- Ability to support methodologies
- Departmental silos
- Alignment with enterprise strategy and priorities

*A copy of this report is available at www.pmhut.com/the-chaos-report-2009-on-it-project-failure, accessed September 2012.

- Effective on-demand tools
- Infrastructure
- Determining requirements
- Financial management
- Communications

The PMO should be capable of successfully delivering on strategic initiatives and achieving strategic results. For most companies, reaching this capability begins with the realization that in addition to managing projects and methodology, the PMO must also manage resources that are shared across projects and help sustain operations.

In order to clarify demand on available resources and define and enable prioritization, the PMO needs to be able to see the impact of sustaining operations on strategic projects. This way, the PMO can compare the results of supporting existing system environments to the results of implementing a new ERP system. Seeing the balance between strategic projects and continuing existing operational projects will help the PMO to understand the total demands on resources, budgets, and the like. This insight is impossible to achieve without using the combined metrics of an integrated resource pool connected to current projects, future projects and operational initiatives. Using a point tool like Microsoft Project Server allows this information to be readily available to the PMO.

As stated earlier, many organizations are missing key elements like resource management, which provides a full overall view of capacity for staffing projects and applications even at the earliest planning stages. Also absent is request management to help the PMO prioritize and align with enterprise goals when projects are coming from all directions. Finally, time and financial management tools are needed to ensure that budgets are met and projects delivered on schedule.

The project manager wears many hats, pays attention to many areas, and, most important, manages the project. In short, the project manager embodies many elements that need to be coordinated and brought together as a single united effort to produce the prescribed deliverable. The skill to do this is the glue that holds together the various functional and management elements of a project to form the project team. Keeping everyone together as a solid team is the greatest challenge facing a project manager.

The project manager has a diverse audience that spans multiple functional areas and management levels. This diversity requires the project manager to invest time in developing a solid communications plan that will provide a continuous means of communication to the entire project team. This diversity also requires the identification of information that is needed by the various project team members and knowledge about how to best communicate this information. An effective communications plan is an integral part of the overall project plan and a critical factor in overall project success. Developing the communications plan must include the implementation project team and principal management and project stakeholders, and must address management of information, information content, communications skill, and accuracy. These elements are the glue behind the communications plan.

Information Management

During the course of a project, a large amount of information is produced and received. Management of this information is the responsibility of the project manager and the PMO as the office of record for all project information. The communications plan addresses the policies and procedures to manage project information. A well-developed report matrix determines how project information is packaged and distributed. This matrix should identify the type of report (formal report, project status reviews, memos, etc.), recipients, frequency of reports, and what action is expected from the recipients. The development of the report matrix must include any contract provisions regarding reports.

Information Content

Once the needed reports are identified, specific data elements, content structure, and level of detail must be documented in templates for each report. The draft templates are provided to the requesting functional or management element for final approval and then made available to the team. The list of the approved project reports is made part of the communications plan with samples in an appendix. In organizations that have a mature PM process, the inclusion of this approved project list into the communications plan has become a standard for all projects. However, these standardized reports and their content still might need to be reviewed for unique data elements and accommodation of specified reports in the contract.

Communications Skills

Project delivery is the first activity that determines how well the information is received. The report matrix indicates whether the information is presented as a stand-alone report or part of a larger report or given as an oral presentation, and in what venue (project status reviews, stakeholder updates, team meetings, etc.).

Clear, concise communication is the essential ingredient to ensuring complete understanding. The information presented must be unambiguous, leaving no room for latent interpretation.

Accuracy

In the presentation, you must be factual and tell the whole story, the good *and* the bad. Do not sacrifice your most important assets: honesty and integrity. Once you have been caught trying to understate a situation or omit it entirely, your reputation as an effective project manager is fatally tarnished forever. A project manager may be forgiven for many missteps but is never forgiven for being found guilty of being less than honest. Honest project managers with a high level of integrity will display enhanced team leadership. By maintaining the standards of communication, one source of truth reporting in their project schedules simplifies and creates good standards and leadership for getting to best practices around project management. Senior management will welcome their presence, even if it does not like what they are saying. Honest messengers are never shot in these situations. We the authors have spent many years consulting and one truth we can rely on is that information is not good or bad, it's just information. Executives and senior management have such a difficult time getting accurate metrics and status, that clear and honest reporting is appreciated (especially if the information comes sooner in the project lifecycle rather than later).

Other Considerations

Over the last couple of decades, managing the delivery of a product or service has been a "company team" effort. All team members worked in the same office and held their meetings in a conference room. It was easy to get the word out or solve project issues, brainstorm, or quickly disseminate project information. The team was also able to mentor others and provide team maturity through knowledge transfer and through working together to solve issues. In many cases this lead to succession planning, enabling junior team members to rapidly step up into senior roles. Advancements in communications technology have spawned global teams that are separated by time zones, cultures, and customs. The acceptance of English as the language of global trade helps overcome some multicultural difficulties but can hinder clear communications when colloquialisms are used. The global delivery team is a reality. With it come greater challenges to the project manager for effective project communications.

The communication tools of the global team include e-mail, video teleconferencing, and virtual conferencing (using both the telephone for conversation and the Internet for graphics). The virtual project team brings a new dimension to ensuring successful delivery. The use of e-mail is the norm for digital communications as it provides almost instantaneous delivery to the far reaches of the globe. However, crossing time zones presents some problems, such as an extended workday. In addition, virtual teams lose the benefits of spontaneous exchanges of ideas and team growth through professional discussions.

If you find yourself explaining "What I said is not necessarily what you heard and may not be what I meant," you have a problem. If your message cannot be received and understood by those you are addressing, your communication attempt is ineffective, resulting in group members not being at the same level of knowledge. Uncoordinated team activities and poor decisions can result. Listed next are some sources of information and suggested management procedures.

- **Correspondence.** Letters, memos, faxes, call records, and the like need to be accounted for and filed either as "in" or "out" correspondence. The project manager controls the document numbers to all project correspondence. A simple document numbering scheme will provide the necessary document accountability (e.g., use the project number as a prefix followed by the initials or code for the customer and then a sequential four-digit number). A spreadsheet log will help you assign document numbers and catalog the type of document (internal, received, or originated)—for example, location, subject, author, date received/sent, and any related documents. Project team members need access to the document log so they can record project documents. If the document was created using a word processing program, the document number can be used to identify the document in the name field.
- **Forms.** All standard forms used in the project lifecycle need to be listed in the project plan along with their purpose and source location. Standardized forms should have a unique number for quick identification.
- **Staffing and coordination procedures.** Activities and decisions must be coordinated via a prescribed process and procedure that follows established corporate guidelines. These procedures should include an escalation matrix that clearly indicates the various levels of decision authority.
- **Trip/meeting reports.** Whenever a project team member meets with the customer or project vendor, this meeting must be documented, numbered, and filed in the project files. These documents detail the meeting agenda, what was discussed, decisions made, and action item assignments. Distribution of these reports needs to be determined and documented in the report matrix.
- **Project and customer status reviews.** These reviews are the project's frontline reports and address its current performance. They include progress reports, expediting and forecasting reports of project resources, and milestone completion reports. These reviews usually are given monthly and in either written or presentation format.
- **Project history.** This report is the collection of all project material into a document that provides the history of the project from inception to lessons learned. It is a compilation of documents that chronicle the project lifecycle.

Holding It Together

Rigorous adherence to a well-developed project communications plan will result in all team members and stakeholders being informed to their level of expectation. Meeting these expectations results in a high level of coordination among team members and fully informed participants and minimizes the risk of misinformation, frustration about the unknown, and ultimate collapse of the project. Effectively applied, project communication is the glue that holds your project together. Stick to it!

Program Collaboration Means Business Results

When the project site (Project SharePoint workspace) is created on the SharePoint Server platform as a common place for project assets, artifacts, and document storage, it provides a platform to support project team and stakeholder communication around project documents, issues, risks, and deliverables. The site can also contain individual collaboration tools, such as workspaces, discussion boards, surveys, wikis, blogs, and contact lists.

Team members should be active in the project site, as their understanding of the project's big picture is valuable in the effort to achieve successful results. The Project site is visible through a browser portal called Project Web App (PWA). PWA allows team members to view the project site, collaborate on documents, share files, contribute to wikis and blogs, participate in team discussions, and more via their Web browser. This Web visibility removes the requirement for every team member to have Project Professional 2010 on their individual machines and allows for remote team participation. PWA can be configured so that each time a user creates a new project, an associated Project site is set up automatically. If you prefer, you can allow team members to manually create a Project site instead and all the while have the benefit of controlling content and user access through permissions.

Microsoft has worked diligently to make a truly collaborative workspace. This has resulted in a singular design with numerous mouse-over features, hidden menus, and other ways to deliver functionality without clutter. It is easy to have a SharePoint Project site for every type of project the organization has, allowing collaboration to be more flexible in the information that should be shared in the project (see Figure 8.1).

Project Manager Empowerment

Project Lifecycle Management

Project lifecycle management (PLM), also known as demand management (DM), is a unified approach that optimizes the consolidation of a significant number of essentially related processes and capabilities. PLM offers a unified view of all work in a central location. Its purpose is to quickly help organizations gain visibility into projects and operational activities, standardize and streamline data collection, enhance decision making, and subject initiatives to the appropriate governance controls throughout the PLM.

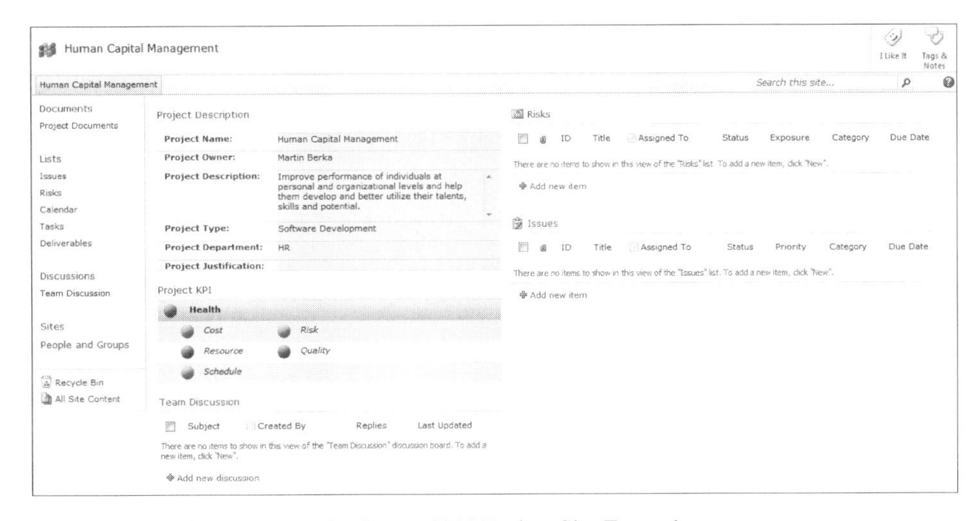

Figure 8.1 SharePoint Server 2010 Project Site Example Source: Advisicon

PLM sets the tone for the evolution and process/organizational improvement of projects to their completion. The information traceability paths, through each phase, also embedded in the project or task tracking system have many different elements. These elements include tasks, assumptions, risks, issues, stakeholder requirements (or functional requirements), time/cost restrictions, quality/regulatory requirements, and so on. Each path has to be both an inbound and an outbound component through these lifecycle states:

- Create
- Select
- Plan
- Manage
- Closure

The advantage of understanding and using the create, select, plan, manage, and closure phases and processes (the governance activities surrounding that phase), is that it provides a simple framework for improving DM capabilities.

Workflows in a PPM Implementation

SharePoint 2010 supports workflows for collaborative applications. Many organizations have benefited by moving manual processes into a SharePoint workflow. The primary reason we use workflows in software design is to manage long-running processes. SharePoint workflows are the ideal way to automate processes that previously required paper or the use of complex software to record and monitor any long-running activities.

With Project Server 2010 now a part of the SharePoint platform, DM now includes workflows that help you manage project proposals and portfolio analyses.

DM is a new concept in Project Server 2010 that integrates project proposals, portfolio analysis, and PM through workflows and project detail pages. The goal of DM is to enable users to propose, view, categorize, prioritize, select, and track projects within their organization. A key component within DM is the workflow governance model implemented in Project Server. Some out-of-the-box workflows available, but new workflows can be customized per your requirements.

Thanks to this feature, all organizations can customize processes, allowing them to have standardized activities for DM and an easy way to authorize documents and projects, making the flow of information among the users more efficient. This level of customization can be achieved by using Visual Studio 2010 and SharePoint Designer 2010.

Having a collaborative environment based on different organizational workflows will make it easy for the organization to make decisions regarding which initiatives should be approved, and what skill set of resources should be hired or outsourced; it also provides the ability to audit project execution and determine whether the efforts should go to the next stages or not.

Business Process and Forms

An essential factor for efficient PM is uniformity. Now all of the processes and PM templates can be easily uploaded to avoid having several versions of a document. This use of templates will allow standardization among the organization, streamline collection of data, and encourage user adoption.

Synchronize the PLM

Prior to and during a PLM, capturing specific information that needs to be accounted for is crucial. Often users cannot leverage a scheduling tool to capture the updates and perform an analysis. Users commonly employ products such as Excel, Word, and Outlook to distribute updates among team members, take relevant notes, and transfer the information to the project schedules. Within SharePoint, people have been able to leverage online tools, such as Excel Services and SharePoint lists.

A project task list in SharePoint displays a collection of tasks that are part of a project (see Figure 8.2). In this sense, a "task" is a discrete work item that a single person can be assigned to. A "project" is typically a series of activities that have a beginning, middle, and end and that produces a product, result, or service (such as producing a product demonstration for a trade show, creating a product proposal for stakeholders, or even putting together a corporate morale event).

ID	Mode	Task Name	Start	Finish	Work	Units Planned	Resource Names
1		**ZCST-100 Mile House**	**1/1/2011**	**12/30/2011**	**358.38h**		
2		January-100 Mile House-ZCST	1/1/2011	1/30/2011	21h	28	100 Mile House-
3		February-100 Mile House-ZCST	2/1/2011	2/27/2011	24.88h	39	100 Mile House-
4		March-100 Mile House-ZCST	3/1/2011	3/30/2011	28.38h	40.5	100 Mile House-
5		April-100 Mile House-ZCST	4/1/2011	4/29/2011	34.13h	56	100 Mile House-
6		May-100 Mile House-ZCST	5/1/2011	5/30/2011	27.5h	42.5	100 Mile House-
7		June-100 Mile House-ZCST	6/1/2011	6/29/2011	31.75h	52	100 Mile House-
8		July-100 Mile House-ZCST	7/1/2011	7/30/2011	40.88h	41	100 Mile House-
9		August-100 Mile House-ZCST	8/1/2011	8/30/2011	26.63h	34.5	100 Mile House-
10		September-100 Mile House-ZCST	9/1/2011	9/29/2011	32.88h	37.5	100 Mile House-
11		October-100 Mile House-ZCST	10/1/2011	10/30/2011	42.13h	52.5	100 Mile House-

Figure 8.2 SharePoint Server 2010 Task List Example Source: Advisicon

After you create a project task list, you can add tasks, assign resources to tasks, update the progress on tasks, and view the task information on bars that are displayed along a timeline.

Although some of the settings for the project task lists differ from those of other lists (such as for contact lists and announcements), you use the same basic procedure for creating a project task list as you do for other types of lists: adding columns, exporting to a spreadsheet, sorting and filtering, or organizing the task list.

Lists: Sync Up SharePoint and Project Data

The ability to capture relevant, critical information in a central repository is essential for organizations. Because PPM is aligned with corporate objectives, any impact to components within the PPM environment has a direct correlation to the overall objectives. As mentioned earlier, stakeholders within a PPM are migrating to using online features such as lists and discussion groups to capture key pieces of data related to the projects and programs. Using SharePoint lists is a great first step, but the ability to integrate those data points directly to a project is a major improvement. Previous versions of Project and SharePoint allowed for list integration, but only on a limited basis.

Project Server 2010 is now tightly coupled with SharePoint task lists. In Project Server 2010, you have the option to create a fully managed portfolio project schedule composed entirely from entries in a SharePoint task list. If you start with a Project 2010 file, you can generate an ongoing two-way sync between the project schedule and a SharePoint task list. Hence, users can now publish a project schedule from Project Server to SharePoint Server and vice versa. Any changes made in Project or SharePoint can be easily updated into the other with the click of a button.

Here is how it works: The project manager creates a simple project schedule in Project Professional 2010, as shown in Figure 8.3:

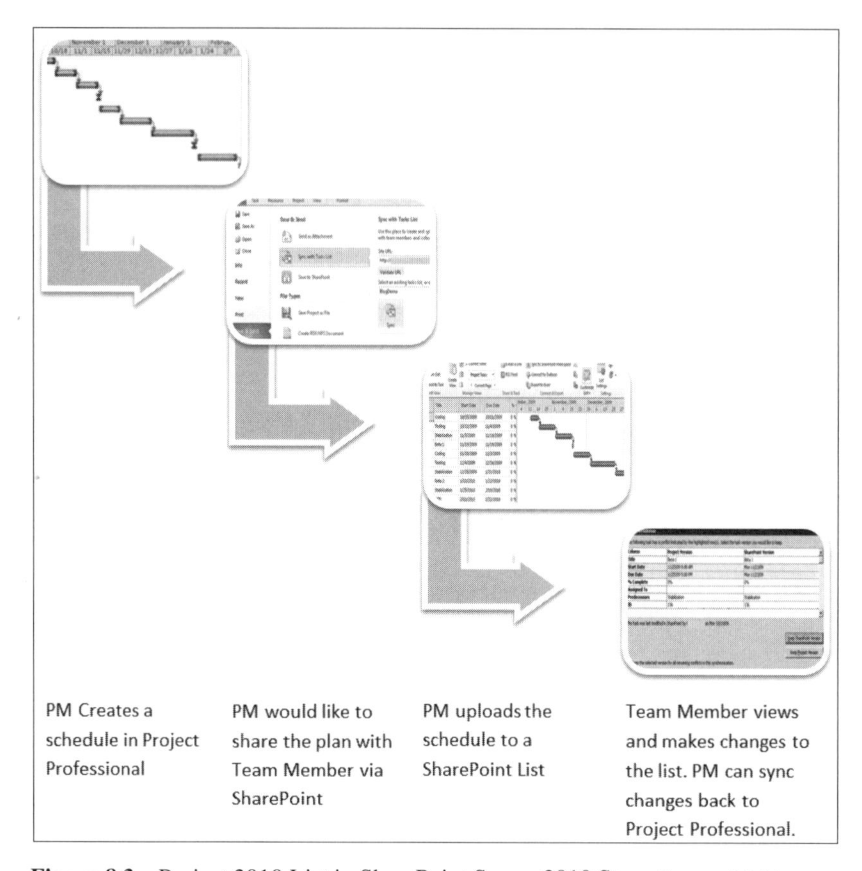

Figure 8.3 Project 2010 List in SharePoint Server 2010 Steps Source: Advisicon

Figure 8.4 shows the process of supporting the PLM if we initiated the schedule in SharePoint by creating a task list and then synchronizing to Project.

As detailed in our whitepaper "Microsoft Project Server 2010: A look at Demand Management"and again mentioned in "Microsoft Project Server 2010: A look at Portfolio Strategy", process definition and use is essential for organizations to gain control over their information and the information traceability paths. The create phase—which the Project Management Institute calls initiation—requires a fluid, agile environment of data gathering in order to organize the demand requests of a proposed project. The SharePoint Server 2010 feature set now enables project managers or key project stakeholders to use a number of options to capture information and synchronize in a Project Server environment.

Sync to SharePoint is a useful feature for those organizations that have SharePoint 2010 but not Project Server 2010 or for those that have both but prefer to use a simple SharePoint task list to publish project plans rather than Project Server 2010.

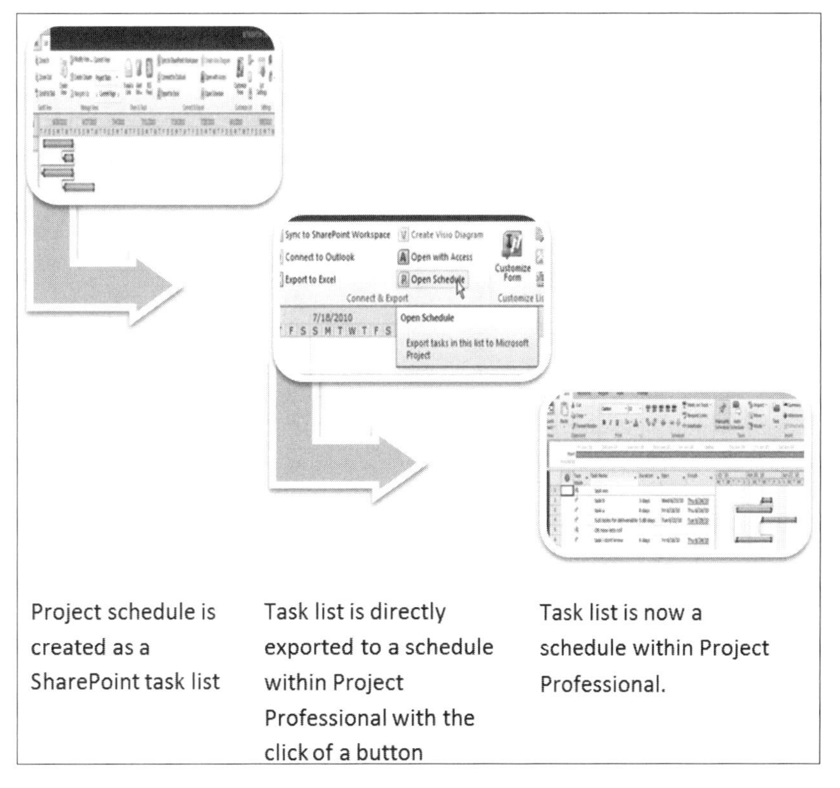

Figure 8.4 Saving Project to SharePoint Source: Advisicon

Using SharePoint and Project 2010, project resources can enter updates in SharePoint, project managers can synchronize updates with Project, and custom field information can be used to generate reports.

Managing Risks and Issues

Thanks to the SharePoint Project sites, the project manager is able to keep track of risks and issues related to the project. Project sites allow capturing, tracking, assigning, and monitoring of risks and issues. Figure 8.5 illustrates the issue and risk lists associated with tasks and projects in Project Server 2010.

Have you ever created a project plan and documented assumptions that need to be true in order for the project to be successful? Assumptions are relied-on, specific circumstances. They are accepted as true, real, and certain without requiring proof. Assumptions are the starting place, as they document what is accepted as "known" to the team and provide a good base for team collection and brainstorming of risk and issue data.

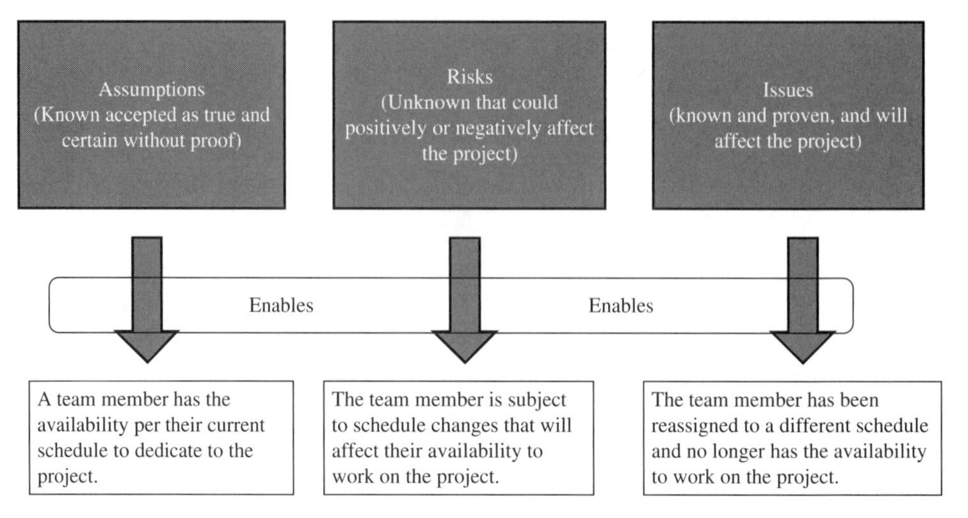

Figure 8.5 Benefits of Better Assumption, Risk, and Issue Management Source: Advisicon

For example, you assume that your customers will be available for a minimum of 20 hours per week throughout the project. Because you've made this assumption, you're relying on the customers to be available, to avoid delays in the project schedule.

Assumptions should be tracked throughout the project lifecycle, to confirm they are still presented as true. If an assumption is deemed no longer valid, it becomes an "unknown" factor within the project. It becomes a risk. Risks are uncertain events that could positively or negatively affect the project. Risks are always connected with uncertainty. Consequently risk management is a set of techniques for controlling the uncertainty in a project. Keep in mind that risks can also arise independently and not be tied to an assumption, but all risks should be managed.

If you assume that your customers will be available for 20 hours per week, a pending management decision may cause their availability to become limited to only 12 to 15 hours a week. Now there is an unknown that can affect the project. Projects that do not track and manage risks do so at their own peril. The most severe risks are those that threaten to delay task, phase, or project end dates; increase the budget; overwhelm; or all three.

Project risks have these attributes:

- Their presence generally is known at the beginning of the project. This is why collecting requirements of risk from the beginning of the project is crucial.
- They can exist only at a specific point in the project, or they can persist throughout the life of the project.

- They can materially affect the outcome of the project if they become reality.
- There is a reasonable likelihood that they could become reality.
- Risks are extraordinary to what normally would be managed on a project.

A SharePoint site created for each project in Project Server 2010 can assist project managers with risk management.

Once collected, risks should be analyzed and prioritized. Although all risks should be managed at some level, some risks will require more management than others. Some risks may not require stringent management due to their low priority, but they still should be monitored in case circumstances change in such a way that they become a higher priority for the project. The high-priority risks need to be proactively managed.

Risk is always connected to uncertainty. If something is certain to occur, it is called an issue, not a risk. Issues are just as important as risks and, like risks, are problems that occur during a project. Issues can arise by themselves or be a result of a risk becoming a certain event instead of remaining an uncertain event. If an issue isn't managed, it can materially affect the successful completion of a project. Where issues typically differ from risks, however, is that they generally don't persist throughout the project, and they may not be known at the outset of a project. Your issue list will not be as persistent as your risk list will be. Issues will open and close as they're identified and resolved.

What's important in identifying and managing issues is this: Dealing with issues is necessary for successful project completion.

Continuing with our example, once the management decision has been made, customer availability is now limited to no more than 12 hours a week. It is now an issue that must be dealt with as it will impact the project schedule.

The keys to identifying these issues are:

- Understand the issue.
- Verify your understanding of the issue.
- Record the issue to keep from forgetting it, which could spell doom to your project.

Project Server 2010 has predefined forms to identify and track risks and issues. Risk can be identified at a project level or linked further to a task, issue, document, or the risk itself. The project team can place risks and issue forms on the Project site. Team members can assign issues, categorize them, and prioritize them. Risks and issues can be linked to specific tasks, documents, and other risks and issues, which then display as icons in the project view. The captured risk and issue data will be stored in the knowledge base and can be used to proactively mitigate similar risks and issues in the future. Figure 8.6 shows the issues and risks that can be both active and future, allowing the project team to work to address these or work to mitigate them.

Risks - New Item

Edit | Custom Commands

Save | Cancel | Paste | Cut | Copy | Attach File | Spelling

Commit | Clipboard | Actions | Spelling

Title *

Owner

Assigned To

Status: (1) Active

Category: (2) Category2

Due Date: 12 AM 00

Probability *: 0 %

Impact *: 5
The magnitude of impact should the risk actually happen

Cost: 0
The cost impact should the risk actually happen

Description
The likely causes and consequences of the risk

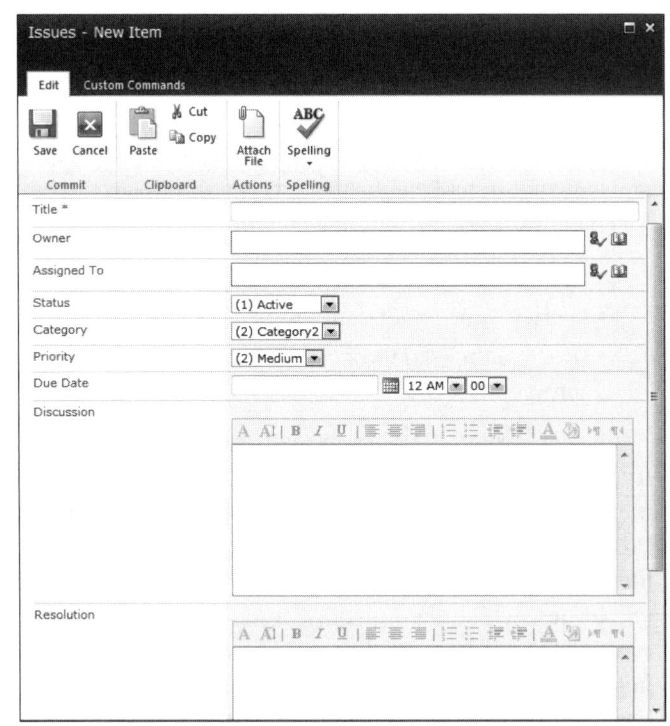

Figure 8.6 SharePoint Server 2010 Risks and Issues Source: Advisicon

Managing Deliverables

It is imperative to manage risks and issues because they are likely to affect the status of the deliverables, which, if delayed or altered, can significantly affect the success of the project. A deliverable is a tangible and measurable result, outcome, or product that must be produced to complete a project or part of a project. Managers of large projects often have several milestones or deliverables to report about. Project Server lets you identify these items in the project plan so you have better control of them and can easily report whether they are being completed on time or not.

Typically, the project team and project stakeholders agree on the project deliverables before the project begins. Clarifying the deliverables before the project work begins can help ensure that the project outcome meets all the stakeholders' expectations and that project goals align with the larger business goals. You can identify deliverables to show an end product of a particular task or of the entire project. Figure 8.7 shows a deliverable that can be added and managed within Project Server 2010 or outside in SharePoint.

As a project evolves, various types of deliverables are produced to support project continuation, measure progress, and to validate plans and assumptions.

Deliverables can be managed using a Project site created for each project. They can be created directly on the Project site or through Project Professional. A deliverable can be independent or associated with a task or phase in project.

Other essential components to deliverables are requirements and quality. Requirements are the characteristics that the deliverable must possess to satisfy an identified need. Success of a project depends on how well the requirements were satisfied. Satisfying the requirements leads to delivering the level of quality desired for the deliverable.

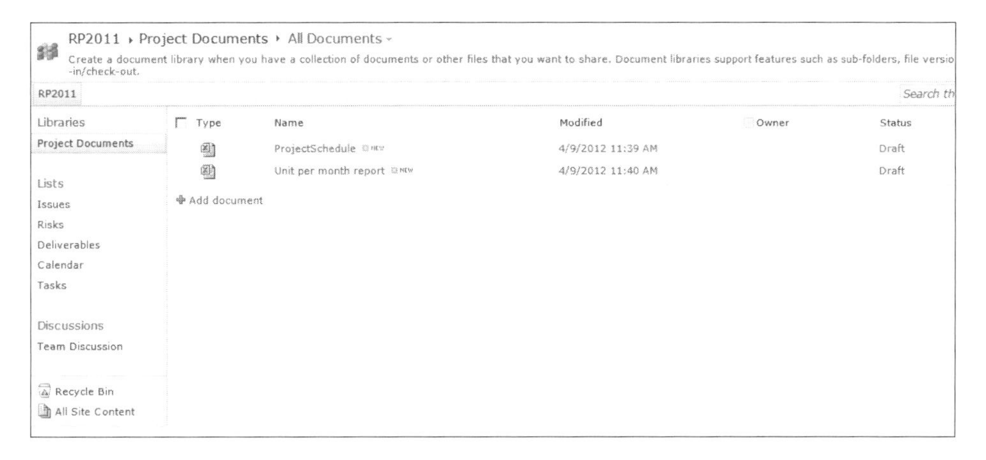

Figure 8.7 Deliverables Can be Added Directly in the Project Site or Through Project Professional 2010 Source: Advisicon

As the management process for deliverables is carried out, Project provides a continuous checking system and views to ensure that requirements and quality standards are being met. If deliverables are not managed efficiently (or, in the worst case, not at all until the end), there are missed opportunities to correlate risks, issues, requirements, and quality and to initiate preventive or corrective actions.

Thanks to the flexibility that SharePoint offers, it is easy to customize a list or workflow to set up the quality assurance process to ensure the deliverables developed meet the quality criteria defined for the project. You also can set up a list that allows you to validate that the requirements defined at the creation phase of the project are being met.

Managing by Exceptions versus Managing Everything

You need to determine where to concentrate your efforts. You can't check every detail, yet there must be a way to monitor operations to find and resolve potential problems while identifying and leveraging new opportunities. You need to ensure that employee and departmental metrics are aligned with overall strategic goals. Your organization may not have the resources of a Fortune 500 company, but your employees are passionate about their jobs and committed to your customers. Maybe your company is relatively small right now, but it's on a high-growth path. You're aware of the power of business intelligence (BI) and know that larger companies—maybe even your direct competitors—are using it to their advantage. At this stage of your company's evolution, isn't it time your analysis capabilities graduated from spreadsheets to more powerful tools?

A major part of any manager's job is to make decisions. If you can improve the overall quality of your organization's decision-making process, you'll improve the overall effectiveness of your organization.

In short, BI helps your organization make smarter decisions, hence its description as "decision support." BI allows organizations to better understand, analyze, and even predict what's occurring in their company. BI helps organizations turn data into useful, meaningful information and then distributes this information to those who need it when they need it, thereby enabling them to make timely, better-informed decisions. It allows organizations to combine data from a wide variety of sources and see an integrated, up-to-date, 360-degree view.

This is especially important for midsize companies that typically can implement business decisions relatively quickly. BI provides a win-win solution for IT and business users. It allows the IT department to be more productive in working with its business users to service special requests while permitting these business users to become more self-sufficient. Operations and analysis are two sides of the business, and BI allows IT to be a valued partner in both.

The BI spectrum is very broad in terms of its tools and functionality. At its core is the traditional functionality of query, reporting, and analysis. This core is complemented by data quality and data integration to accurately and consistently consolidate data from multiple sources. Dashboards and other visualization techniques can help users quickly understand analysis results and are often considered part of the BI spectrum. BI can encompass these functionalities:

- Search functionality for locating information and reports
- Predictive analysis to discover hidden patterns and enable what-if analysis
- Scorecards and performance management to help monitor business metrics and key performance indicators (KPIs)

These KPIs might include customer satisfaction, profitability, and sales per employee. Additionally, KPIs can be monitored to support alignment of individual and departmental metrics as well as the organization's strategic goals. A simple query might access your company's data to ask "What total sales to XYZ Corporation were last December?", "What is the current salary of employee Tim Runcie?", or even "How many training manuals do we have in inventory?" Most query tools also provide simple reporting functionality and can also be used to generate a simple report listing the accrued vacation of all employees, sorted and totaled by department.

Enterprise reporting or production reporting typically involves high-volume, high-resolution reports that are run on a regular basis. An example might be a sales manager report showing monthly sales and associated commissions sorted by salesperson and then by customer. The report distribution likely would be controlled so that each sales manager could see entries only for his or her salespeople. The report might be e-mailed to them or viewed through a Web browser. Enterprise reports can also be used to generate customer statements, invoices, or individualized benefit summaries for each employee.

With advanced analysis functionality, users can view data across multiple classifications or dimensions (e.g., product, customer, location, time period, and salesperson). They can slice and dice the data to look at various combinations, such as the sales in each region for December or which products each customer purchased last year. Advanced analysis functionality also permits organizations to define hierarchies so that users can first view sales for each region and then drill down to view the sales in each country, of each product in each store, or for each salesperson. These advanced analytics make it easy to compare the results from one time period with those of another—say, this July's product sales compared to last July's—while performing year-over-year comparisons by store, customer, or salesperson.

Other advanced functionality, such as filtering, can be used to include or exclude specific stores, regions, products, salespeople, or time periods and look at the top or bottom best- or worst-performing products, stores, or salespeople. Combined with drilldown, slice and dice, and filtering functionality, this is a powerful multidimensional analysis.

In the past, interactive analysis and simple views and reports were designed for passive viewing (static without the ability to drill down). SharePoint and some of the SQL analytical tools that come with Project Server have provided advanced abilities to perform interactive analyses and drilldown capabilities. Many of these advanced functionalities were once available only in specialized Online Analytical Processing (OLAP) products that involved the use of proprietary databases and highly skilled technical specialists. Now OLAP functionality often is incorporated into query and analysis tools, such as Performance Point and PowerPivot. Thus, business users can perform interactive analysis and click on a number in a report to drill down and analyze the underlying details.

Effective BI should be an interactive process. Query and analysis tools with embedded OLAP functionality permit business users to perform dynamic analyses on their data. As most IT practitioners will attest, users requesting a static report typically ask for additional modifications and details. Query and analysis tools allow business users to formulate a high-level query and then immediately explore the underlying details on their own.

Graphical Techniques

Dashboards strongly complement the other members of the BI spectrum. With graphical gauges analogous to an automobile dashboard and color-coded indications as exemplified by traffic lights (red represents an alert condition and yellow a warning), users can quickly identify exception conditions.

It has often been said that if you can't measure it, you can't manage it. Scorecards and other performance management tools enable you to establish business metrics, update and monitor the results, and communicate these metrics as appropriate so that minor problems can be identified early on and corrective action can be taken quickly.

Dashboards frequently are used to display performance metrics and can allow users to drill down from the visual image to view the underlying detail. Other visualization techniques include slider bars that allow a user to perform what-if analyses to execute such tasks as showing that potential profit margins increase when maintenance revenues are increased or distribution expenses are reduced.

Distribution and Control

BI is not just about tools and their applications; it's also concerned with distribution and control. Companies should be able to publish reports to the Web and deliver them to a user's preferred mobile device. However, not every employee should have access to every report or analysis. Administration, monitoring, security, and control are also part of the BI environment. Furthermore, the use of commercial BI products does not necessarily mean the elimination of spreadsheets. Rather, BI can provide controlled linkage of spreadsheets to up-to-date data while enforcing proper distribution and control. This way, spreadsheet chaos is no longer an issue, and trying to determine whose spreadsheet is "more correct" is no longer part of every company meeting. The ability to locate and search out relevant reports is also part of the BI landscape, as a report is of little value if no one knows it exists or how to find it.

Data quality is of paramount importance in both operational systems and data warehouses. In an operational environment, no one wants to ship the wrong order to the wrong address, provide a patient with the wrong medication, or transfer funds to the wrong bank account. In a data warehouse environment, no one wants to make decisions based on incomplete, incorrect, or inconsistent data. The deployment of data quality tools can help ensure that this does not happen.

By using BI against both operational systems and data warehouses, a company can improve its daily operations and compare current results with historic values to identify trends and head off problems before they become more serious.

Rinse and Repeat Key Processes

Taking a repeatable process and making it a company standard is a common practice. Sometimes we can take a large component and customize it a bit to work in different situations across the enterprise; other times we can take apart an operation by breaking it into smaller pieces that then can be fit together as needed to help define commonly shared modules. As we work to prevent reinventing the wheel in our business processes, the methods by which we manage our projects also requires a standardization of techniques and practices.

To increase project effectiveness, a PM workforce has the ability to support and impose standardization on many levels over how a company functions. By directing project staff to use processes that focus on scheduling, expense, and cost-to-quality relationships, a more successful portfolio is possible. Establishing these strategies with a sponsored PMO dedicated to overseeing their implementation can generate a stronger, more refined and professional business environment that is reliable and dependable—and thus more attractive to customers.

Why Standards?

Higher levels of specific PM standardization have demonstrated that increased project effectiveness can encompass a variety of industries. For example, organizations will highly tailor a PM lifecycle to their needs, like manufacturing over construction. Bringing in standardization to an enterprise does not guarantee, however, that there will be universal implementation success. Even when an inclusive and integrated PM culture, construct set, and system is developed, there is no surefire winning formula that will bring about project effectiveness.

There is always the challenge of trying to meet stakeholder requirements or project deliverables and completing them as fast and efficiently as possible. It is a consistent demand from the beginning to the end of the project and may even extend beyond the normal project parameters. Keeping deliveries within agreed-on time frames, costs, and standards of excellence is required in order to remain competitive and support a professional identity in the industry. No matter how technology or business needs may change, leaders must be adaptable to remain valid.

The increasing speed of delivering a PPM system is a strong force in the aggressive technology markets that we are currently experiencing. Organizations typically decide what types of products or solutions to prioritize and engage in by comparing the applications or products with how rapidly they can be distributed. The pace of industry changes and their accompanying needs accelerates at a rate that puts demands on projects designed to support this growth. Both hardware and software development efforts are affected; they are also pushed to have shorter and shorter cycle times.

A key feature in leveraging a project system is the ability to manage, track, and ensure that projects are focused on goals that organizations are motivated by. In many organizations, planners and senior management are highly focused on cost factors, so the ability of Microsoft's PPM system to ingrate with enterprise resource planning (ERP) systems, including dynamics is an example of where working in the Microsoft stack provides a higher value. Other organizations, including our company, Advisicon, have built connectors to other ERP systems to move key actuals over and integrate them with Project data. While we may yearn for the latest gizmos, gadgets, and other kinds of improvements, in the end, cost is more of a factor in the decision on whether to purchase and support new offerings. Cost-driven competition has pushed us to be more economical in our development processes and more conservative in our builds, thereby pursuing more refined efforts that may result in less product diversity.

Yet another concern may lie in focusing efforts on product quality. In an organization's desire to shorten time to delivery and keep costs down, the quality of project deliverables can suffer. It is important to balance producing high-quality products while simultaneously incorporating the need for reduced cost or speed to market. In many cases, this balance between quality and timely delivery is the very core of why technology-based products or solutions are started. Many sales and marketing divisions have sold quality and feature sets that have enticed customers to purchase; now the need for the organization to deliver, and do so within a timeline or at the most efficient cost, creates the need to track the time, work, and resources for that project. Focusing on quality is important and may be a deciding factor in making a niche in an industry, but quality is difficult to accommodate if it is influenced by the need to aggressively follow delivery schedules and keep costs down.

This question comes up quite frequently: How do you track quality in a product, project, or deliverable? In many cases, the identification value of quality tracking isn't addressed by just adding a separate field to enter a value, as it is the actual physical tasks that are being included in a schedule that ensure that quality is achieved. Being able to monitor the time or effort as well as the features being delivered according to a baseline estimate helps all parts of the project team ensure they aren't straying or burning valuable time.

In Project 2010, you can instantly turn on or off different baselines as well as compare one baseline to another or even planned and actuals. This capability helps planners and senior management to quickly spot check views, details, and key trends to see if they are at risk.

Promoting Standards for Maximum Value Proposition

Time, cost, and quality—each places a burden on the projects so that an organization can better compete in their industry. The idea of collaborating with business units to incorporate all of these components in a standardized and repeatable fashion requires a managerial octopus; hence the need for a PMO.

Understandably, when a standard has few steps or components, it is easy to implement, replicate, and enforce. Conversely, the more complex a standard is, the more difficult it is to support and impose. While we can conduct extensive project postmortems and publish all our best practices data on a site for everyone to access, it is another matter to take that information and turn it into practicable business standards.

The use of a PMO helps reinforce repeatable processes that will not differ no matter what the project. Through the PMO, the development of predictable, regular actions creates functionality that operates regardless of the convoluted parameters of client wish lists and the turbulent environment of the competition. Maintaining and supporting the implementation of standardization makes a more effective use of time (an expense that cannot be reclaimed once it is expended). When the reinvention principle is suppressed in each project group, standardized processes can be put in place, thereby saving not just time but resources and money as well.

Art and Science of Technology Delivery

Different project approaches (design, test, build), software development projects present unique challenges. Complexities arise around managing and delivering due to the intangible nature of software, the lack of a software standard development process, and the rapid pace of technological advancement. With these types of projects, end users, stakeholders, and customers often don't know what they want ahead of time. Once they visualize what has been created stakeholders sometimes decide it won't meet their needs. This tumultuous project environment compound the complexity of software delivery. These are some of the reasons having a prototype or a mockup of the solution, helps to expedite and create clarity around the solution or end result product.

There are similar complexities with engineering projects. The same criteria used to determine the successful outcome of engineering projects can be applied to software development: Both types of projects should be delivered on time, on budget, and on quality. Software development projects and engineering projects should both meet the end users' expectations. Unfortunately, software projects have one of the poorest track records of delivery; research by the Standish Group in 2011 says that only 28 percent of projects succeed, while 23 percent are canceled and 49 percent are challenged (very late, over budget, or missing features . . . or all of these issues combined).

Approaches to Improving Successful Delivery

A great approach in software development is to build the solution in an iterative fashion. This is highly helpful in reducing the risks inherent in development and provides early views and reviews with customers/stakeholders who can provide insight, scope changes and ensure there are no surprises at the end of the project. Complex or high-risk items should be addressed early in the development process. Establishing a solid foundation or architecture is critical to a solid system, especially if the idea is to continue to scale or build upon the initial solution. Starting with a solid architecture is even more important for large systems. It is a good practice to create and document an architecture baseline early in the development process. When developing iteratively, this architecture should be produced within the early iterations, enabling the team to ensure that the foundational tables, structure, code base, and other key architecture points are leveraged so that the final solution will not outgrow it's platform. In software development, the architecture should support the customer requirements, however the developers should not start "goldplating" their solutions with features they think users or systems might need in the future, which usually adds unnecessary complexity. Keep it simple and basic; scale later as budget, scope or new features/functionality are requested and approved.

Successful application development uses component-based development. This process helps you design and develop applications that handle change better, allows the reuse of code, functions and other modular functionality, and can reduce the maintenance costs associated with the project and environment. Generally speaking, in software development projects 70 percent of the costs of a system implementation are incurred after the project has been completed. This means that applications built using component-based modules allow for a more compartmentalized and streamlined design, which in turn allows for easier troubleshooting, simpler maintenance, and are more cost effective. Pair programming is very helpful and creating a process of code reviews can dramatically reduce the number of defects that appear after the initial implementation. Establishing these and other best practices will help ensure that programmers write better code; they know it will be reviewed. These reviews should take place regularly and after a feature is coded. Reviewers should not only help to improve the code, but they should take responsibility for the code byadding their name in the comments when the code is checked in. Reviews also improve knowledge sharing within the team and improve performance and efficiency in downstream development.

In traditional waterfall development, the integration is done late in the development cycle. This presents a high level of risk due to the potential issues may be nested and difficult to address. One way of reducing this risk is to integrate and test continually, not at a single point in the project (such as a "big bang" integration), as done in traditional development. Martin Fowler (2000) defines continuous integration (CI) as "a fully automated and reproducible build, including testing, that runs many times a day. This allows each developer to integrate daily, thus reducing integration problems."

CI gives you rapid feedback and enables mistakes to be corrected quickly. The three main steps for CI are listed next.

1. **All code is checked in daily** into a single source code repository.
2. **Automatic build,** runs compiler scripts and reports on success or failure of the code.
3. **Automatic testing** executes unit tests for the code and, if possible, runs regression tests.

In traditional development, testing is usually an afterthought. In using a more Agile development process, it stresses the importance of testing by the concept of test-driven development (TDD). In TDD, an automated unit test is written before any code is written. The developer will continually add more test cases and make them pass by writing the required code. TDD provides superior quality software and ensures that testing is at the heart of development, not something done at the final stages of the development process.

Another approach is to leverage Refactoring. Refactoring improves the structure of components without changing their functionality. This practice improves quality and also makes the software being developed easier to maintain. By having automated tests it will ensure you have not inadvertently changed component behavior. Refactoring is particularly useful to do on critical or complex pieces of code.

Planning Level Approach

In reviewing initial requirements in software solutions developed and delivered, studies have shown roughly that 40 to 50 percent of features created from early specifications were never used and an additional 15 to 20 percent were rarely used (Khurana 2009). These statistics suggest that we should do iterative development, which will allow us to

- **React to changing requirements.** Developing for customer and future systems will require planning for changes. The development process and management required needs to be able to have flexibility to adjust and adapt for these changes. Traditional waterfall approaches expect stable and unchanging requirements. In software development, this is rare.

- **Optimize & Prioritize Work.** Focus on addressing high-value, high-risk items first; high-value, low-risk items second; and low-value, low-risk items last. Avoid low-value, high-risk items as a priority.
- **Focus on continually delivering business value and executable software.** Remember that technology is there to make the business more successful; fast and incremental delivery of value solutions is the key.

Note that iterative development provides shorter feedback cycles, which provides the following benefits:

- Testing is done much sooner than in traditional waterfall development.
- Continuous process and learning improvement through the project. (In traditional waterfall development, you often don't learn until post implementation or the closeout of the project)
- Risks are identified and addressed sooner.
- Management or the customer can make strategic and tactical adjustments along the way.

In the initial stages of planning, the estimates often are made initially without the benefit of understanding the full requirements of the solution being requested.. A best practice is to establish an initial requirements mapping session and follow-up by reviewing and revisiing these requirements throughout the development phases. When estimation is done at the beginning of the lifecycle before the requirements are fully defined, how can you accurately estimate without knowing the problem in it's entirety? To make this even more complex, the requirements may become obsolete or the needs of the stakeholder/customer may change as the application is being developed. Early estimations are sometimes guestimates (yes, a known form of estimating according to PMI), however how many of us cringe when asked for an estimate? We hear those famous worlds, "We won't hold you to it …", but we know that the initial number sometimes becomes etched in marble. The team is held to deliver on the estimate, even though the features and functionality hadn't been fully defined when it was made. A best practice is to revisit the estimates as the project progresses.

A good example of this is used at NASA. They advocate re-estimation at defined points in the lifecycle. Unfortunately, many project managers know that they should revise their estimates as they get a better understanding of the requirements, but they don't because they feel they will be seen as failing in their role as a PM. Project managers need to be honest with themselves, their team, and their stakeholders.

Another example is that management often reacts to problems or issues that are raised, rather than proactively reducing the likelihood of such problems occurring. While this reactive approach may be a product of the high level, busy executive, this can easily be addressed by building a proactive mind-set or review process with senior stakeholders and the project team that is focused on asking and addressing what should be done at each task or stage, to prevent your project from slipping in the future.

It is important that the project manager is not bullied into enforcing a totally unrealistic completion date that has been defined by a superior manager. Project managers often commit to unachievable dates, leading to failure and disappointment. In such situations, managers need to explain the key dimensions in project management: scope, time, quality, cost, and risk. Altering these variables might allow the project manager to achieve the date specified by the "bully" sponsor (through scope or quality reduction).

Good PM is about managing expectations. You must not surprise your customer. Many projects are considered big successes even though they have delivered only a small percentage of what was specified when they began, but they managed expectations obsessively.

Ultimately, a significant factor for project success is to have executive support. Typically, large projects should have a steering committee (or Project Board in Prince 2 methodology). This committee is chaired by an executive or project sponsor and consists of a senior user, a supplier, and a project manager. Essentially, this steering committee maintains commitment and provides a business involvement looking and addressing and making decisions regarding project scope and direction, and resolves issues in a timely manner, ultimately weighing strategic interests versus tactical project only decisions.

IMPORTANT CONCEPTS COVERED IN THIS CHAPTER

In this chapter, we focused on these key topics and learning points:

- The value of PMOs and processes supported in PPM
- The integration of issues, risks, and deliverables in the SharePoint/Project server Environment
- Leveraging lists to help project collaboration, "not just schedules anymore"
- Synchronized project lifecycle management in Project Server and its benefit to reinforcing processes
- Technical solutions or needs and best practice approaches

REFERENCES

Fowler, Martin. 2000. "Continuous Integration (original version)." Last modified September 10, 2000. http://martinfowler.com/articles/originalContinuousIntegration.html.

Khurana, Sanjeev. 2009. "The Art of Software Delivery." Last modified September 21. 2009. www.gantthead.com/content/articles/251635.cfm.

CHAPTER 9

INTELLIGENT BUSINESS PLANNING AND CONTROLLING USING MICROSOFT PROJECT 2010

IN THIS CHAPTER

In this chapter, we describe how strategic planning with Project Server facilitates the linkage between the strategy and program/project execution for the organization. Microsoft Project 2010 is one of the best project management (PM) and collaboration tools that supports organizational linkage between strategy and execution.

What You Will Learn

- How Project Server enables strategic planning
- The value of linking strategy to performance
- The shape of the Portfolio Lifecycle with Project Server
- How to leverage business drivers for success in projects
- How to use Project Server to master demand management
- Best practices for determining portfolio selection criteria
- How to create portfolio views and link dependent projects
- How to view or address constraints in a portfolio
- How to creating multiple scenarios for portfolio planning
- How to commit new work through portfolio planning
- Project Server's optimization of governance for project management offices (PMOs).

UNDERSTANDING STRATEGIC PLANNING WITH PROJECT SERVER

The content in this section has been structured around key portfolio management knowledge constructs illustrated in figure 9.1, and describes how Project Server 2010 can be used to provide a complete end-to-end integrated solution to address the portfolio management needs of a department or enterprise. Figure 9.1 illustrates the topic map for key knowledge concepts and components.

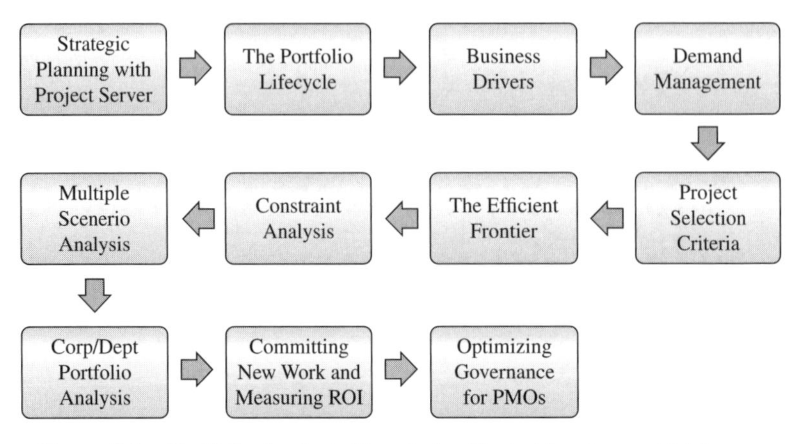

Figure 9.1 Portfolio Analysis Knowledge Topic Map Source: Advisicon

What Is Strategy and Strategic Planning?

The online Bing.com dictionary defines "strategy" as "a carefully devised plan of action to achieve a goal, or the art of developing or carrying out such a plan." Strategic planning is the process by which leaders of an organization determine what the organization intends to be in the future and how it will get there. In other words, leadership develops a vision for the organization's future and determines the necessary priorities, procedures, and strategies to achieve that vision.

Although most organizations have a strategic plan, it is often in the form of a mission or purpose comprising a set of products and services, a set of quantifiable goals, or just some narrative or spreadsheets describing how the company will achieve the plan. It is uncommon for organizations to have a set of actionable strategies and plans that detail how organizational resources will meet their objectives. Even less common is the presence of a corporate-wide project portfolio management (PPM) system that provides visibility, insight, and control into the execution of the organization's strategic plan.

Project Server 2010 facilitates this critical linkage by providing a fully integrated PPM solution that spans the vision of the executives: through project management, through the realization of benefits to the enterprise, and through the successful delivery of products and services in the marketplace.

Importance of Strategic Planning

Strategic planning helps ensure that an organization remains relevant and responsive to the needs of its customers and contributes to organizational stability and growth. It provides a basis for monitoring progress and for assessing performance. It facilitates new program development. It enables an organization to look into the future in an orderly and systematic way. Figure 9.2 shows an example of

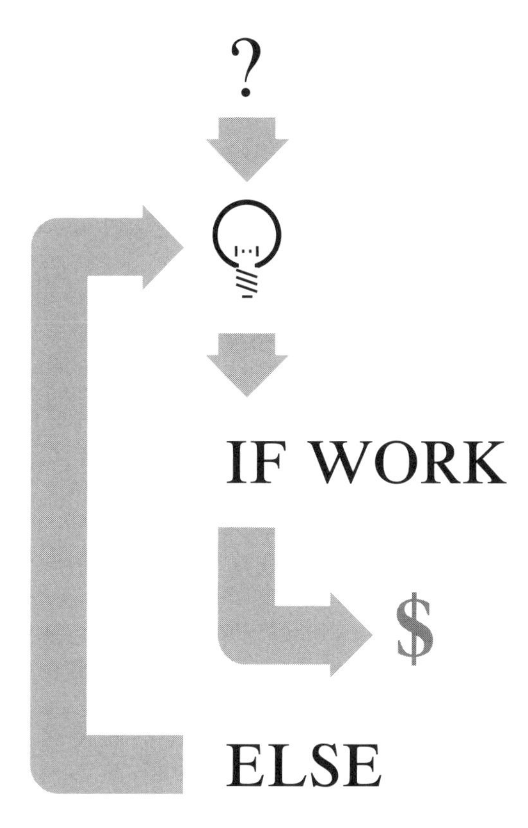

Figure 9.2 Business Plan Source: Advisicon

a business plan. In thinking about creating new proposals or projects, addressing the business need is a key element that increases the likelihood of acceptance and selection of a project.

From a governance perspective, strategic planning enables executive management to establish goals and strategies to guide the organization and provides a clear focus to the managers and staff for program implementation and PM.

From a performance perspective, strategic planning helps organizations define, prioritize, and deliver those initiatives, programs, and projects that are important to the overall success of the business. Doing the right things versus doing things right!

Linking Strategy to Performance

Measuring performance provides an organization and its key leadership resources to assess the overall health and vitality of the enterprise. Most important, however, performance measurement enables an organization to assess, monitor, and course-correct performance to align all employees with key business objectives. This clears the way for the company to implement its business strategy.

Key performance factors are quantifiable metrics that reflect the critical elements of the enterprise, department, or project. Some examples of key performance factors include:

- Increased market share
- Improved employee performance
- Increased return on investment (ROI)
- Reduced time to market

A metric-based scorecard approach is certainly one way of measuring performance; however, this approach is akin to flying an airplane using a rearview mirror. We need to pay more attention to where we are going (so we don't fly the aircraft into a mountain) and pay less attention to our instruments and flight plan. Many organizations focus on past trends or metrics of project performance without spending the necessary time to integrate project metrics based on future expectations. By combining future results with existing work, an organization gains the ability to see where it is going in the context of where it expected to be.

Our flight plan starts with the ultimate goal and mission of the organization that eventually must align with all the business units and departments. Teams need to align with the means (our methods to achieve our desired outcomes).

Figure 9.3 illustrates how high-level strategic objectives, processes, and systems help align business units, departments, teams and individuals. Key metrics are used to assess performance toward the goals and mission of the organization.

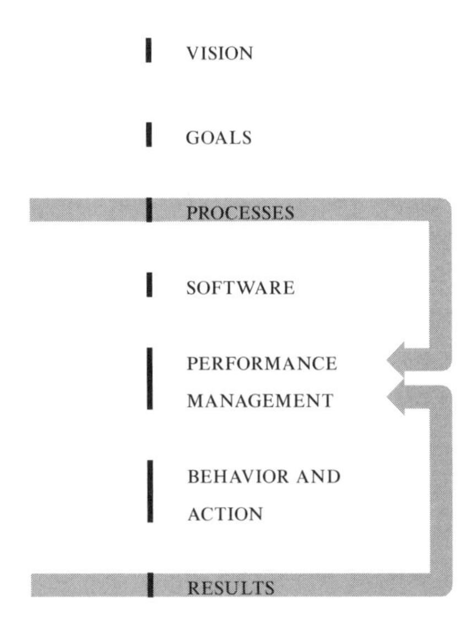

Figure 9.3 Mapping Strategic Objectives to Organization Performance Source: Advisicon

A simplified view depicting the strategic linkage in the project portfolio is presented in Figure 9.4. Here we can visualize how the goals and strategies act in concert to help align the work (the programs and projects) to the mission of the organization. Now we are on course with an approach that we can dynamically adjust as strategic intent of the organization changes due to external and internal conditions. We are in control of our flight.

In the next section, we show how mapping organizational strategy to programs and projects is accomplished by defining business drivers. Utilizing Project Server allows us to support executive decision making to better understand and manage priorities, resources, and the work of the organization. We introduce the concept of the Portfolio of Projects (often shortened to Portfolio). The Portfolio is used to evaluate the priority of one project versus another by determining the total demand of the organization and then evaluating competing project or proposal candidates for the limited resources. The Portfolio is the critical control point in the enterprise. If properly implemented, it can become the single source of truth for all work demand and resource utilization. However, work managed

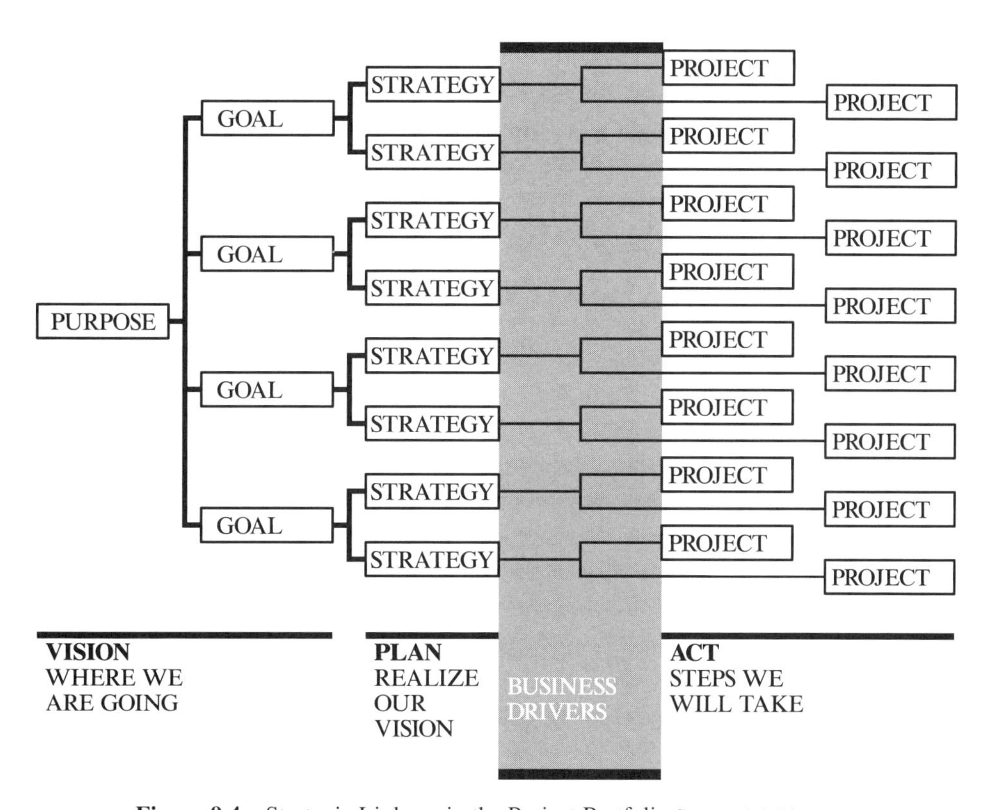

Figure 9.4 Strategic Linkage in the Project Portfolio Source: Advisicon

outside the Portfolio results in an over-allocation of resources because their actual availability will not be accounted for accurately. When this happens, it is extremely difficult (if not impossible) to reflect true demand versus capacity in the rollup reporting to management. Without an accurate depiction of this relationship, management cannot make informed decisions about work priority or assignment. This is a common problem in most enterprises; often it is considered a resource management problem.

CREATING AND MANAGING PORTFOLIO LIFECYCLE FOR PROJECT SERVER

Portfolio management is the art and science of balancing an organization's PM skills and resources to achieve optimum strategic, financial, and operational impact across all product lines in all lifecycle phases.

Project Server 2010 enhances the overall support of the Portfolio lifecycle by streamlining DM from idea creation to successful execution and close-out. The best-of-breed portfolio management techniques are included in Microsoft Project Server 2010, providing a single server with end-to-end PPM capabilities.

This section describes how Project Server supports the entire lifecycle of projects in an organization.

Portfolio Lifecycle

The Portfolio lifecycle follows a project from the initial concept to business case analysis and project initiation to tracking the project through various stage gates of progress with workflow to the final measurement of success using tools that provide accurate and timely reporting of project and portfolio results.

Figure 9.5 is an example Portfolio lifecycle that illustrates four key phases: create, select, plan, and manage. A phase represents a collection of stages grouped to identify a common set of activities in the project lifecycle. A stage represents one step within a project lifecycle (e.g., propose idea, deliver project).

Phases and stages are managed in Project Server 2010 through the use of enterprise project templates that help guide projects through each stage through the use of workflows.

Portfolio Lifecycle Governance and Workflow

Portfolio governance and lifecycle management enable organizations to define processes that synchronize the efforts of distributed teams to consistently create the best possible products, capture greater market share, and increase customer satisfaction.

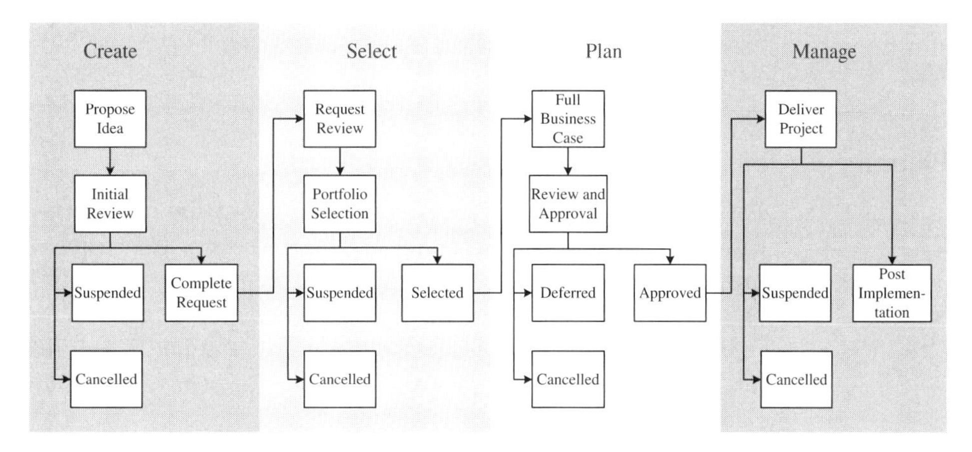

Figure 9.5 Example Portfolio Lifecycle Source: O'Cull 2009

The project workflow models the organizational governance processes to provide a structured way for projects to proceed through the various Portfolio phases and stages. Workflows, along with other project attribute data in project detail pages (PDPs), are captured and integrated within Project Server 2010.

To better understand workflows, it is important to understand the relationships and roles of the key Project Server 2010 components. These components are illustrated in Figure 9.6.

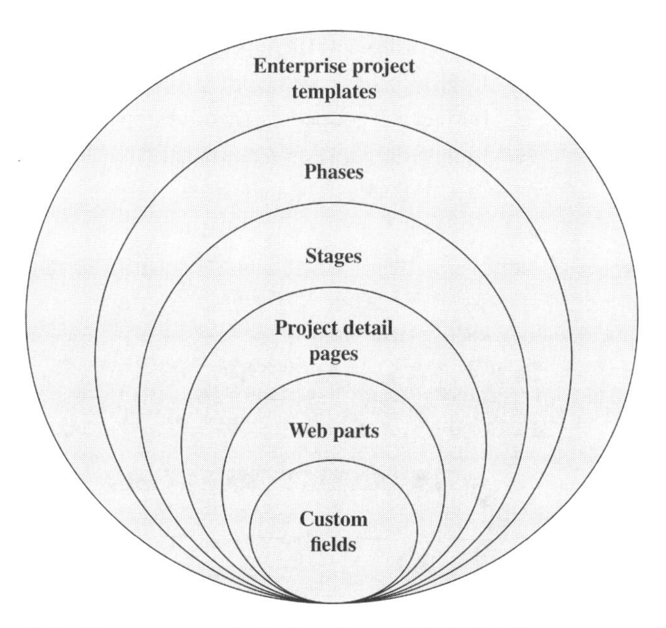

Figure 9.6 Components of a Project Server 2010 Workflow Source: Advisicon

Every project is associated with an Enterprise Project Template, which governs it through the Portfolio lifecycle.

Planning for the Portfolio Lifecycle

Portfolio management is focused primarily on defining the needs of the business: capturing total demand, aligning to the priorities of the business, and then selecting projects for detailed planning and execution—essentially ensuring that the organization is focusing on the right projects.

PM then entails the detailed planning, execution, monitoring, reporting, and final closure at the completion of the project (in other words, doing projects right).

Figure 9.7 provides an overview of the Portfolio Project lifecycle.

The remainder of this chapter addresses how Project Server 2010 supports each of the next key aspects of the portfolio management lifecycle:

- **Defining business drivers,** the interface between the business strategy and the Portfolio of Projects.
- **Capturing demand** to provide a detailed understanding of the entire work and resource demands. Doing this helps to align all of the requests for work and resources to the business priorities and capabilities of the organization. We use portfolio management methods, techniques, and analytics.
- **Aligning to business priorities.** How work (programs and projects) aligns to the mission (or purpose) of the organization.
- **Performing constraint analysis.** The review of restrictions of resources, budget, or proposals that are dependent on each other.
- **Selecting an optimal mix of projects,** by applying strategic portfolio analysis and selection criteria to a group of proposed projects.

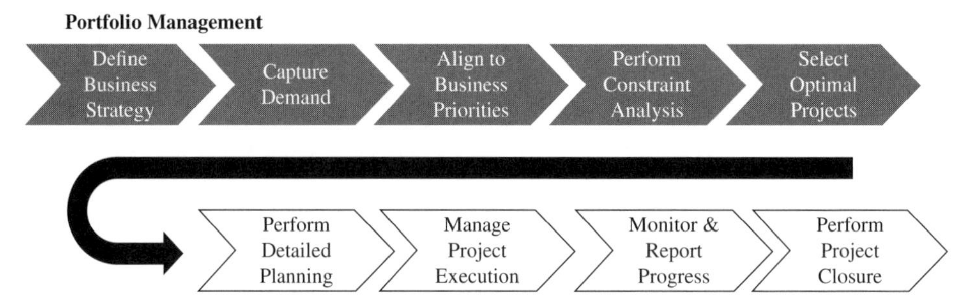

Figure 9.7 Overview of the Portfolio Project Lifecycle Source: Advisicon

These key project management aspects are shown here to provide a complete picture of the integrated PPM lifecycle.

UNDERSTANDING AND BUILDING BUSINESS DRIVERS

What are business drivers? They are the main factors and resources that provide the essential products, services, marketing, sales, and operational functions of a business. Stated another way, business drivers are the people, information, and tasks that support the fulfillment of a business objective. Business drivers include the people, knowledge, and conditions (e.g., market forces) that initiate and support activities for which the business was designed.

Common key business drivers are listed next.

- **People and their competencies** help define the long-term intent, objectives, and strategies of an organization. They manage the execution of critical decisions and provide constant innovation to move the business forward.
- **Products and services** are developed and delivered by the organization in response to critical customer pains. These pains result from changing market forces and customers' ability to assimilate new methodologies.
- **Technology and innovation** are also key differentiators in today's competitive business environment. Innovative technology is a key enabler that can assist businesses with everything from managing large volumes of data, to analysis of trends, to the support of organizational planning through the use of decision support systems.

Most business drivers are specific to a particular industry or aspect of an organization (e.g., construction, aerospace, consumer packaged goods, or information technology [IT]). Some business drivers, however, are common to all organizations in one respect or another (e.g., customer experience and satisfaction, quality of goods and services, and ROI).

Value of Using Business Drivers

The intended purpose of business drivers is to help us define the priorities of the business. They are derived by having a clear definition of the mission, vision, goals, metrics, and strategies of the organization. Figure 9.8 illustrates this relationship.

Business drivers act as the interface between the business strategy and the Portfolio of Projects to define key business priorities and to assist with the prioritization and scheduling of work and resources.

Too many tools and too many organizations fail to combine their portfolio and work planning with the existing set of approved projects, which results in an

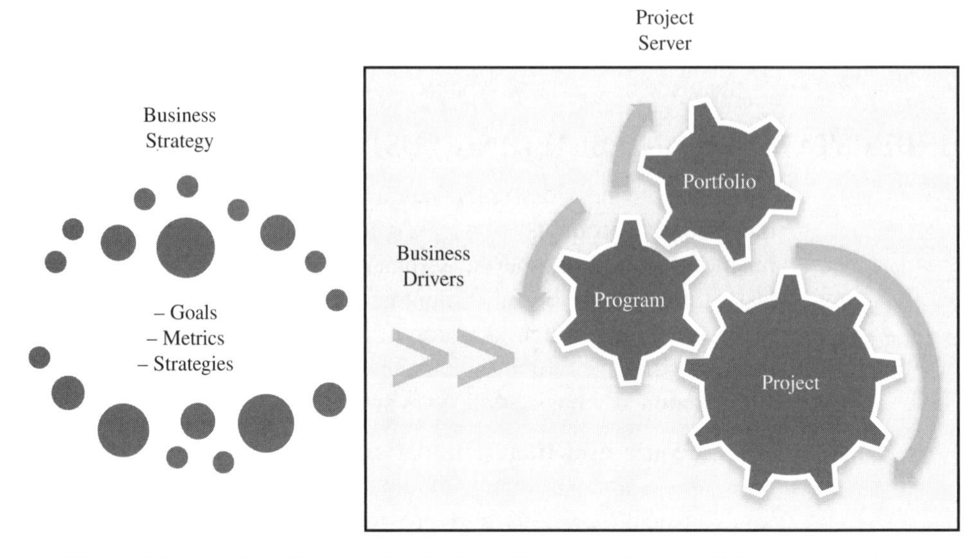

Figure 9.8 Interface Between the Business Strategy and the Portfolio Source: Advisicon

incomplete and inaccurate understanding of the organization's total work demand and capacity.

Understanding and properly defining business drivers is a key step in ensuring the success of a portfolio management system. Business drivers are an effective way to ensure alignment between strategy and execution as they:

- Provide the linkage between the business strategy and the Portfolio of Projects.
- Ensure a consistent way for key stakeholders to agree on cross-organization business objectives.
- Help establish a basis for mapping projects back to business priorities.
- Offer a central single source of truth for all ongoing and planned work so that prioritization can be properly performed.

Defining Business Drivers

Here is where we separate the approach of defining business drivers from creating the business and technical aspects (keeping in mind they are both necessary elements). The business approach to deriving business drivers is usually done in a facilitated workshop. Defining business drivers in Project Server involves succinct steps in the definition process with the goal of building a business driver with a quantifiable metric (enabling specific and measurable information gathering for business intelligence (BI) analysis both before and after the project).

Art of Defining Business Drivers

A facilitated-workshop approach can be used to determine the key priorities of the business. This approach is an important first step, as it solicits key input from the business leaders, knowledgeable subject matter experts, and other key stakeholders in the organization to obtain strategic-level and goal-oriented input that will be used in defining the business drivers.

It is important to focus on the definition of the business drivers without putting too much emphasis on how to configure the Project Server environment. The initial focus should be on the key business aspects, including the long-term strategic intent of the organization, its vision, mission, goals, metrics, and proposed strategies.

As stated earlier, business drivers are somewhat specific to particular industries or aspects of an organization, so it is difficult to list a complete set as they will in turn be heavily influenced by the goal set, metrics, and strategies of the organization's business strategy.

Technology of Business Drivers

Project Server 2010 provides an integrated PPM platform to help organizations break down their strategy into actionable, measurable, and discrete business drivers. This solution helps define and effectively communicate the business strategy in actionable terms to provide a blueprint that can be understood and implemented by departments throughout the organization.

Now that you have derived your business drivers, you can start to consider how you are going to define them in Project Server. This process includes:

- Creating business drivers in the driver library.
- Selecting those departments containing projects that will be measured against a given business driver.
- Establishing driver impact statements that describe how specific projects support a given business driver.
- Objectively prioritizing business drivers to drive consensus within the executive team.

Creating New Drivers

The steps for creating or removing business drivers are fairly straightforward and easy to manage from Project Server's interface.

Typically, after you have worked with your executive team and subdivided your strategic goals and objectives into succinct business drivers, you will need to enter each driver into PWA in the driver library.

To get started with drivers, it is recommended that you first enter the driver name and description. The description provides a more detailed explanation of the driver, which includes how it maps to strategic goals. It should also provide an overall goal for the driver, such an increase in repeat business.

An organization can review the business drivers periodically and respond to changes in industry trends or business directional changes. The referential information will help to link the original drivers and what their focus or intent was at the time of their creation.

It is best practice to keep the name for your business driver somewhere between two and six words on average; doing so makes the name easier to read in the views and rating and ranking screens provided by Project Server.

To create a new driver (see Figure 9.9):

1. From the Quick Launch, click Driver Library.
2. On the ribbon in the Driver group, click New.
3. In the Name and Description section, enter the Name and Description.
4. On the ribbon, click Save & Close.

Deleting Drivers

Drivers that are no longer used or needed should be deleted. Doing so will reduce the list of drivers that you have to pick from and may help simplify the creation of a new driver prioritization.

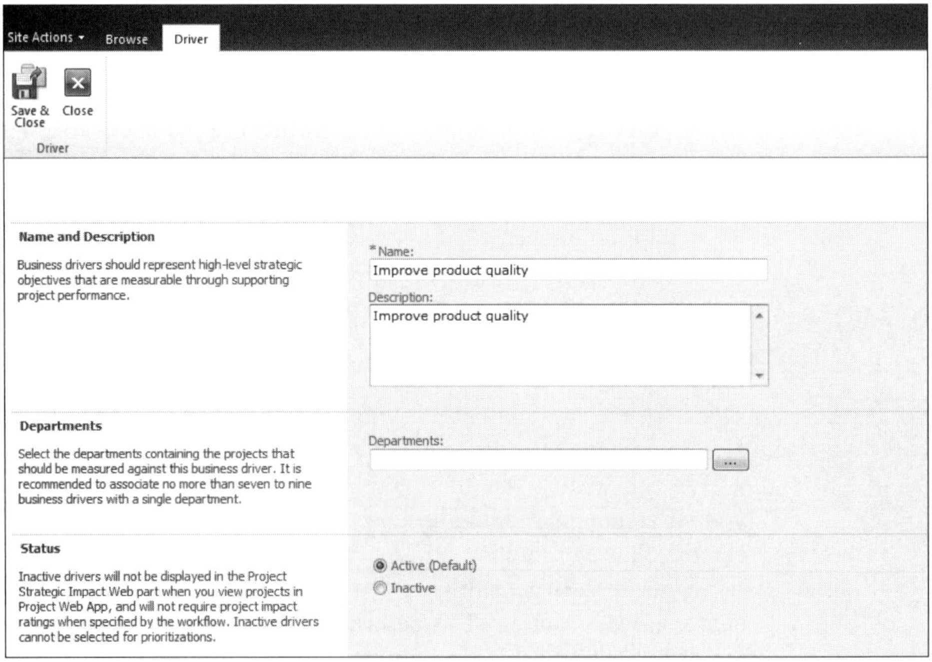

Figure 9.9 Creating a New Driver Source: Advisicon

Be careful in removing or deleting drivers. Deleting a driver is permanent and cannot be undone. You do have an option to make a driver inactive, which we will address later in this chapter.

To delete a driver (see Figure 9.10):

1. From the Quick Launch, click Driver Library.
2. Select the driver you want to delete by clicking on any portion of the row that does not have hyperlinked text.
3. On the ribbon on the Driver tab in the Driver group, click Delete.
4. In the Message from webpage dialog box, click OK to complete the deletion process.

Understanding Project Impact Statements

It is important to establish a Project impact statement for every proposal or project being evaluated in the Portfolio Management module. Without a project impact rating scale, each project that affects a business driver would be given the exact same rating against that driver. If so, it would be virtually impossible to assess which project might be better to select over another, especially if multiple projects are connected with the same business driver.

The advantage of the business driver feature in Project Server is that each driver is affiliated with both an impact rating scale and an affiliated impact

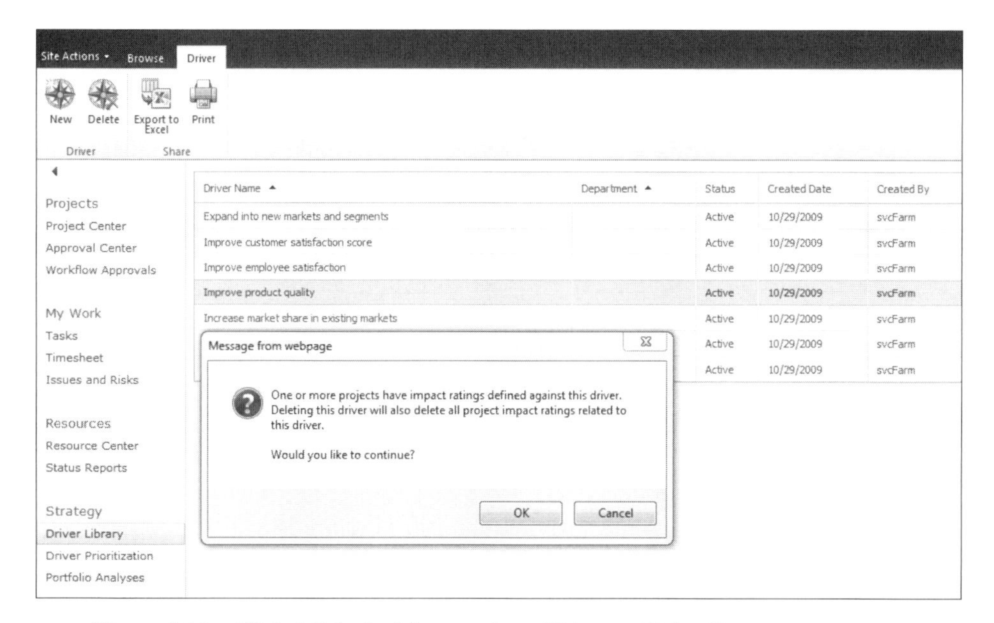

Figure 9.10 Click OK in the Message from Webpage Dialog Box Source: Advisicon

statement. This ensures that you can differentiate between projects of various impacts against a strategic objective. Also, a very important reason for completing the impact statement is it identifies a standard for each level that everyone must follow so there is common understanding of what it means. For example, to rank a project in a low or moderate value, there should be a definition that is understandable for the person creating the project/proposal. Failure to fill in the project impact statement leaves a lot of room for individual interpretation and reduces consistency between project and portfolio managers.

To enter a project impact statement:

1. From the Quick Launch, click Driver Library.
2. Click the name of the driver you want to edit.
3. On the Edit Business Driver: name page, in the Project Impact Statements section, enter information that will clearly describe each level.
4. On the ribbon, click Save & Close.

Types of Driver Prioritization

Since not all organizational objectives have equal importance, you will need to complete driver prioritization so Project Server can make appropriate suggestions when doing a portfolio analysis. There are two types of driver prioritizations: calculated and manual.

Two different types of driver prioritizations are necessary because in some cases you will want to compare driver to driver and let the software calculate a scale of importance. In other cases you may want to enforce an executive preference. Should you choose to enforce an executive preference, you will need to use the manual driver prioritization type. Both of these types are explained further in upcoming topics.

Creating New Manual Prioritization

Project Server allows you to establish your own unique process for prioritizing business drivers by creating a manual process. When you create a manual prioritization, you specify a priority value for each driver. Doing this allows you to create a prioritization that meets the needs of an executive, team, or department. Be sure that whatever driver you give the highest value to is actually the most important business driver in relation to the other drivers selected for the prioritization.

Priority values must total 100 percent. If you fail to achieve 100 percent, the system will make adjustments to result in a total of 100 percent. The overall concept is very simple: Each driver prioritization is compiled to score up to 100 percent. This gives an overall ranking for all drivers that equal the maximum rating or ranking.

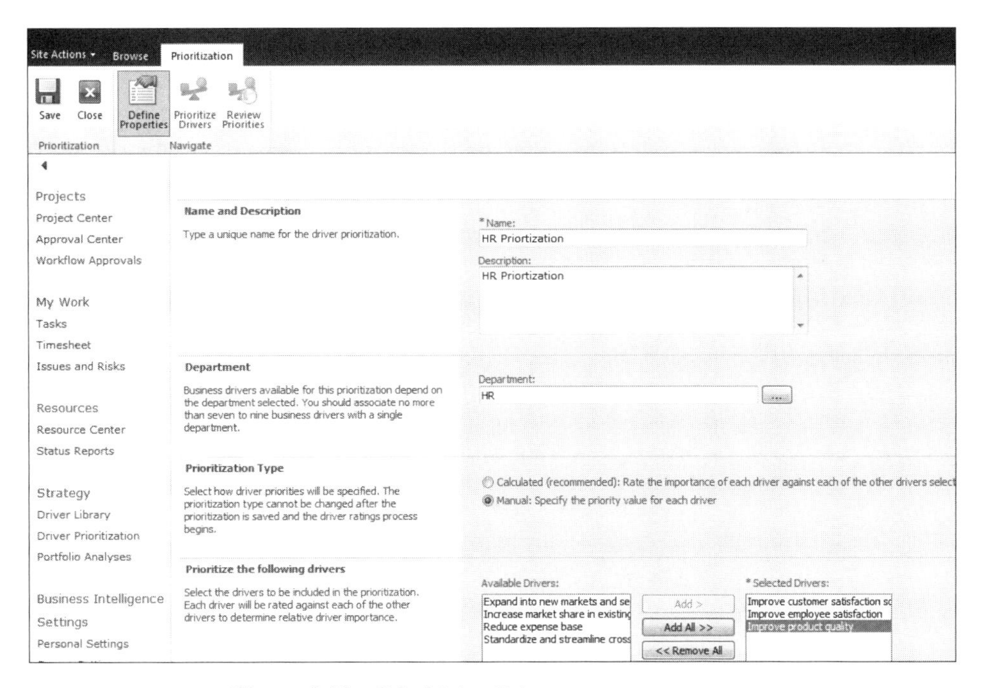

Figure 9.11 Prioritizing Drivers Source: Advisicon

In many cases, organizations will return and fine-tune the percentages either up or down as the portfolio of work is evaluated in order to achieve a more realistic ranking system for new project proposals. Figure 9.11 shows the prioritization of drivers. It is good to note that as the Portfolio rating and ranking process is going on, many organizations come back in and tune their drivers over and over again until each scenario and driver is giving a good overall ranking of projects.

To create a new manual prioritization:

1. On the Quick Launch, click Driver Prioritization.
2. On the ribbon, in the Prioritizations group, click New.
3. In the Name and Description section, enter the name and description.
4. Optional: In the Department section, select a Department from the Select Value button.
5. In the Prioritization Type section, select Manual.
6. In the Prioritize the Following Drivers section, select each driver and click Add. Alternately, you can click Add All to choose all the drivers.
7. On the ribbon in the Prioritization group, click Save. When complete, click Close.

Creating New Calculated Prioritization

The default process is to utilize Project Server's prioritization engine. This fits most organizations that are using the out-of-the-box features. When you choose to do a calculated prioritization, you compare each driver to every other driver that was selected for the prioritization. The comparison is based on this fixed seven-point scale:

1. Is extremely more important than
2. Is much more important than
3. Is more important than
4. Is as important as
5. Is less important than
6. Is much less important than
7. Is extremely less important than

This process is also called a pair-wise comparison and will yield driver priority values that will total 100%.

To create a new calculated prioritization:

1. On the Quick Launch, click Driver Prioritization.
2. On the ribbon in the Prioritizations group, click New.
3. In the Name and Description section, enter the name and description.
4. Optional: In the Department section, select a department from the Select Value button.
5. In the Prioritization Type section, select Calculated.
6. In the Prioritize the Following Drivers section, select each driver and click Add. Alternatively, click Add All to choose all the drivers.
7. Click the button Next: Prioritize Drivers.
8. Using the Select a Rating list, choose the appropriate scale when presented with each driver comparison.
9. Click the Next Driver button to complete the next series of comparisons. Repeat until finished.

Reviewing Driver Priorities

After creating business drivers, it is best practice to review them and their weighted prioritization ranking. In many organizations, fine-tuning the ranking process occurs even during the portfolio evaluation process as the results sometimes don't match expectations.

Following completion of a new driver prioritization, you will have the opportunity to review the list of driver priorities. This list is in order of highest priority at the top and lowest priority at the bottom (see Figure 9.12).

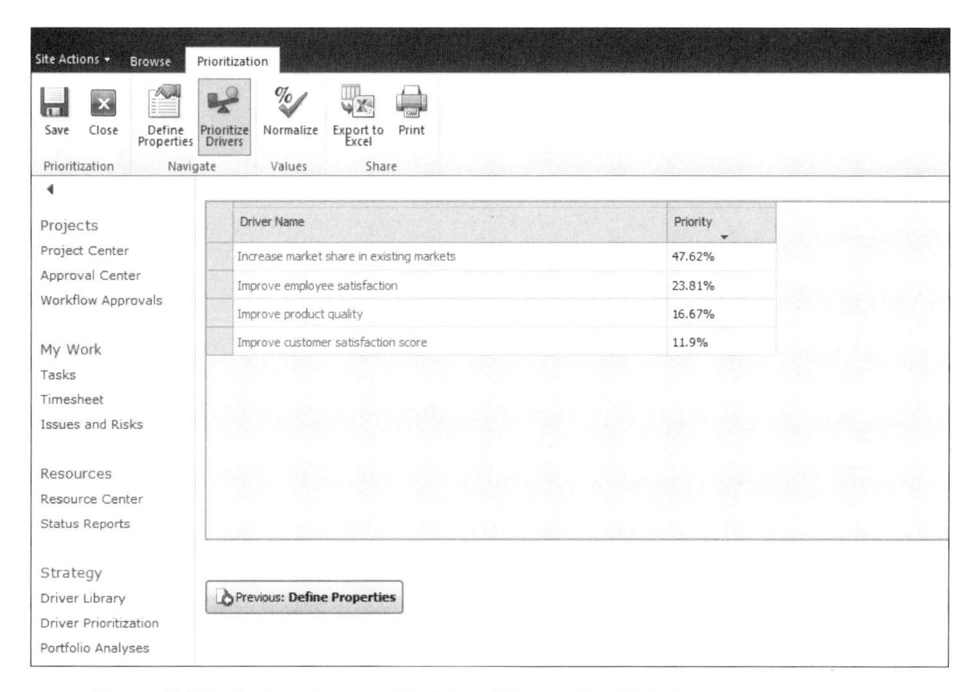

Figure 9.12 Drivers Arranged in List of Descending Priority Source: Advisicon

If you have selected manual prioritization, you are free to make additional changes to the priority values until you are satisfied with the results. You can use the normalize feature to adjust the values to total 100 percent, or you can manually make adjustments to reach that total.

If you have selected calculated prioritization, it is recommended that you go back and review your pair-wise comparison if the priority values seem to be unexpected.

To normalize a manual prioritization:

1. On the Driver Priorities page, review the priority values entered and make changes as desired.
2. On the ribbon in the Values group, click Normalize. Notice the values will be recalculated to total 100.

Significance of Consistency Ratio

Just below the view showing the prioritization list is a consistency ratio rating scale graphic (see Figure 9.13). It measures how many logical conflicts occur in the driver prioritization. If one driver is less important than all the other drivers in one instance, it should not be marked as the most important driver when compared again against the same drivers. This would be a logical conflict.

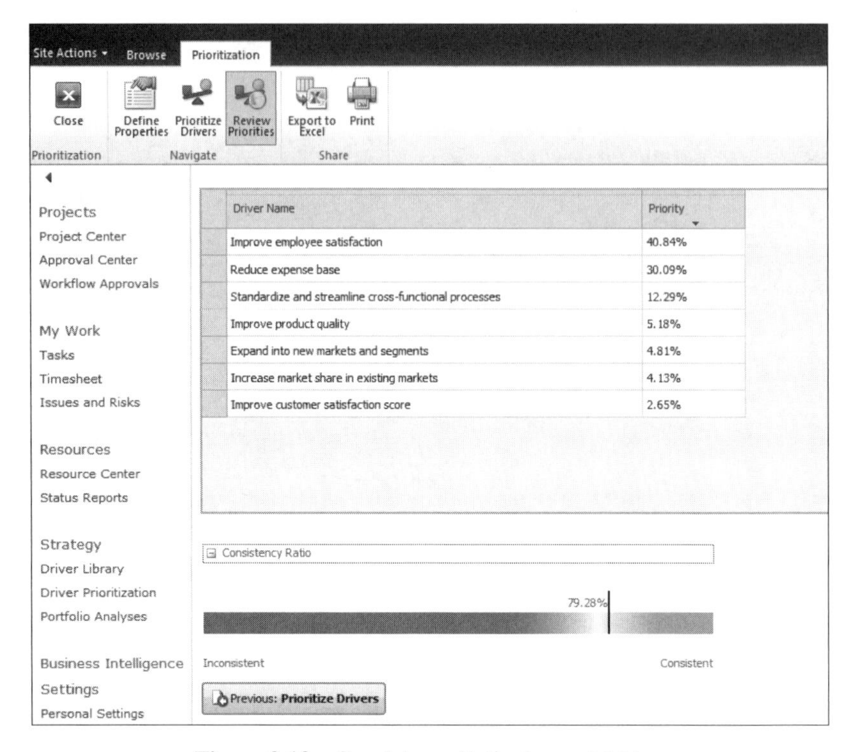

Figure 9.13 Consistency Ratio Source: Advisicon

Normally the consistency ratio is less than 100 percent; a number around 75 percent or higher is considered to be a good ratio. Very low numbers typically indicate mistakes or inaccuracies; you may want to consider completing a new driver prioritization before accepting a low ratio.

USING PROJECT SERVER TO MASTER DEMAND MANAGEMENT

The overall aim of portfolio management is to provide an organization with the ability and flexibility to quickly adapt to changing market conditions, make better investment choices for projects and initiatives, and in general make better-informed decisions. In order to do that, we need to centrally manage requests for all work and account for the planned and in-flight allocation of all resources.

In most situations, demand generally exceeds the ability of an organization's resources to deliver. As discussed in the introduction to this chapter, there are many strategies and priorities for the different business units and departments of the enterprise. These requirements eventually result in project requests that need to be reflected in a total demand picture. We need an agreed-on mechanism by which we can objectively track and manage demand.

Value of Demand Management

The starting point to good governance and planning is to have a clear and complete understanding of the total work picture and demand for resources in an organization, including both new and in-progress projects. Demand management (DM) provides a consolidated view of the total work and resource demand picture across the entire organization.

Demand management provides us with a detailed understanding of the entire work and resource demand to help align all of the requests for work and resources to the business priorities and capabilities of the organization, using portfolio management methods, techniques, and analytics.

DM provides the ability to:

- Have a single collection point for the capture of work and resource requests.
- Have a starting point for end-to-end insight for stakeholders, eliminating the black hole of requests.
- Make smarter decisions faster (based on the latest demand), collaborating better, and attain optimal ROI.
- Align investment decisions across the organization by having a total view of all work requests.

Role of Demand Management in an Organization

Often there may seem to be an opaque ceiling blocking your view into items resulting in limited organization's success in regard to project execution. Or worse yet, there is no opaqueness as the organization chooses to not keep limitations in sight at all.

Successful DM often requires organizational change. Executives and senior management must incorporate the balance of projects (demand) and constraints (supply) to make the best project decisions for the organization. Doing this requires clear identification of business goals and objectives as well as a transparent definition of organizational constraints. Streamlined communication is required from the team member to the chief executive officer.

Through openly visible criteria defined for project selection and measurement, paired with consistent governance and thorough analysis of project data and requests, organizations will have created the best internal environment to have the support of the entire team in driving project success.

Planning for Demand Management

It is important to leverage the key pieces in Project Server to get the most out of establishing a good demand and work forecasting and management system.

Some of those elements are:

- Enterprise project types
- Custom fields and lookup tables
- Project detail pages
- Project phases
- Project stages

In this section, we provide some straightforward tips, tricks, and building techniques for creating these Project Server entities. Building some or all of these objects rapidly increases the adoption, simplification, and ease of use for project creation, management, and reporting around demand management.

Enterprise Project Types

Every project or proposal that is placed in Project Server is started with an enterprise project type (EPT). EPTs are associated with specific types of projects, such as new product development or software upgrade. Different EPTs may be needed for various departments that may have stages or processes that do not follow the organization's typical stages or phases.

One of the best ways to automate the creation of projects and help speed up the process of getting an accurate and consistent schedule is to leverage EPTs. Doing so sets the stage for decreasing time for project managers or planners to create schedules and to see the upcoming or new demand of work in the existing or future work portfolio.

What Makes Up Enterprise Project Types?

Every project that is created in Project Server from the ribbon begins with an EPT. An EPT is a template for a specific kind of project that may or may not be associated with a workflow. Some areas where you may need different EPTs include internal versus external projects, departmental-versus enterprise-level projects, or for-profit versus nonprofit projects. An EPT can include all or some of these components:

- Phases
- Stages
- Workflow
- Project detail pages
- Project, task, resource regular, and custom fields

Default Project Server Enterprise Project Type

Since the EPT is a building block to project creation, it is important to begin with a selection of initial EPTs. Default EPTs provide a great place to get started.

However, creating schedules that have the same building blocks of tasks will rapidly expedite the schedule creation process for organizations looking to streamline project types.

New projects are not necessarily just project schedules. They may include a series of enterprise fields necessary for keeping information about a proposal or a project through is project lifecycle.

Microsoft has created some sample default EPTs for every deployment. These are the ones that appear in the Project Center when you click on the New button on the Projects tab on the ribbon. Clicking the New button displays the EPTs in a list format. Only two default EPTs are provided initially:

1. **Sample Proposal.** This is affiliated with the sample proposal workflow and is located right above the Basic Project Plan in the list of EPTs.
2. **Basic Project Plan.** No workflow is associated with this EPT. It contains only the schedule and Project information detail pages. It is the default EPT during project creation.

A nice feature is that EPTs without a workflow have a schedule icon next to the name, while EPTs with a workflow display an icon with a curved arrow on them.

To display the default EPTs:

1. On the Quick Launch, click Project Center.
2. In the ribbon on the Projects tab, click New to display the list of default EPTs.

Creating an Enterprise Project Type

Project Server allows you to create Project types and link them to specific departments. This method enables a multiple-use scenario for different parts of an organization to use Project Server, without having to wade through a large list of choices.

To satisfy the needs of different departments or to account for differences between various types of projects that your organization creates, you will need to create new EPTs.

To create an EPT (see Figure 9.14):

1. On the Quick Launch, click Server Settings.
2. In Workflow and Project Detail Pages, click Enterprise Project Types.
3. On the Enterprise Project Types page, click New Enterprise Project Type.
4. In the Name section, type a name for this EPT.
5. In the Description section, type a description (optional).
6. In the Site Workflow Association section, in the list, click the desired site workflow association.

   If you click No Workflow, the options in the next section will change.

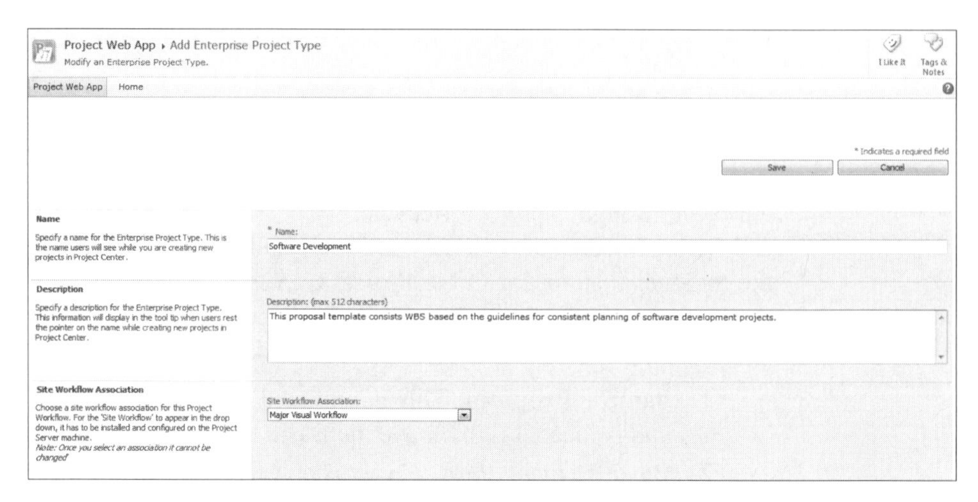

Figure 9.14 Adding an Enterprise Project Type Source: Advisicon

7. In the New Project Page/Project Detail Pages section, click the desired New Project Page.
8. In the Default section, check the box "Use this as the default Enterprise Project Type during Project Creation" if you want this to replace the existing default EPT (optional).
9. In the Departments section, select the desired department from the Select Value button (optional).
10. In the Image section, type the URL for the desired image (optional).
11. In the Order section, check the option to Position this type at the end (if desired) or click the EPT item in the list that you want to appear before your new EPT.
12. In the Project Plan Template section, click the desired Project Plan Template (optional).
13. In the Project Site Template section, click the desired Project Site Template.
14. Click Save.

An important step in building an EPT is the taking time to map and plan out exactly what you are looking for. Doing so saves time and rework later.

Modifying or Deleting Enterprise Project Types

EPTs allow you to organize the fields and what is displayed for an end user. Essentially, they are for organizing and presenting the metadata associated with a project to a specific page. These PDP pages are SharePoint pages tied to the project they are created with.

The creation, modification, and removal of EPTs are relatively simple processes. To modify an EPT:

1. On the Quick Launch, click Server Settings.
2. In Workflow and Project Detail Page, click Enterprise Project Types.
3. Click the name of the EPT you want to modify.
4. Make the desired changes and click Save.

To delete an EPT:

1. On the Quick Launch, click Server Settings.
2. In Workflow and Project Detail Page, click Enterprise Project Types.
3. Select the EPT(s) you want to remove and click Delete Enterprise Project Type.
4. When you receive the message: Are you sure that you want to delete the selected enterprise project type(s), click OK.

If possible, plan your custom fields out in advance of creating projects to reduce or eliminate the need to open and modify existing projects.

When you create a custom field, you will have to fill in key informational fields as described later the Entity and Type field selections. Proper selection of the Entity and Type is required to ensure your ability to report on these fields and enter desired information later on.

Enterprise Custom Field Entities

Project Server allows you to quickly and efficiently create additional fields to help enable BI reporting, viewing, sorting, filtering, and comparison analysis.

In demand and strategic analysis, these field entities are broken into different categories related to the major databases behind the scenes.

- **Project.** Used for a field that will be applied at the overall project level (e.g., project sponsor, or project industry type).
- **Resource.** Used for a field that further describes a resource (e.g., level of college, or speaks Russian).
- **Task.** Used for a field that will be applied at the task level (e.g., accounting code, task priority level).

Enterprise Custom Field Types

There are different custom field types to choose from. Listed next is a breakdown of what you can leverage for project to portfolio usage. Remember that if you are going to do manual portfolio evaluation, you will definitely want to leverage these different custom field types.

- **Cost.** Accepts numbers and is formatted using the currency settings configured for Project Web App.
- **Date.** Accepts only a date.
- **Duration.** Accepts a numeric value that represents a time span.
- **Flag.** Accepts only two options, yes or no.
- **Number.** Accepts only numbers.
- **Text.** Accepts any combination of letters and numbers as well as some symbols and special characters.

Some other important options when creating a field include:

- **Custom attributes.** "Single line of text" is recommended.
- **Department.** Choose a department only if this field will be limited to individuals in a specific department.
- **Values to display.** "Data" is recommended.

Figure 9.15 illustrates the additional options that are important to enabling customization or different types of portfolio rating raking and dashboards.

To create a new custom field:

1. From the Quick Launch, click Server Settings.
2. Under the Enterprise Data heading, click Enterprise Custom Fields and Lookup Tables.
3. In the Enterprise Custom Fields section, click New Field.

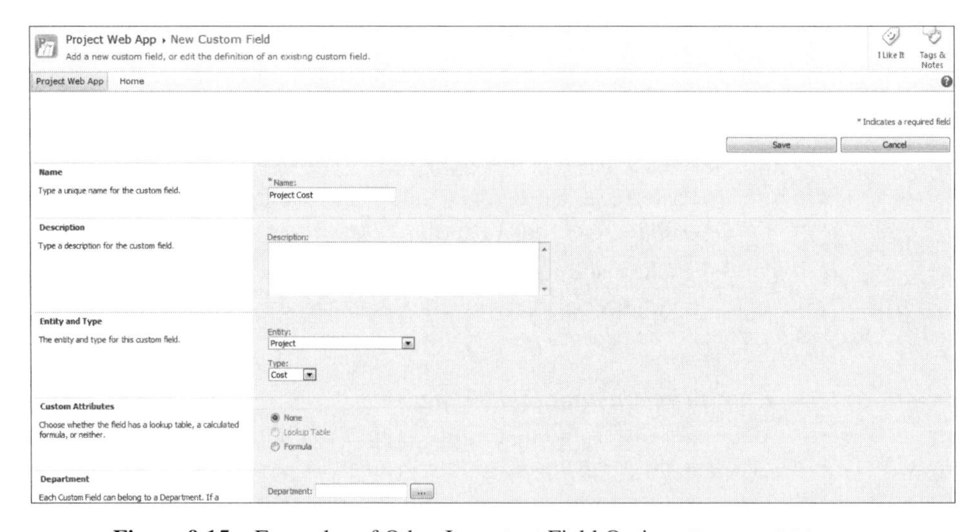

Figure 9.15 Examples of Other Important Field Options Source: Advisicon

4. Complete the fields as desired.

   Any field with an asterisk symbol is required.

5. Click Save.

Custom Fields and Lookup Tables

Project Server offers flexibility and customization so organizations can incorporate their own way of doing business into the software. Most organizations need to add information that does not currently match with an existing Project Server field.

The use of custom fields and optional lookup tables is a very simple way to capture that information. There are many advantages to using custom fields, such as the ability to create views and reports tailed to the information in the fields. It is possible to use custom fields to integrate with third-party applications.

In this section, we walk through the process of creating custom fields for the three different entities in Project Server. We show how to create a lookup table that can be attached to a custom field to provide a pick list and discuss how to control user behavior with various custom field options.

Creating New Custom Fields

Each custom field is a designated space you are adding to the Project Server database for capturing additional information. Custom fields are needed when your organization wants to include information in each project that existing fields do not account for. These fields typically are created to facilitate custom views online or to facilitate reports.

Creating New Lookup Tables

A lookup table contains a list of values. Lookup tables can be assigned to custom fields so an individual can select an option from the list instead of having to type in a value. A lookup table can be attached to a field to ensure that individuals always pick from a limited list of choices. This attribute simplifies data entry and the options when this field is used for reporting purposes.

When creating a lookup table, you must differentiate between two choices:

1. **Code mask.** This is used to define the structure of the lookup table. A different line is created in the code mask for each level you desire. Levels can contain information that matches one of four types: numbers, characters, uppercase letters, or lowercase letters.
2. **Lookup table.** This is where you enter the actual values that will be displayed in the field. You also designate the level at which each item belongs.

You do not change levels by typing the level; instead, you indent the item that you want to move to the next level. Doing this automatically changes the level number.

To create a new lookup table:

1. From the Quick Launch, click Server Settings.
2. Under the Enterprise Data heading, click Enterprise Custom Fields and Lookup Tables.
3. In the Lookup Tables for Custom Fields section, click New Lookup Table.
4. Complete the fields as desired.

   Any field with an asterisk symbol is required. Figure 9.16 illustrates the required field. Remember: if you make a field required, it will have to be filled out. In many instances, especially early on, the information may not be known, so take that into consideration.
5. Click Save.

To assign a lookup table to a custom field:

1. From the Quick Launch, click Server Settings.
2. Under the Enterprise Data heading, click Enterprise Custom Fields and Lookup Tables.
3. In the Enterprise Custom Fields section, click the name of the field you want to assign a lookup table to.
4. In the Custom Attributes section, select Lookup Table and choose the name of the lookup table you created in the list.

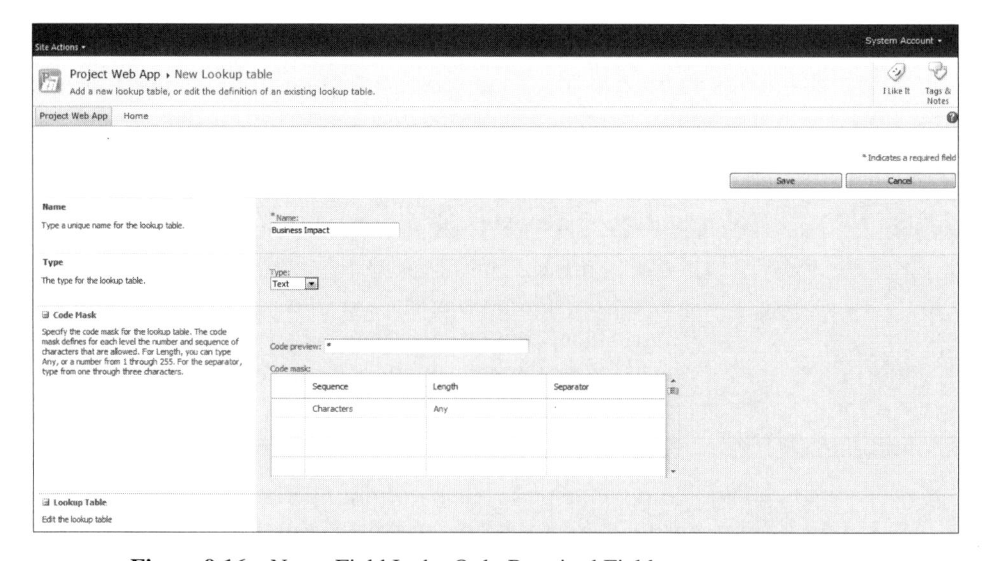

Figure 9.16 Name Field Is the Only Required Field Source: Advisicon

5. Choose any additional options as desired with the Custom Attributes section.
6. Click Save.

Determining Behavior of Custom Fields

The organization may further tailor custom fields so the use of that field is mandatory during specific phases of the project. Requiring a field during a workflow stage or before information can be saved means the individual is forced either to enter a value or to choose from the field list (if available). This is a way for an organization to enforce a corporate standard and to ensure that specific types of reports are able to be run because the information was required.

A drawback to requiring fields to be filled out is that individuals may cancel their work on a specific page when they don't know what information goes in that field. Instead of having a project with some partial information, you may end up with a project with even less information provided due to the restriction of the required custom field.

The behavior of a custom field can be controlled in these ways:

- **Behavior controlled by workflow.** Choose this option if you want a specific stage of the workflow to require this field to be filled in before moving to the next stage.
- **Require that this field has information.** Select "Yes" on this option only if you want to prevent information on a specific page from being saved until that field is filled in.

When the behavior is controlled by a workflow, you do not have the option to choose whether or field has information.

To modify the behavior of a custom field:

1. From the Quick Launch, click Server Settings.
2. Under the Enterprise Data heading, click Enterprise Custom Fields and Lookup Tables.
3. In the Enterprise Custom Fields section, click the name of the field you want to change the behavior of.
4. In the Behavior section, choose the options you desire.
5. Click Save.

Project Detail Pages

When you work with Project Web App, you may not realize that you are using a series of PDPs presenting a collection of Web parts. With the functionality available in Project Server 2010, it is possible to create, customize, and even trigger these PDPs based on actions in Project Web App. Since so much functionality is available, it is important that you understand the PDP.

Throughout this section, we want to explore the use of PDP pages and how they are integrated with features in Project Web App.

What Is a Project Detail Page?

A project detail page is another way of describing a Web page available in Project Web App. These pages consist of Web parts, and each PDP page can be used for either collecting information or displaying information. Many PDP pages are already available, but you can create additional ones as the need arises. Furthermore, PDPs can be associated with stages in a workflow.

Default Project Detail Pages Available in Project Server

Before making any modifications to PDPs, it is useful to have an understanding of what pages are already available and how they are integrated in Project Web App.

If your organization does not use all of the features of Project Web App, some PDPs will not be needed.

The titles for the default PDPs are listed next. You can review the description for these items on the same page as the PDP list to gain a general understanding of how each page integrates with Project Web App.

- Post Implementation Review
- Project Details
- Project Information
- Proposal Details
- Proposal Start and End Dates
- Proposal Stage Status
- Proposal Summary
- Schedule
- Strategic Impact

Each PDP has a page type associated with it. Three page types are available within the PDPs. You can see the page type next to the display name of the PDP page when you review the PDP list.

1. **Project.** This type is used when working with an existing project.
2. **Workflow Status.** This type is used to display the status and stage of a project.
3. **New Project.** This type is used when creating a new project and typically is connected with an enterprise template.

Not all PDPs will display information when you click on them; some require specific project information. The ones that do display will look like an online form.

To review the list of PDP pages:

1. On the Quick Launch, click Server Settings.
2. Under Workflow and Project Detail Pages, click Project Detail Pages.

Components of PDPs

A PDP is made up of one or multiple Web parts. Some PDP default pages include Web parts specifically designed for that page.

It is possible to combine multiple Web Parts from various categories to make a more complex PDP page. Web Parts can be from both Project Server and SharePoint Server.

This is a list of all the Web Part categories for reference:

- Lists and Libraries
- Business Data
- Content Rollup
- Filters
- Forms
- Media and Content
- Outlook Web App
- PerformancePoint
- Project Web App
- Search
- Social Collaboration
- SQL Server Reporting

Creating a PDP

In working with PDPs, you may discover that the available ones do not meet your needs, and you may want to create a new one. Building a new PDP is a way to fully customize a piece of Project Web App without requiring a developer. It is also a way to help enforce organizational processes or data entry requirements.

To create a new PDP:

1. On the Quick Launch, click Server Settings.
2. Under Workflow and Project Detail Pages, click Project Detail Pages.
3. Click the Documents tab on the ribbon.
4. Click New Document.
5. Enter a Name for your new PDP, then choose the desired layout and click Create.
6. In each portion of your PDP, you can choose the desired Web Part. Click one of the Add a Web Part hyperlinks.

7. Click the desired Category, click the desired Web Part, click the desired Field and click Add.

The Project Web App category is a popular place to start when building PDPs.

1. Your Web Part will be previewed in the area you previously selected.
2. Click another Add a Web Part hyperlink and repeat step 7 until all the Web Parts are filled in.
3. On the Page tab, click Stop Editing to finish the PDP.
4. On the Navigate up Arrow, click Project Detail Pages and see your new page listed.

New PDPs are automatically assigned the Project type. You can change this by editing the properties of the PDP.

Modifying and Deleting PDPs

As your organization's use of Project Server evolves, you probably will discover a need to further refine your PDPs. This could involve modifying a page or deleting one that is no longer needed. Becoming skilled in both of these areas will allow you to make changes relatively quickly and will free up your Project Server administrator to perform other tasks.

To modify a PDP:

1. On the Quick Launch, click Server Settings.
2. Under Workflow and Project Detail Pages, click Project Detail Pages.
3. Click the Documents tab on the ribbon.
4. Select the check box next to the PDP you want to modify, and on the Document tab, click Edit Document.

To delete a PDP:

1. On the Quick Launch, click Server Settings.
2. Under Workflow and Project Detail Pages, click Project Detail Pages.
3. Click the Documents tab on the ribbon.
4. Select the check box next to the PDP you want to modify, and on the Document tab, click Delete Document.
5. When prompted to confirm this action, click OK.

Deleted PDPs go to the recycle bin on the main PDP. Your administrator determines how long items are available in the recycle bin.

Workflow Stages

This section will help you learn how to create and modify stages and phases as part of a preinstalled Project Server workflow. A workflow is simply a process for achieving a goal. In terms of Project Server, a workflow is used to support your project lifecycle.

If you have difficulty completing these steps in your own environment, you will need to contact your Project Server administrator.

Developing a custom workflow is best supported by your Project Server administrator and is beyond the scope of this book.

Within each workflow, individuals must go through a number of steps (also called stages) to complete each phase. Project Server supports a sequential series of stages to ensure that every project is created in the same way and important business information is captured at the exact time it is needed. Since stages are building blocks toward a phase, they are addressed first. Throughout this section, we explore the existing concept of stages and how they relate to PDPs to address a business need. While working with stages, you will create, modify, and deletePDPs.

What Is a Stage?

A stage is a step within a project lifecycle or project workflow. Stages are subcomponents of phases. Each stage has a minimum of one project detail page associated with it. Using multiple stages can enforce a business process. Some things that can be controlled on the PDP include being required to enter information in certain fields and submit the page before you can continue. Some stages require another party to approve your submission before moving to the next stage.

As you move from stage to stage, additional PDPs can be shown while other PDPs can be hidden to prevent changes to submitted information. Hiding select PDPs also prevents individuals from jumping ahead and creating a schedule when the high-level proposal for the project has not been approved yet.

Default Stages Available in Project Server

To get started with a workflow in Project Server, Microsoft has provided a five-phase workflow containing multiple stages. You can choose to accept the stages provided, modify them, or delete them as desired. Since each stage represents a step in the workflow, you can compare your current business process with the listed stages and look for the deltas. Pending internal discussion, you can then make modifications as needed. The rest of this section focuses on changes to stages; the next section focuses on changes to phases.

The next list presents the default phases with their associated default stages.+

- Phase—Create
  - Automated Rejection
  - Initial Proposal Details
  - Initial Review
  - Proposal Details

  - Rejected
  - Selection Review
- Phase—Select
  - Not Selected
  - Proposal Selection
- Phase—Plan
  - Canceled
  - Resource Planning
  - Scheduling
- Phase—Manage
  - Execution
- Phase—Finished
  - Completed
  - Final Assessment

You can review affiliated PDPs and a description of each stage on the Workflow Stages page.

Creating a New Stage

After reviewing a stage, you may notice that it needs to be divided into two steps to facilitate different approvals, or you may notice that a step in a particular phase is missing. In both of these cases, you will need to create a new stage.

Here is a checklist of things you will need to have available before creating a new stage:

Information to Obtain

- Name for the stage
- Description (optional)
- Description/tool tip for the individual (optional)
- Phase the stage belongs in
- Introduction PDP—typically Project stage status
- List of additional PDPs individuals should be able to see at this stage (will be displayed on the quick launch)
- Additional descriptions for the visible PDPs—may be used to help remind an individual about the purpose of each page during this stage (optional)
- Required custom fields—fields that were required before entering this stage (Note: These fields must have been added previously to a PDP.) (optional)

- Read-only custom fields—fields you want individuals to see but not change, such as budget (optional)
- Strategic impact behavior—if required, you will need to determine a strategic impact value for every business driver (for portfolio analysis)
- Project check-in required—if required, the project will need to be checked in before completing this stage and submitting it to advance the project to the next stage (optional)

To create a new stage:

1. On the Quick Launch, click Server Settings.
2. Under Workflow and Project Detail Pages, click Workflow Stages.
3. On the Workflow Stages Page, click New Workflow Stage.
4. On the New Workflow Stage Page, enter or complete the information as needed and click Save.

Modifying and Deleting Stages

As your business processes evolve, you may choose to modify steps in the workflow or even delete outdated steps. This is done by either modifying or deleting stages.

To modify or delete a stage:

1. On the Quick Launch, click Server Settings.
2. Under Workflow and Project Detail Pages, click Workflow Stages.
3. On the Workflow Stages Page, click the name of the stage you want to modify.
4. Make any necessary modifications on the stage name details page and click Save.

To delete a stage:

1. On the Quick Launch, click Server Settings.
2. Under Workflow and Project Detail Pages, click Workflow Stages.
3. On the Workflow Stages Page, select the row of the item you want to delete.

   To select the row, click the gray box to the left of the phase name or click anywhere on the row that does not have hyperlinked text.
4. Click Delete Workflow Stages.

Workflow Phases

An important next step to managing stages is managing phases, which are a higher level than stages in the PM lifecycle.

If you have difficulty completing these steps in your own environment, you will need to contact your Project Server administrator.

Throughout this section, we explore the concept of phases, including the phases available instantly when Project Server is fully installed. You will make a series of changes to phases, which include creation, modification, and deletion. As part of the discussion in this section, you will explore the benefits of using a naming convention when working with phases.

What Is a Phase?

A phase is a collection of stages that are grouped together for a common purpose, such as project selection or project scheduling. By grouping your entire enterprise list of projects into phases, you can easily identify trends and make decisions based on where the majority of projects fall.

For example, a majority of projects still in the project selection phase might indicate a backlog with the individuals or teams responsible for approving projects or assigning projects to a project manager. In the next few sections, we cover creating, modifying, and deleting phases and the importance of naming conventions.

Default Phases Available in Project Server

There are five default phases in Project Server. These phases support many PM lifecycles and business processes but can be easily tailored to your organization's needs. The available default phases are listed next, in workflow order:

1. Create
2. Select
3. Plan
4. Manage
5. Finish

Creating a New Phase

The existing phases available from Project Server may not completely cover your organization's PM lifecycle. It is relatively simple to create a new phase. After that phase is created, you can either create new stages or modify stages so they become part of the new phase. Doing this will support your project management lifecycle.

To create a new phase:

1. On the Quick Launch, click Server Settings.
2. Under Workflow and Project Detail Pages, click Workflow Phases.

3. On the Workflow Phases Page, click New Workflow Phase.
4. On the New Workflow Phase Page, enter the name and description of the new phase and click Save.

Modifying and Deleting Phases

As your business lifecycle evolves, you may choose to make changes to phases in the workflow or remove outdated phases. This is done by either modifying or deleting phases and takes place after you have already modified the related stages that may be affected.

To modify an existing phase:

1. On the Quick Launch, click Server Settings.
2. Under Workflow and Project Detail Pages, click Workflow Phases.
3. On the Workflow Phases Page, click the name of the phase you want to modify.
4. Make the changes as desired and click Save.

To delete an existing phase:

1. On the Quick Launch, click Server Settings.
2. Under Workflow and Project Detail Pages, click Workflow Phases.
3. On the Workflow Phases Page, select the row of the phase you want to modify.

   To select the row, click the gray box to the left of the phase name or click anywhere on the row that does not have hyperlinked text.
4. Click Delete Workflow Phases.
5. When prompted to confirm the action, click OK.

Before deleting a phase, it is a good idea to verify that no stages are connected to the phase to avoid problems with your workflow.

Deciding Naming Convention for Phases

A naming convention is a standard that you create to apply structure to your phases. There are three reasons why you might want to consider a naming convention for phases.

1. You may want to differentiate between organizational phases and those provided by Project Server.
2. You may want the phases to be listed in a particular order when you sort by phase.
3. You may want to differentiate which phases belong to which workflow.

It is recommended that you plan out your naming convention structure and, if necessary, explain it to Project Server users to ensure complete understanding of how these phases will illustrate your organizational lifecycle.

BUILDING PROJECT SELECTION CRITERIA

In order to help manage the new work being evaluated with an organization or a PMO, Project Server has an opportunity to create selection criteria options that will help an organization to rate, rank, and select the right projects rather than just selecting based on discussion. By allowing the user to rate, rank and score a project, it helps to provide better selection criteria when evaluating and choosing projects over each other.

It is important to have a weighted or metrics-driven selection or ranking criteria as it helps to drive out the emotion of project prioritization and lends a system of balanced scoring based on quantifiable metrics. Of course, in most organizations, there will be selection of projects, regardless of whether they score higher or lower. These may be pet projects or projects that have a relationship with an already started or approved project. In this case, Project Server enables users to create dependencies between projects or during the selection process to force a project in.

If a project is forced in, other higher-rated projects may be forced out, but it the status of each project is shown within the tool so that those who are in the review and selection process can see the impact of their choices.

In Project Server, different selection criteria can be specified to help prioritize the new work being reviewed. These selection criteria are made up of elements within the Project Portfolio Server module within Project Server and include business drivers (associated to a project), impact statements, and the risk score. Later in this chapter, we detail these further.

Our intent is to showcase the fact that if an organization is interested in leveraging Project Server as a scalable solution, these prioritization and ranking features do not require you to have all of the local details or tactical activities built in Project Server. For example, in a fully built schedule the idea is to put proposals on the table that can be further detailed out after selections have been made. Some organizations use this flexibility to help map and plan the work that will be approved and then managed either in or outside of Project Server. The key here is that you don't have to do all of the rating, ranking and prioritizing, and building detailed schedules from the outset. You can start with a portfolio approach and then grow up your detailed schedule and resource planning capabilities. Project Server can act as a scalable top-down planning tool as well as a bottom-up detailed scheduling solution.

Project Dependencies

Figure 9.17 outlines the impact that project dependencies have on the portfolio selection criteria and Figure 9.18 showcases the dependencies drop down that

Dependency Type	Description
Dependency	If *Project A* is selected, *Project B* must also be selected, or *Project A* depends on *Project B*
Mutual Inclusion	Within a set of projects, if one project is selected then all projects must be selected. If all projects cannot be selected, none of the projects are selected
Mutual Exclusion	Within a set of projects, only one of the projects can be selected. Used to model alternative options
Finish to start	Finish to Start dependencies are used only in Resource Constraint Analysis for scheduling precedence

Figure 9.17 Project Dependencies

allows you to link projects together. Note some projects are dependent upon other projects, so if one is cancelled the other should not be undertaken

Creating New Portfolio Analysis Views

The power of a portfolio analysis view is the ability to control the information you see across the enterprise list of projects selected for a specific portfolio. Creating a new view is a way to display fields of information that would meet the needs of specific departments or could be mapped to specific executive needs, such as the chief information officer. It is easier to switch between existing views than to configure the fields of information displayed.

To create a new portfolio analysis view:

1. On the Quick Launch, click Server Settings.
2. In Look and Feel, click Manage Views.
3. On the Manage Views page, click New View.

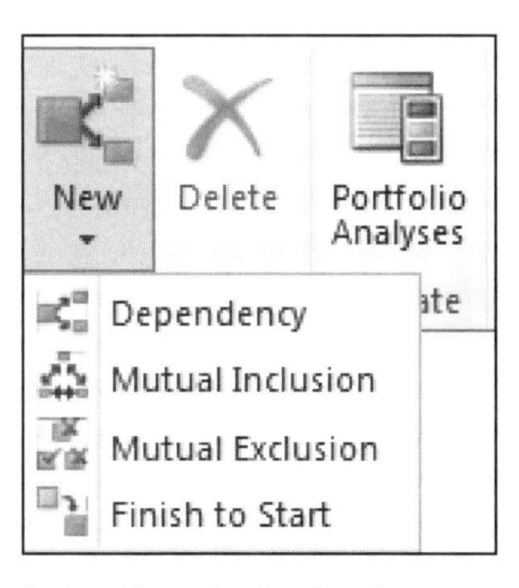

Figure 9.18 Project Dependencies Dropdown Source: Advisicon

4. On the New View page, in the Name and Type section in the View Type list, click Portfolio Analyses.
5. In the Name box, enter a name for the new view.
6. In the Table and Fields section, click the desired table(s) and field(s) and click Add. Use CTRL to select more than one field at a time.
7. In the Security Categories section, click the desired category (or categories) and click Add.
8. Click Save to save your new view.

Working with Driver Prioritizations

Working with business drivers as independent entities does not represent how an organization typically operates. There are relationships between drivers, and these relationships help influence decisions in an organization. As the preferences of executives vary and as the needs of different areas of the business vary, it is possible that the priorities and collection of related drivers will vary too. This section focuses on setting up two types of driver prioritizations, essentially allowing for multiple scoring and ranking criteria to be used based upon different business priorities. This allows the PMO or the portfolio team to be able to evaluate and adjusting the results until the consistency ratio is acceptable in the business driver planning scenario.

As in building business drivers, the driver prioritization, whether automatic or manual, values must always total 100 percent. If you fail to achieve 100 percent, the system will make adjustments to result in a total of 100 percent.

If you decide to create a new manual prioritization, follow these steps:

1. On the Quick Launch click Driver Prioritization.
2. On the Ribbon, in the Prioritizations group click New.
3. In the Name and Description section, enter the Name and Description.
4. (Optional) In the Department section, select a Department from the Select Value button.
5. In the Prioritization Type section, select Manual.
6. In the Prioritize the Following Drivers section, select each driver and click Add or click Add All to choose all the drivers.
7. On the ribbon in the Prioritization group click Save, and when complete, click Close.

Defining Portfolio Analysis Properties

When creating a new portfolio analysis, there are several decisions you need to make that will affect options available during the analysis and will affect how the portfolio analysis behaves. After you apply these options, you can enable:

- A portfolio analysis for a specific department.
- A portfolio analysis that maps to a driver prioritization.
- A designated value that can be used for the total portfolio budget.
- An identified value that resources will be categorized on.
- Designated reasons to include and exclude projects.
- Necessary project relationships.

Complete understanding of how and when to apply these options is critical since most of these options cannot be changed after the portfolio analysis is saved, and the setup of these options will impact how you are able to work with an existing portfolio analysis.

Prioritization Types

There are two prioritization types you can pick from when doing a portfolio analysis. If your organization has already subdivided your strategic objectives into business drivers and also created a driver prioritization, you should select the first type, "Prioritize projects using business drivers." If instead your organization has not created business drivers or is still evaluating the capabilities of Project Server, you should choose the second type, "Prioritize using custom fields." This option generates an additional step during the portfolio analysis that enables you to select the custom fields and their weights.

The only way the software can create a recommendation on which projects to include or exclude from your portfolio is to have some basis for prioritization. That is why this option is important. The first option is recommended since it can be applied easily to other portfolios, and the option represents work by your executive team in figuring out what key variables projects must be measured on. The custom fields option is more of an on-the-fly prioritization that may not be replicated easily and may not be fully tied to strategic objectives.

To specify the prioritization type:

1. On the Quick Launch, click Portfolio Analyses.
2. On the ribbon in the Analysis group, click New.
3. In the Prioritization Type section, select Prioritize projects using business drivers.
4. In the Driver Prioritization list, select the desired option.
5. Continue with the Portfolio Analyses.

Configuring Force-in and Force-out Options

The force-in and force-out feature of portfolio analysis gives you the option of manually designating a project that must be included or excluded from the portfolio, ignoring Project Server recommendation. If you choose to use the default settings for these options, the software simply displays force-in or force-out. Instead, you could choose a custom lookup field for force-in and another custom lookup field for force-out. If you do this, you can choose from a list of reasons why a project must be included or excluded. For example, you could indicate "executive preference" as a force-in reason and "postponed" as a force-out reason.

To configure force-in and force-out options:

1. On the Quick Launch, click Server Settings.
2. In Enterprise Data, click Enterprise Custom Fields and Lookup Tables.
3. In Lookup Tables for Custom Fields, click New Lookup Table.
4. In the Name section, enter Force-in.
5. (Optional) Modify the Code Mask section.
6. In the Lookup Table section, enter the reasons for forcing in a project.
7. Click Save.

To use the lookup tables during a portfolio analysis:

1. On the Quick Launch, click Portfolio Analyses.
2. On the ribbon in the Analysis group, click New.
3. Expand the Alias project Force-in and Force-out options section.
4. Check the Alias Force-in option, and select the Force-in lookup table from the list.
5. (Optional) Repeat this for the Force-out option.
6. Continue with the Portfolio Analyses.

Configuring Project Dependencies

After a portfolio analysis is created and listed on the Portfolio Analyses page, you will be able to set Project Dependencies. A project dependency is a relationship you want to include between projects that will influence the portfolio. This relationship may result in additional projects moving during a force-in or force-out situation or moving as the portfolio changes. Available dependencies are listed next.

- **Dependency.** Unless a specific list of dependent projects are included in the portfolio, this specific project will not be selected.
- **Mutual Inclusion.** All or nothing. Either all of these projects are selected or none is.
- **Mutual Exclusion.** Alternative projects. Selecting one project will exclude another project.
- **Finish to Start.** Designates a priority of projects. In other words, one project must finish before another project starts. Note: This does not guarantee that the successor project will be in the portfolio.

To create project dependencies:

1. On the Quick Launch, click Portfolio Analyses.
2. On the ribbon in the Navigate group, click Project Dependencies.
3. On the Project Dependencies page in the Dependencies group, click New and choose the desired dependency option.
4. In the Name section, type the Name for the dependency.
5. Following the options provided with each dependency type, complete the requested information including the selected projects.
6. On the ribbon in the Dependency group, click Save, and when complete, click Close.

Prioritizing Projects and Reviewing Priorities

An important next step in defining the properties for a portfolio analysis is mapping each project and its relative importance to each business driver. Doing this will result in a highest to lowest priority listing of projects. Project Server has enabled a six-point prioritization scale, for review and to allow the planner to adjust factors that drive project priorities. For further analysis, Project Server allows you to export the information from these areas to Excel. Many organizations use this Excel document to further test and adjust prioritization between drivers before coming back and making adjustments in Project Server.

The seven-point scale comes out of the box with Project Server. Each driver can have these rankings against other drivers:

1. Is Extremely Less Important Than
2. Is Much Less Important Than
3. Is Less Important Than
4. Is Important As
5. Is More Important Than
6. Is Much More Important Than
7. Is Extremely More Important Than

Exporting Data to Excel

If you decide that you want to share your project to driver prioritizations and the resulting levels with individuals who may not have access to Project Server, you can export this information to Excel. You will have this option during the Prioritize Projects step or the Review Priorities step.

To export data to Excel:

1. Complete the prioritization of projects and review the resulting priorities.
2. On the ribbon in the Share group, click Export to Excel.

WHAT THE EFFICIENT FRONTIER IS AND HOW TO USE IT

Efficient frontier analysis was originally popularized by Dr. Harry Markowitz, a professor of finance at the Rady School of Management at the University of California, San Diego. He won the Nobel Memorial Prize in Economic Sciences in 1990 for his work on the foundations of portfolio theory.

Simply stated, the efficient frontier is the combinations of investments that produce the highest return for the lowest possible risk. Figure 9.19 illustrates the concept of the efficient frontier curve.

Each point on the curve represents a portfolio comprised of multiple projects. Those projects that show up first on the left side of the curve provide the best ROI. There is a point in the curve (just as it starts to flatten out) where there is a declining return in the value obtained with each additional increment of cost.

Project Server 2010 utilizes efficient frontier analysis tailored to PPM analysis. There are some key differences from efficient frontier's initial use with financial applications and the way it is used in Project Server:

1. Projects address a range of strategic goals instead of having a single focus on financial ROI.

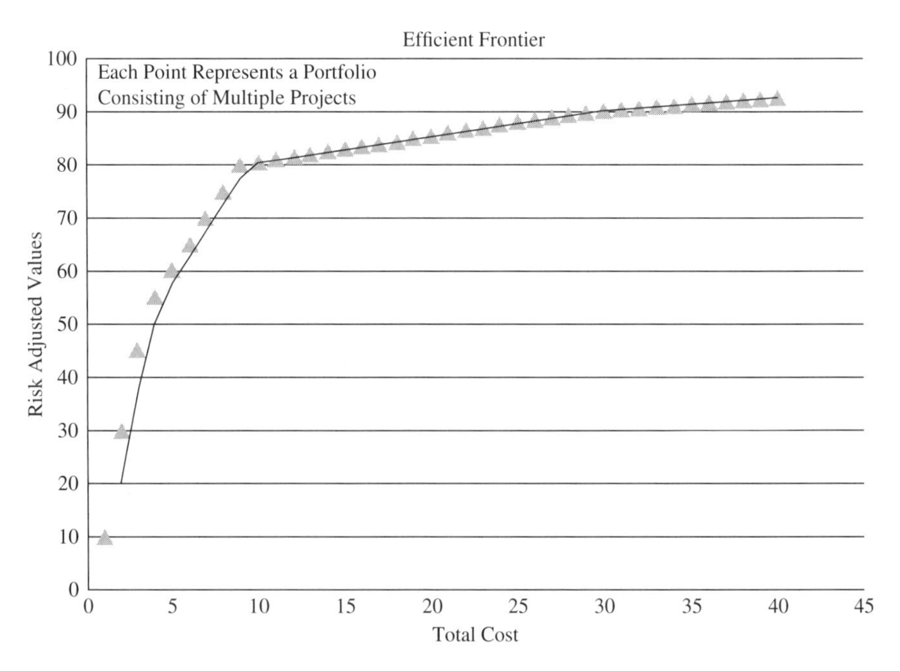

Figure 9.19 Efficient Frontier Curve Source: Advisicon

2. Programs and projects are not merely traded like financial transactions. There can be significant costs to cancel a project and move its resources over to another project.
3. Projects are complex and constrained by factors other than just cost (e.g., resources requiring specific knowledge or skills).

The value of the efficient frontier in Microsoft Project is that:

- It can be used in conjunction with other portfolio analytics to provide what-if scenario analysis and derive a set of projects that yield a desired risk/return value.
- Efficient frontier also supports the modeling of many combinations of projects in a variety of prioritizations and combinations of primary constraints. This allows the efficient frontier to predictively determine the outcome of the selection of a specific portfolio combination.
- It allows an organization to obtain the greatest possible value from any specified available budget.

Understanding and Using the Efficient Frontier

In Project Server's Portfolio Analysis tool, there is a view with the label "The Efficient Frontier" (see Figure 9.20). This view is comprised of an *x*-axis that represents total budget from zero to N and a *y*-axis of strategic value percentages based on the selection of projects that meet the business drivers. This overall line chart shows the efficient frontier with a 100 percent being all project selected. As fewer projects are selected, it showcases what percentage of strategic value is ultimately delivered. The overall goal is to get the highest percentage with the budget or capacity available to deliver the projects.

In general, the efficient frontier view is to help guide the parties using the tool in managing budgets of revenue or resources to get the highest strategic value in selecting the right projects. As a user or organization makes changes, choices, forces projects in or out, or reduces the overall budget for projects, the tool recalculates which projects make the selection and the efficient frontier line shows a maroon square where the actual value is ranked at with the remaining selected projects.

Analyzing Cost Scenarios

There are two options when analyzing a portfolio: evaluate costs or evaluate resources. This section covers costs, and the next section covers resources. Since portfolio analysis is about choosing the right mix of projects, when you are evaluating costs, you are considering the budgeted cost of a proposed group of projects. These projects may or may not be selected for further planning and schedule development.

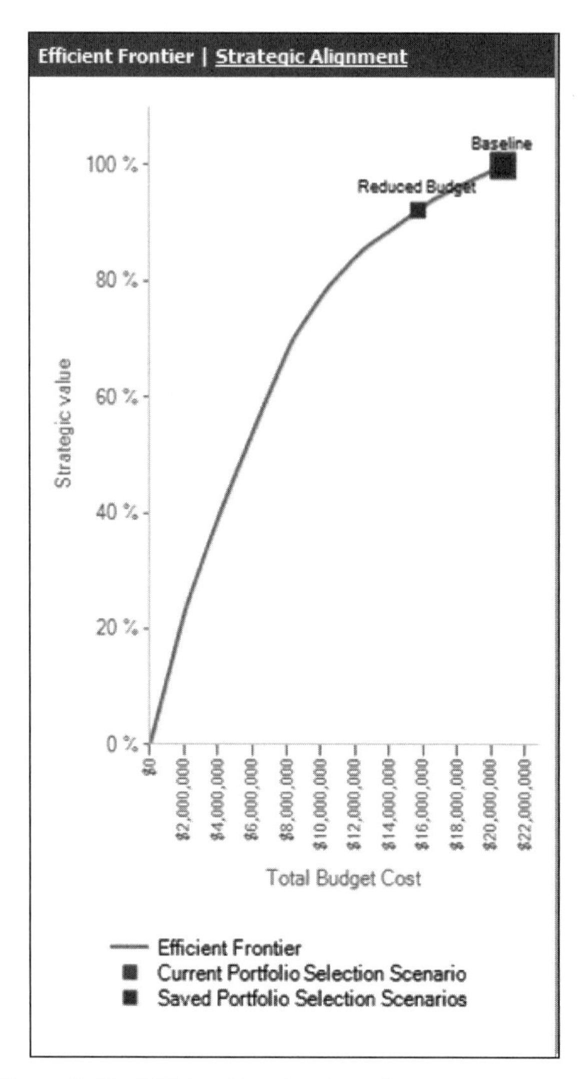

Figure 9.20 Efficient Frontier Example Source: Advisicon

In this section, we review the efficient frontier of projects against cost and strategic alignment. Project displays different ways to view and edit the selected projects, including in a scatter chart where you can also adjust the budget and review the impact to the portfolio.

Baseline Cost Scenario

The first time you create and save your portfolio analysis, that portfolio analysis becomes the baseline scenario. This scenario is automatically maintained so you can compare the baseline with your desired scenario (if you choose to create one). The baseline cost scenario is especially useful if you review the portfolio later when projects are in progress or finished to evaluate how your budget compares with actual costs or costs to date.

For this baseline cost scenario to be useful, you must have values entered in your primary cost constraint field.

To create a baseline cost scenario:

1. After you have accepted the driver priority levels, click "Next: Analyze Cost."
2. After the Cost Constraint Analysis page displays, on the ribbon in the Analysis tab, click Close.

Notice that the name of the portfolio analysis is listed on one row and the baseline scenario is listed directly below.

Comparing Projects with Business Drivers

The second step after setting up a portfolio analysis is prioritizing projects. When considering a Portfolio of Projects, you will designate how projects relate to each business driver in terms of importance, and this will further refine the results in the portfolio. This step is important because it will help determine the strategic value for the overall portfolio and will indicate a priority level for each individual project (if desired). The overall goal is that each project is rated against the business drivers in a 7-point scale ranging from "no rating" to "extreme."

To compare projects with business drivers:

1. Complete the Define Properties portion of the portfolio analysis.
2. Click the Next: Prioritize Projects button.
3. For each project, under each business driver, select the appropriate rate from the 6-point scale.

Using the Scatter Chart for Analysis

The scatter chart is the plot of all projects in the portfolio, including both selected and not selected projects. You can identify forced-in or forced-out projects as well. The x-axis for this chart is total cost, while the y-axis is strategic value. The information in the scatter chart is the same information in grid, just represented visually with a diagram showing the value delivered by the selected projects, but it is organized in a different way, which may be more visually appealing and easier to explain to some executives.

You can pause on each bubble in the scatter chart to display information about that project.

To display the scatter chart, click Scatter Chart on the ribbon in the Projects group of the Cost Constraint Analysis page.

Forcing Projects In and Out

When you force a project in or out, you are overriding the portfolio and choosing specific projects to include or exclude. The force-in or force-out dropdown list will pull up the custom lookup field you selected when you defined the portfolio analysis properties. The purpose of this feature is to override the portfolio when necessary and to evaluate the impact on costs, resource requirements, and strategic values.

To force a project in or out:

1. On the Cost Constraint Analysis page in the Projects section, locate the Force in/out column.
2. For the project you want to force in or force out, click the force in/out cell and choose the appropriate reason from the select value button that appears to the right.

Reviewing Project Priorities

Through the information you specified in the driver to project comparison, Project Server will calculate the project priority levels. The levels will total 100 percent. You should review this information to ensure that the priority matches what you expected since this information helps drive project selection in your portfolio in the next step. If necessary, you can return to the prior step to make changes that will adjust the priority levels.

To review the list of project priorities:

1. Complete the prioritizing of projects in the portfolio analysis (previous topic).
2. Click the Next: Review Priorities button.
3. Review the priority levels listed, and either accept the results or click the Previous: Prioritize Projects button to go back and make changes.

WORKING WITH CONSTRAINTS IN PORTFOLIO PLANNING

Portfolio management is all about establishing strategic goals and a governance model and then reaching an understanding of organizational resource capacity, including financial, cost, and other means. It is important to understand what constrained resources are and how they affect resource planning and project portfolio planning.

Microsoft Project Server 2010 includes new features to help evaluate whether project proposals comply with financial and resource constraints. These new Portfolio analysis capabilities provide the ability to prioritize projects and make selection decisions based on cost and resource constraint analysis.

Cost Constraints

Following the creation of the portfolio analysis and review of the relative priority values of the proposals, the next step is to analyze the proposals based on high-level cost constraints. Cost limits can be set that help narrow the list of proposals that can reasonably be approved as projects, based on available funding.

Figure 9.21 depicts the cost constraint analyses capabilities of Project Server 2010, where multiple scenarios can be created against a baseline set of projects. Additional cost limits can be added to the analyses from a predefined list of available constraints (e.g., including enterprise custom fields). Fields can be added to the view that potentially can be used as portfolio selection scenario totals. The aggregated values for all selected projects will be shown for each selected field.

Resource Constraints

Once we have analyzed the portfolio based on high-level cost constraints, the next step is to review resource requirements to determine if proposed projects in the portfolio can be executed in the time frame specified. After examining the project schedule and timephased resource requirements of a portfolio, many of the proposals cannot go forward without significant schedule or resource plan modifications. Resource constraint analysis identifies resources for projects by using role-based availability and project requirements (calculated from the resource oool availability data).

Figure 9.22 illustrates the resource constraint analysis view of Project Server 2010 that support multiple scenarios from a single portfolio baseline, including

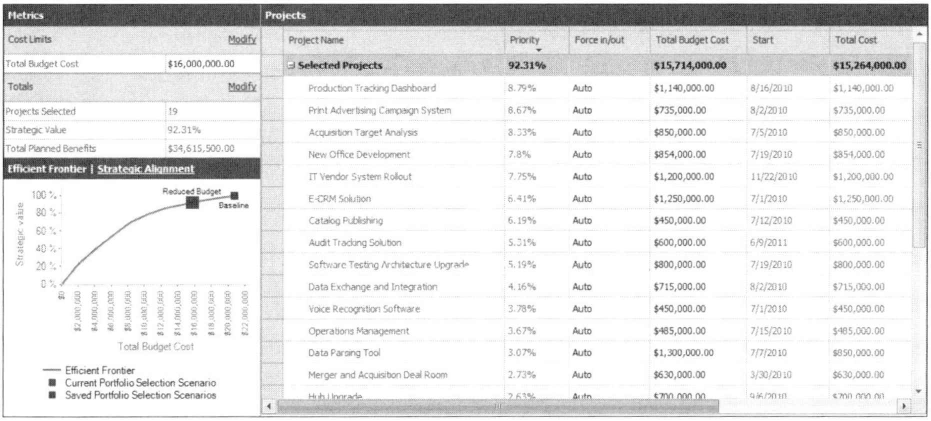

Metrics	
Cost Limits	Modify
Total Budget Cost	$16,000,000.00
Totals	Modify
Projects Selected	19
Strategic Value	92.31%
Total Planned Benefits	$34,615,500.00

Project Name	Priority	Force in/out	Total Budget Cost	Start	Total Cost
Selected Projects	**92.31%**		**$15,714,000.00**		**$15,264,000.00**
Production Tracking Dashboard	8.79%	Auto	$1,140,000.00	8/16/2010	$1,140,000.00
Print Advertising Campaign System	8.67%	Auto	$735,000.00	8/2/2010	$735,000.00
Acquisition Target Analysis	8.33%	Auto	$850,000.00	7/5/2010	$850,000.00
New Office Development	7.8%	Auto	$854,000.00	7/19/2010	$854,000.00
IT Vendor System Rollout	7.75%	Auto	$1,200,000.00	11/22/2010	$1,200,000.00
E-CRM Solution	6.41%	Auto	$1,250,000.00	7/1/2010	$1,250,000.00
Catalog Publishing	6.19%	Auto	$450,000.00	7/12/2010	$450,000.00
Audit Tracking Solution	5.31%	Auto	$600,000.00	6/9/2011	$600,000.00
Software Testing Architecture Upgrade	5.19%	Auto	$800,000.00	7/19/2010	$800,000.00
Data Exchange and Integration	4.16%	Auto	$715,000.00	8/2/2010	$715,000.00
Voice Recognition Software	3.78%	Auto	$450,000.00	7/1/2010	$450,000.00
Operations Management	3.67%	Auto	$485,000.00	7/15/2010	$485,000.00
Data Parsing Tool	3.07%	Auto	$1,300,000.00	7/7/2010	$850,000.00
Merger and Acquisition Deal Room	2.73%	Auto	$630,000.00	3/30/2010	$630,000.00
Hub Upgrade	2.63%	Auto	$700,000.00	9/6/2010	$700,000.00

Figure 9.21 Cost Constraint Analysis Cost Constraint View Source: Advisicon

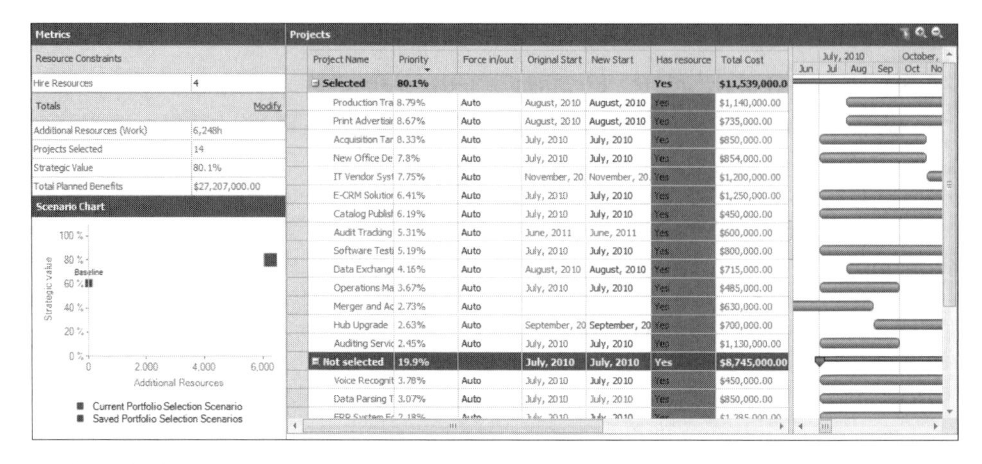

Figure 9.22 Resource Constraint Analysis Resource Constraint View Source: Advisicon

manual adjustment of project start dates, hiring of additional resources, and the ability to force projects in or out of the portfolio analysis.

Baseline Resource Scenario

As indicated earlier, the baseline cost scenario is created automatically, as is the baseline resource scenario. Both cost and resource scenarios are listed with the baseline portfolio scenario. It is very useful to have a baseline resource scenario for a later analysis against a change in resources, including the number available in each resource role.

If a resource plan was not created and if you have not designed a specific Resource Role field for both the portfolio and the resources in the resource plan, you will not be able to use the resource scenario.

To create a baseline resource scenario, evaluate the baseline cost scenario and then click Next: Analyze Resources. The baseline scenario will be created automatically.

If you previously closed the cost scenario, you can simply open the baseline scenario and display the resource information by choosing Analyze Resources.

Analysis Primary Cost Constraint

The option for the primary cost constraint is where you designate a specific field that Project Server will use to represent the budget for the analysis. Some organizations use calculated cost values for this purpose. However, when you are in the process of selecting projects, this level of cost detail may not exist.

In those cases, it is suggested you that create a field that can be used during initial planning of the proposal that will represent the anticipated budget (e.g., a high-level budget). You will be able to view the overall totals of this field during the display of the portfolio of projects.

The default field for this purpose is Sample Proposal Cost.

To specify the primary cost constraint:

1. On the Quick Launch, click Portfolio Analyses.
2. On the ribbon in the Analysis group, click New.
3. In the Analysis Primary Cost Constraint section, select the desired field from the list.
4. Continue with the Portfolio Analyses.

Efficient Frontier and Strategic Alignment

Project Server calculates the efficient frontier by factoring in the cost of each project against its strategic value and plots the list of projects. Given the cost value (either generated by the baseline scenario or entered by the portfolio manager), the efficient frontier of projects is identified. Essentially this means that given an amount of available money, Project Server can determine the highest strategic value that can be achieved with the mix of projects in the portfolio. The efficient frontier is automatically displayed when you display a cost constraint analysis. You can switch the display to the strategic alignment list instead.

To display the efficient frontier or strategic alignment:

1. On the Cost Constraint Analysis page, you will see the efficient frontier chart.
2. (Optional) Click the strategic alignment link to display the list of drivers.

Modifying Cost Limits

One of the benefits of cost constraint analysis is the flexibility to run various scenarios. By simply changing the total cost for the portfolio and clicking on the recalculation button, Project will do a recalculation and refresh the screen with the newest changes, Project Server will re-rank the highest strategic value projects and the portfolio selection view will showcase the newly prioritized projects with the new cost budget applied. This flexibility allows you to run scenarios with slightly higher budget values, slightly lower budget values, or very large changes in budget values. The resulting project list can help you determine if a large increase to the budget will generate a large increase in the strategic value of the portfolio.

To modify the cost limits:

1. On the Cost Constraint Analysis page, under Cost Limits, change the Total Cost value to a new number.
2. On the ribbon in the Portfolio Selection group, click Recalculate.

Resource Role Custom Field

If as part of your proposal planning you created a high-level set of your needs in a resource plan, you will be able to expand your portfolio analysis with an added resource analysis. Since the resource plan feature is time oriented instead of task oriented, you will also be specifying time-related options as part of your resource options. One required field when including resource information is the resource role. This field is where you designate something specific about each resource that can be used for high-level grouping.

To clarify, a portfolio analysis is done at a role level, not a resource level. For example, let's say you have a resource named Penelope Coventry whom you planned you would need for eight days in March and another resource named Mahmoud Magdy whom you would need for three days in March. If both resources are assigned to the role of Lab, the portfolio analysis would display Lab at 11 days in March. Since the goal of a portfolio is to make high-level decisions about which projects to include and exclude, resource role information is all that is required at this level.

To designate the resource role custom field:

1. On the Quick Launch, click Portfolio Analyses.
2. On the ribbon in the Analysis group, click New.
3. In the timephased Resource Planning section, select Analyze timephased project resource requirements against organizational resource capacity. This displays some additional options.
4. In the Resource role custom field section in the Role Custom Field list, select the desired field.
5. Continue with the portfolio analyses.

Resource Assignments

The resource assignments page allows you to review all of the project work assigned to each of the resources in a centralized view. This can be useful when project priorities are shuffled and decisions about who might be available when a task in one project needs to be reassigned changes. It can also identify trends in resource assignments that might show when an organization continuously overassigns or underassigns work to specific resources.

Two options are available on the resource assignments page:

1. **Gantt Chart** displays a Gantt chart for each resource that includes a grouping by project within each resource's section.
2. **Timephased Data** displays a resource usage view for each resource that includes a grouping by project within each resource's section. This option can be further modified with the Set Date Range option, which allows you to adjust the way assignments are filtered in or out on this page by changing the assignment date range.

To check resource assignments:

1. In the Quick Launch menu, click Resource Center.
2. In the Resource Center page, click the check box next to several resources and click Resource Assignments.
3. In the Assignments tab, click Timephased Data.
4. In the Assignments tab, click Set Date Range, modify the From and To dates and click OK.
5. In the Assignments tab, click Resource Center.

Requirements Details versus Gantt Chart View

There are two useful views to use when considering resources in the portfolio. The Gantt Chart view is similar to the Project Center view except that it illustrates a grouping of selected and unselected projects. One very useful column in this view is the "Has Resource Requirements" column, which will help you determine if a project was planned out with resources or without. How the project was planned may influence your resource decisions based on which projects will be impacted.

The other view is the Requirements Details view, which is a timephased breakdown of resource needs by the Resource Role you specified earlier and by the selected versus unselected list of projects. An advantage to this view is that time periods where there is a shortage of resources will be highlighted.

To display different resource display options, click either the Gantt Chart button or the Requirements Details button in the Projects group (on the Resource Constraint Analysis page).

General Settings

After populating the enterprise resource pool, it will be necessary to fill in other fields of information about each resource. We are collectively calling these other fields of information "general settings." The importance of addressing all of these settings now is that they will help to drive the features that use resources, such as time sheets. Completing these general settings will also complete the process of planning how each resource will be used throughout a collection of projects. To enhance the power of resource management in Project Server, we are going to explain some optional features and their uses.

Optional features include:

- **Resource Booking Type.** When resources are assigned, you have the choice to make each assignment a committed or proposed assignment. Each organization can specify guidelines for the correct use of committed or proposed assignments but, in general, a committed resource is an assignment that is confirmed while a proposed assignment requires some type of approval process. Proposed assignments are not visible to the resource in PWA. Unless you choose to turn on proposed, only committed assignments display in PWA on the Resource Assignments or Resource Availability page available on the ribbon in the Resource Center. Even though a resource may be predefined with one option, you can change the resource booking type when the resource is assigned.

In the Assignment Attributes section, under Default Booking Type, choose either Committed or Proposed.

- **Timesheet Manager.** By default, a resource will also be the timesheet manager. This means that unless this option is changed, resources can approve their own time sheets. If your organization wants to require a review of a time sheet before it is approved, change the timesheet manager. Reasons for a review may include simply to spot errors or to make sure individuals are charging their time against the correct project. A timesheet manager is also critical when you require approval for administrative time, such as sick time and holiday or vacation time.

  In the Assignment Attributes section, under Timesheet manager, click Browse to select the specific resource.
- **Resource Assignment Owner.** By default, a resource will also be the assignment owner. This means that a resource is responsible for updates on the task and will be the person logging into PWA to make changes. Reasons for changing the assignment owner may be if a personal assistant will be making the changes on PWA or if a resource is going to be unavailable because of a personal holiday or vacation.

  In the Assignment Attributes section, under Default Assignment Owner, click Browse to select the specific resource,
- **Account Status.** Changes in the list of resources are common with many organizations. For example, a resource may leave the organization or no longer be available for project work due to an extended leave of absence. While you may think that you should delete a resource who is no longer needed, that is not the best approach. Doing so would remove all historical data about that specific resource. A better method is to make a resource inactive and reassign all future work to someone else.

  In the Identification Information section, under Account Status, choose either Active or Inactive.

All use any of these items:

1. On the Quick Launch menu, click Resource Center.
2. Select a resource and then click Edit Resource. The Edit Resource Web Page appears.

Hiring Resources

Careful evaluation using the Requirements Details view will highlight specific projects or time periods where you will need to hire resources to meet project needs and remove any deficits. There are two options in terms of hiring resources.

1. Return to the proposal and adjust the resource plan.
2. Hire resources in the portfolio and allow Project Server to specify where the hired resources should go.

To hire resources in the portfolio:

1. On the Resource Constraint Analysis page in the Portfolio Selection group in the Scenario list, select [Current].
2. In the Metrics section under Resource Constraints, change the value in the Hire Resources field to a number greater than 0.
3. On the ribbon in the Portfolio Selection group, click Recalculate.

Viewing Deficit and Surplus Reports and Hired Resource Reports

Two reports are available in the Resource Constraint Analysis to help you make decisions about hiring resources.

1. **Deficit and Surplus.** What is displayed represents either a surplus or a deficit. Positive numbers indicate a surplus, negative numbers are a deficit.
2. **Hired Resources Report.** This report displays resource roles that were hired after changing the hired resources from 0 to another number.

The Hired Resources Report is available only after the process of hiring resources (explained in the previous topic) has been completed.

To display one of the resource reports, select the desired report from the Reports button in the Portfolio Section of the Resource Constraint Analysis page.

CREATING AND RUNNING MULTIPLE SCENARIOS FOR PORTFOLIO PLANNING

In 1998, the U.S. Government General Accounting Office released "An Executive Guide: Measuring Performance and Demonstrating Results of Information Technology Investments" (U.S. GAO 1998), which describes portfolio management as one of four strategic enterprise objectives including:

1. Enterprise strategic planning and goal accomplishment
2. Enterprise management of the portfolio of IT applications
3. IT financial and investment performance
4. Use of IT resources across the enterprise

Project Server 2010 aids both governmental organizations and private sector enterprises by providing the ability to prioritize and assess projects using multiple scenarios for portfolio planning. It does this by utilizing key portfolio modeling capabilities, including:

- Strategic value, financial value, and risk score analysis, to provide objective comparisons of portfolio scenarios
- The use of the cost constraint analysis view to help model varying budget constraints that employ an optimization algorithm to recommend the project portfolio that best aligns to the business strategy of the enterprise
- The use of efficient frontier, strategic alignment, and compare scenario views to provide key information to help executives identify trade-offs and evaluate and refine portfolio selection

Creating and running multiple scenarios for planning portfolios of projects provides the opportunity for PMOs and key decision makers throughout the enterprise to have rich (yet objective) conversations regarding capital investment decisions and resource planning.

Creating Multiple Scenarios

One of the important new features of Project Server 2010 is the ability to analyze data from projects and resources and build multiple portfolio scenarios. The process of providing an optimized and balanced portfolio is facilitated through the use of the Compare Portfolio Selection Scenarios view illustrated in Figure 9.23.

It is important to understand that Project Server 2010 does not automatically specify an optimal portfolio solution set. Instead, it provides the analyst with the opportunity to adjust portfolio settings to identify specific scenarios until an acceptable scenario is determined.

Scenarios can be run and compared to other scenarios as well as to the baseline portfolio calculation. Force-in and force out options are available in the Gantt chart view of the Resource Analysis screen. Other incremental resources options can be found on the Options tab of the Resource Analysis view.

Portfolio Baseline

The portfolio baseline is the starting point where typically all of the revenue or resources are available and all projects are included. This includes a working starting point for making changes or returning to a starting point.

In doing what-if planning, you can make changes to project, start periods, and budget available, and see what the overall impact is. At any point, you can save that view and return to a standard baseline to continue doing comparison or prioritization planning.

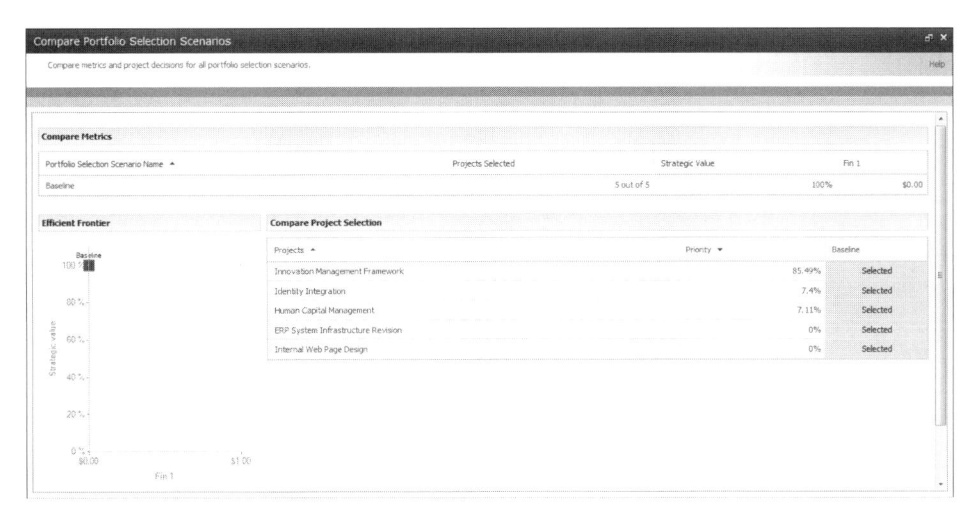

Figure 9.23 Compare Portfolio Selection Scenarios Source: Advisicon

Portfolio Scenarios

With Project Server you can create different scenarios for reviewing different business drivers and optimizing and selecting project. Each scenario typically will include a set prioritization ranking against a set of business drivers.

What is useful about having scenarios is that different parts of an organization can create their own optimized portfolio selection for proposals or projects. These different scenarios can then be compared against each other. Organizations with different funding or fiscal budgets to plan with can apply different scenarios depending on their budgets.

In general, you can do a tremendous amount of what-if planning, fine-tuning business drivers and prioritization of selection criteria quickly and easily, and save them for review or for comparison. This feature is one that many planning organizations use to save and review planning periods against each other (like quarterly selection). Each portfolio selection period can evaluate if projects are no longer relevant and that project needs to be cancelled or delayed as the organization uses this feature to queue up and prioritize new projects that may have a higher relevance or are needed to respond to market demands.

Saving New Scenarios

After you experiment with changing various options in your portfolio view, you may want to keep the results for future discussions or for comparison against the baseline. Each recalculation that you perform can be saved as a scenario within the same portfolio.

This ability allows you to test different options and save those options for later review. Portfolio decision teams can come back to where they left off after validating or reviewing options for changes or additions to the portfolio being reviewed.

To save a scenario:

1. Complete the changes and be sure that the portfolio has been recalculated.
2. In the ribbon in the Portfolio Selection group, click the Save As button.
3. Enter the desired scenario name and click OK. The scenario will be available in the Scenario dropdown list in the Portfolio Selection group of the ribbon.

Comparing Scenarios

After you have at least one scenario saved in addition to the baseline, you can perform a comparison. A comparison allows you to view the similarities and differences between each scenario. The main advantage to the Compare option in Project Server is that it compares all the scenarios at once, allowing strategic reviewers to review the variance of one scenario versus another. This compound view will showcase which projects tend to be higher across all scenarios and any resource or cost differences that can be expected in gaining the most value from selecting projects in the Portfolio Analysis process.

A very useful portion of the comparison is that it creates a list of projects and displays which ones are in which scenario. You can use this information to identify trends. For example, if one project is not selected for any of the scenarios, perhaps it does not belong in the portfolio.

To simplify the comparison window, save only a limited number of scenarios.

To perform a comparison:

1. In the ribbon in the Portfolio Selection group, click the Compare button.
2. To exit a comparison, click the Close button in the upper right corner.

APPLYING STRATEGIC ANALYSIS FOR CORPORATE TO DEPARTMENTAL NEEDS

As discussed earlier, corporate strategy defines the goals and priorities of the business. Business strategy defines how organizations compete to gain market share and maintain a competitive advantage. Effective strategic planning results in making informed decisions that enable an organization to achieve its goals. Key metrics measure the performance of a company's goal set and help us understand how much progress has been made toward achieving those goals. Defining the corporate portfolio and striving to improve worker performance enables corporate success.

Project Server 2010 provides a level of autonomy while maintaining enterprise standardization and control. Using the Department field, Enterprise Project Types, Resources, and Custom fields can be associated with specific departments. Doing this eliminates unnecessary clutter and allows departments to focus on their data while supporting the needs of an enterprise rollup for executive reporting and decision making.

Think of the "Department" field as a grouping mechanism within Project Server 2010. If a company has a Sales and Marketing Department that has three separate functions (e.g., a Sales Order Desk, a Field Sales team, and a Marketing organization), even though these groups are all within a single "department," they will likely have separate work processes, business requirements, and reporting and analysis needs. The primary purpose of the Department field is to be a filter for:

- **Projects,** which can be assigned to none, one, or multiple departments. Projects associated with different departments will show only their associated custom fields.
- **Resources and Resource Custom Fields.** Different groups can have their own custom fields. (Note: Custom fields that are not assigned to a department are available for global use.)
- **Enterprise Project Types.** Different groups can see different user interfaces based the department field, the pivot field new in Project 2010, as it can be used to filter Enterprise Project Templates and Custom Fields.

The value of using the Department field is that it:

- Helps manage the custom field list and helps define (at a resource, task, or project level) which fields are required or not required.
- Can be used for Online Analytical Processing (OLAP) database data filtering from any of the 13 OLAP cubes created from Project Server's reporting database or a hypercube built by an end user against Project Server data.
- Allows for enterprise-wide consistency and support different processes, forms, and fields for different areas and groups.
- Supports the assignment of users to none, one, or multiple departments. Note that Departments are not tied directly to the resource breakdown structure.

Portfolio drivers and driver prioritization can be associated with departments. Driver prioritization can be associated with specific departments, or the prioritization settings can be open for use with any analysis.

Using Custom Fields in Strategic Analysis

Project stakeholders and management often have difficulty identifying relevant project information across a portfolio of projects. Project Server 2010 offers six types of custom enterprise fields at the task, resource, and project level (including an unlimited number of Cost, Date, Duration, Flag, Number, and Text fields).

Enterprise custom fields can use lookup tables, including hierarchical selection (for text) with code-masks. Formulas and graphical indicators can also be used with enterprise custom fields. Microsoft Project Server 2010 calculates the values of formulas for project, task, and resource custom fields when a project is published.

The following is an example of using a custom calculated field to derive a common project "value" score:

$$\frac{\text{Benefits} - \text{Total Cost Ownership (TCO)} - \text{Taxes}}{\text{Project Risk Score}} = \text{Value}$$

The value of using a formula to derive a common currency for each project is that the resulting number indicates how much net new cash you can expect from the project (i.e. the project's value to the bottom line).

Departmental Association of Business Drivers

As part of a portfolio analysis, you map a collection of projects against a collection of business drivers. To simplify the process of selecting business drivers, each driver can be associated with a department. Then, as an individual who is a member of that department logs in, the list of drivers will be filtered to display only those relevant to his or her department. This will simplify the PWA page and provide a tailored list for each department.

To associate a driver with one or more departments:

1. From the Quick Launch, click Driver Library.
2. Click the name of the driver you want to associate with a department.
3. On the Edit Business Driver: name page, in the Department section, click the Select Value button and click the department(s) that you want this driver to be associated with.
4. On the ribbon, click Save & Close.

Creating New Portfolio Selection Views

When you are choosing projects for your portfolio analysis, you have the ability to create a new view that will list different fields of information for each project. This is extremely important since it will help you identify which projects should be considered in the analysis by your executive team. If you do not create a new view, you will only have the following basic information available about each project in the summary view:

- Project Name
- Project Departments
- EPT Name
- Workflow Stage Name

The Portfolio Selection View is available only from the Prioritize these projects section when you create a new portfolio analysis.

To create a new portfolio analysis project selection view:

1. On the Quick Launch, click Server Settings.
2. In Look and Feel, click Manage Views.
3. On the Manage Views page, click New View.
4. On the New View page in the Name and Type section in the View Type list, click Project Analysis Project Selection.
5. In the Name box, enter a name for the new view.
6. In the Table and Fields section, click the desired table(s) and field(s) and click Add. Use CTRL to select more than one field at a time.
7. In the Format View section, click a Sort by and Order option (optional).
8. In the Filter section, click Filter to create a new filter, select the filter values, and click OK (optional).
9. In the Security Categories section, click the desired category (or categories) and click Add.
10. Click Save to save your new view.

COMMITTING NEW WORK PORTFOLIOS AND MEASURING FOR ROI

An old management adage states "You can't manage what you don't measure." The small percentage of people who do set goals, however, do so incorrectly. If you really want to bring about change, you must measure it. Another way of stating this is "Whatever you measure, you will focus on, and what you focus on you will change."

Portfolio analysis involves creating analyses and prioritizing projects, analyzing the portfolio based on high-level cost constraints, and assessing timephased resource requirements. When an analysis is complete and agreement on the proposed projects has been reached, the next step is to commit the portfolio selection and communicate it to the stakeholders.

In the last section, you learned how to create a driver prioritization (a collection of drivers with priority levels tailored based on either calculated or manual settings). The driver prioritization is selected when you first define the portfolio analysis. By reading through this section, you will understand how to create a portfolio analysis and explore various properties you can set and how they impact the portfolio, and you will map selected projects to business drivers.

Two main constraint scenarios are generated with a portfolio analysis; cost and resource constraints will be explored as options to assist with selecting the optimum mix of projects. You will learn how to generate and compare various scenarios and make additional decisions for the purpose of forecasting, such as hiring resources and increasing the budget. This section wraps up with selecting the most appropriate scenario to commit, which will drive projects to the next step in the workflow.

Managing Scenarios

This final section is about fine-tuning the scenarios and finally choosing the scenario that will be committed. Several options will be explored regarding this fine-tuning, including overriding the selection of projects and designating a preference to include or exclude specific projects. As part of the evaluation process, it is important to create one or several scenarios and compare them against each other to ensure that the most appropriate scenario is selected.

After a full evaluation of scenarios is complete, you will learn how to commit the Portfolio of Projects for future detailed planning and to advance the projects through additional phases and stages in the workflow.

Commit a Portfolio Analysis

The Commit button in the Portfolio Analysis tool signals to Project Server 2010 that a portfolio has been committed and Project Server can now move to the next phase in the project lifecycle. Once the Commit button is selected, a warning dialog box will displayed indicating that the portfolio selection is about to be committed. (See Figure 9.24.)

The only other indication that the portfolio selection has occurred is that, after pressing the OK button illustrated in Figure 9.24, there will be a message that "the Portfolio Selection Scenario has been committed."

The Commit button triggers the population of a total of six project-level fields:

1. **Committed Planned End Date.** The finish date of the project as committed to in a Portfolio Selection Scenario during resource constraint analysis.
2. **Committed Planned Start Date.** The start date of the project as committed to in a Portfolio Selection Scenario during resource constraint analysis.
3. **Committed Portfolio Selection Decision (Cost).** The result of a cost constraint analysis on a project. You can choose Selected, Unselected, Forced-In/Out, or Custom Forced-In/Out.

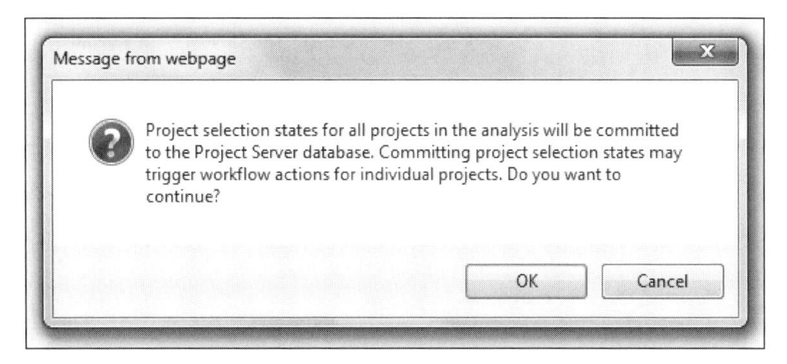

Figure 9.24 Commit Portfolio Selection Dialog Box Source: Advisicon

4. **Committed Portfolio Selection Decision (Schedule).** The result of a resource constraint analysis on a project. You can choose Selected, Unselected, Forced-In/Out, or Custom Forced-In/Out.
5. **Committed Portfolio Selection Decision Date (Cost).** The commitment date of a Portfolio Selection Scenario as determined during cost constraint analysis.
6. **Committed Portfolio Selection Decision Date (Schedule).** The commitment date of a Portfolio Selection Scenario as determined during resource constraint analysis.

These fields can also be added to the Portfolio Analysis and Portfolio Analysis Project Selection views.

It is important to understand that there are two independent Commit Selection Portfolio decisions. The next enterprise fields are updated when the portfolio selection is committed independently for each of the Analyze Cost and Analyze Resources selections:

1. **Analyze Cost view.** Where the Committed Portfolio Selection Decision (Cost) and the Committed Portfolio Selection Decision Date (Cost) fields are populated
2. **Analyze Resources view.** Where the Committed Portfolio Selection Decision (Schedule) and the Committed Portfolio Selection Decision Date (Schedule) fields are populated

Internal to Project Server 2010, the next conditions also occur:

- If there is a workflow attached to the various projects, the Commit button will fire the OnProjectCommit event to allow a workflow to be triggered.
- The Commit button will also write to the Project Server ReportingDB table called MSP_EPMProjectCommit. This table contains all projects within

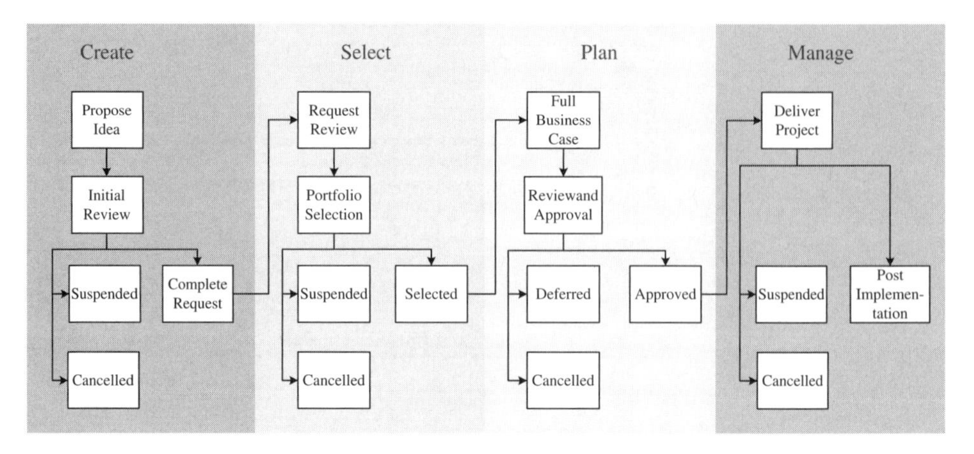

Figure 9.25 Project Portfolio Management Lifecycle

the Portfolio Analysis and their relative decisions (Selected, NotSelected, Forced in/out, etc.).

Once committed, the projects in the portfolio analysis now move to the Select phase of the Project portfolio lifecycle, as illustrated in Figure 9.25. In this generic PPM model, these selected projects are now authorized for detailed planning and execution in the Plan and Manage phases of the PPM lifecycle.

Measuring for ROI

Monitoring and reviewing the performance of a portfolio is highly recommended. It can be performed by the use of dashboards and other reporting formats supported by the Project Server 2010 solution.

Before we start configuring the tool, we must have a good understanding of what we plan to measure. As we indicated earlier in this chapter, performance measurement enables an organization to assess, monitor, and course-correct performance and align all employees with key business objectives. We need to be aware of the impact that our selection of metrics can have on performance in an organization and be mindful of the impact on motivation and behavior.

Let's first examine some popular financial measures, starting with some of the more common investment analysis techniques.

Once you have created an optimized mix of projects and fully evaluated this mix against costs and resources, you are ready to commit the portfolio and move the projects to the next collection of stages and phases for detailed planning and executing of projects. The process of committing takes field information from a saved scenario and applies it to views and fields in other project detail pages. Many of the fields are locked as baseline fields so you have a baseline value for later comparison against actual results.

Return on Investment Analysis

According to Investopedia,[1] ROI can be defined as "a performance measure used to evaluate the efficiency of an investment or to compare the efficiency of a number of different investments. To calculate ROI, the benefit (return) of an investment is divided by the cost of the investment; the result is expressed as a percentage or a ratio." The following formula illustrates the ROI ratio:

$$\text{ROI} = \frac{(\text{Gain from Investment} - \text{Cost of Investment})}{\text{Cost of Investment}}$$

Furthermore, the definition states that "if an investment does not have a positive ROI, or if there are other opportunities with a higher ROI, then the investment should be not be undertaken."

When comparing one potential project with another, the term "ROI" applies to cash flow analysis. In this case "ROI" simply means the "return" (incremental gain) from an action, divided by the cost of that action, over time. Three ways to maximize ROI are represented in Figure 9.26: reduce costs, increase benefits, and accelerate the returns. These are ideal opportunities which will produce deliverables in the Project Portfolio ahead of schedule.

Several factors can complicate ROI calculation or interpretation. For that reason, many organizations do not attempt to present ROI as a quantitative result. Instead, they incorporate financial metrics into their portfolios such as present value (PV), discounted cash flow, internal rate of return, and Payback Period.

As we have seen throughout this chapter, the Project Server 2010 solution provides the basis for tracking initiatives, including their proposed benefits. Using Enterprise Custom fields, it is easy to calculate the ROI using the investment formula. What is not so obvious is how to calculate the "gain" or "benefit" of an investment.

Cost of Ownership

Sometimes called total cost of ownership, cost of ownership is the total cost of acquiring, installing, configuring, developing, and training over an extended period of time. The cost of ownership can be the cost side of a cost/benefit analysis. However, cost analysis does not take into consideration the benefits of the initiative.

Cost/Benefit Analysis

This analysis is used for planning, decision support, program evaluation, proposal evaluation, and other purposes in a variety of ways. The term itself has no precise

[1]*Investopedia*. s.v. "Return on Investment—ROI," accessed April 5, 2012, www.investopedia.com/terms/r/returnoninvestment.asp; see also s.v "Return on Investment (ROI)" in Marty J. Schmidt, *Encyclopedia of Business Terms and Methods*, www.solutionmatrix.com/business-encyclopedia.html.

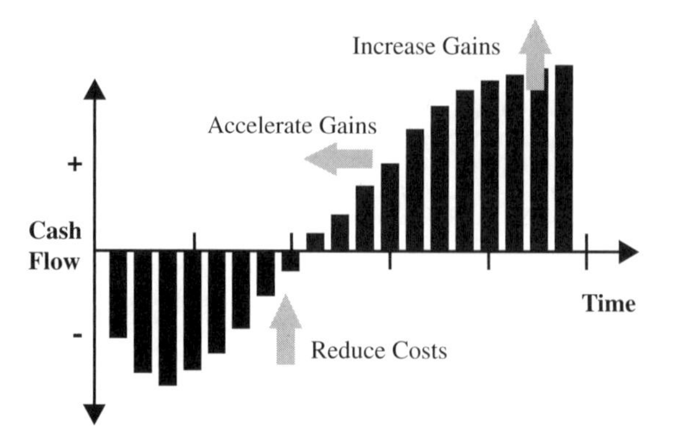

Figure 9.26 Cash Flow Investment Curve Source: Advisicon

definition other than that both positive and negative impacts are going to be analyzed and weighed against each other. The upside to this approach is that both the investment and the payback are analyzed to ensure that a net value can be derived for any given investment.

Present Value Analysis

Using PV and future cash flows of an investment, an interest rate can be calculated by using the PV calculation. The PV formula is useful for predicting future expected future cash flows for a given level of risk. The PV formula is shown in the following formula, where PV is the present value, FV is the future value, r is the discount rate (measure of risk), and n is the number of periods:

$$PV = FV \times \frac{1}{(1 + r)^n}$$

Balanced Scorecard

Measuring for financial ROI is most certainly top of mind for senior management, especially during the challenging economic period at the time this book was published. Financial analysis is, however, only one aspect of scoring or ranking projects for viability. Other business drivers, such as future expansion or optimizing efficiency should also be taken into consideration when doing portfolio analysis.

Drs. Robert Kaplan and David Norton[2] introduced the performance measurement framework in the mid-1990s. It added nonfinancial performance strategic

[2]For the basics of the balanced scorecard approach and a concise summary of Kaplan and Norton, see "Balanced Scorecard Basics" at www.balancedscorecard.org/BSCResources/AbouttheBalanced Scorecard/tabid/55/Default.aspx.

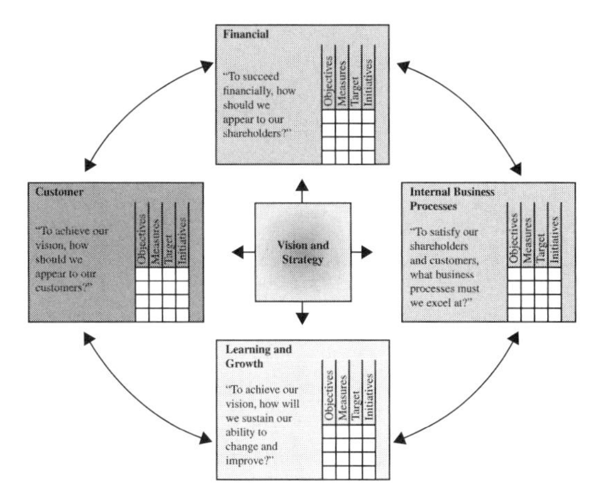

Figure 9.27 Using the Balanced Scorecard as a Strategic Management System Source: Advisicon

indicators to traditional financial metrics to give managers and executives a more "balanced scorecard" of organizational performance.

Figure 9.27 illustrates a view that the organization should consider from four perspectives to develop performance metrics, collect data, and analyze it relative to each of the next perspectives:

- **Financial.** The traditional aspect of financial data
- **Customer.** The importance of customer focus and satisfaction
- **Learning and Growth.** Employee training and corporate culture
- **Internal Business Processes.** Internal business processes or metrics

The balanced scorecard is not a piece of software. Unfortunately, many people believe that implementing software amounts to implementing a balanced scorecard. Once a scorecard has been designed and implemented for the organization, however, Project Server 2010 BI and reporting capabilities can provide a powerful set of key performance indicators and dashboards with visual representation of the necessary performance indicators to enable a strategic management and planning system.

Figure 9.28 illustrates an example of a Project Server 2010 dashboard.

PROJECT SERVER OPTIMIZING GOVERNANCE FOR PMOS

First come, first-serve project management is an ineffective way for organizations to analyze, prioritize, select, and assess a portfolio of current and future projects. Most organizations continue to experience significant challenges.

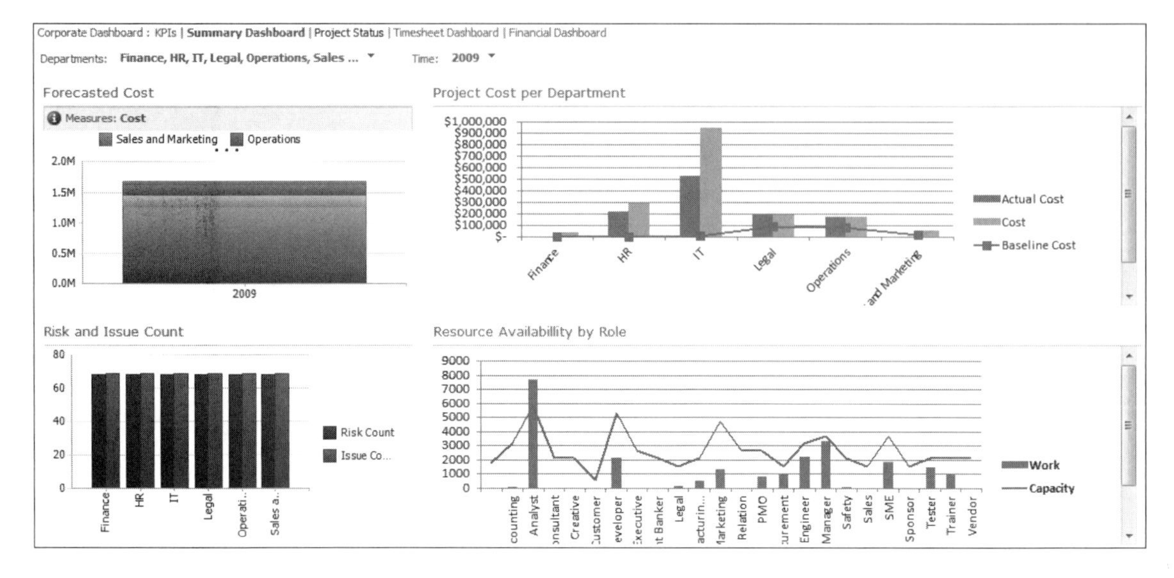

Figure 9.28 Project Server 2010 Dashboard Source: Advisicon

The Gartner Group (www.gartner.com) provides many different statistics on the percentages of projects canceled due to inadequate risk management as well as the percentage of projects that actually will succeed. While we cannot provide the actual statistics here, and the studies from change year to year, we recommend that you check this source of reference statistics on the success and failure percentages of projects.

Investments in a PMO as a work management discipline can provide common planning and reporting processes and bring structure and support to evaluating, justifying, defining, planning, tracking, and executing projects.

Enable the Strategic Role of a PMO

A number of key steps need to be undertaken to properly enable the strategic role of a PMO:

- Implement a PPM office, including appropriate governance, methods, roles and responsibilities, and oversight of the program and project management.
- Institutionalize disciplines and processes to help build, refine, and prioritize programs and project portfolios.
- Align organizational strategies and plans, business cases for investment proposals and performance, and success metrics for enterprise programs and initiatives.

There are three basic organizational styles for a PMO. Which one you select will determine the role of the office within the project development to management lifecycle:

1. **Project repository.** The project office simply serves as a source of information on project methodology and standards. This method is often used as a first step to consolidating or sharing management practices. However, it falls short of direct project oversight within the business. Project managers continue to report to their respective business areas.
2. **Project coach model.** This model assumes a willingness to share some project management practices across business functions and uses the PMO to coordinate the communication with the various stakeholders. The PMO in this model is a permanent structure with staff and has PM responsibility for all projects. A dotted-line reporting relationship exists between business-staffed project managers and the PMO for performance and reporting.
3. **Enterprise Project Management Office (EPMO).** This is the most consolidated organizational model. The EPMO will have direct management and oversight over larger enterprise projects within the organization.

PMO Governance

The PMO is assigned the key roles of assessing and validating project estimates as well as staffing the project manager function. Five key roles must be incorporated into the design of a PMO, although implementations vary based on business structure, the degree of dysfunction, and the sense of urgency across business divisions that a need exists for a shared solution to project control.

1. **Standard methodology.** A consistent set of tools and processes for projects is necessary for clear communication.
2. **Resource evaluation.** The initial assessment of resources (i.e., people, money, and time) is critical to organizational capacity planning.
3. **Project planning.** The project plan is a cooperative effort coordinated by the PMO, which serves as a PM competency center and as an archive for previous project plans.
4. **Project management.** Consistent practices, frequent reviews, and a governance responsibility are the baseline roles for management within the PMO.
5. **Project review and analysis.** Enterprises need to know if project deliverables are achieved on time, on budget, and deliver the required functionality.

Collaboration Infrastructure

Although governance is a critical component of the PMO, a second major enabler is the provision of tools and best practices for an organization to support collaboration across the enterprise. Most organizations are running their businesses today on e-mail, spreadsheets, and PowerPoint. This is not only costly, complex, and does not scale, but it is becoming increasingly apparent that it is an ineffective way to manage the increasing work and resource demands, given today's ever-changing business environments and requirements.

Project Server 2010 enables a PMO to step into a more strategic role and begin to help lead organizational initiatives to increase the likelihood of success for the projects it is responsible for staffing, managing, tracking, reporting on, and evaluating whether they have hit their goals.

Project Server 2010 provides the key collaborative infrastructure elements to support the information demands of the divisional or enterprise PMO:

- Demand management
- Portfolio planning and scenario analysis
- Capacity planning
- Scheduling planning and tracking
- Cost management
- Resource planning and management
- Risk and issue management
- Team collaboration
- Management reporting

It is increasingly critical to organizations that project data is accessible and reportable. These requirements typically are key for effective delivery to the strategic objectives, which makes Project 2010 much more inviting to enterprise decision makers.

PMO Competencies

Certain critical competencies also are required for the successful implementation of a PMO. These competencies include technology skills, domain expertise, business process aptitude, communication skills, and other related "soft" skills.

To function at a highly competent level in the modern PMO environment, there is significant need for competency development beyond the baseline PM knowledge areas. Much of this can be gained through a PMI certification (such as the Project Management Professional or PMI Scheduling Professional or PMI Risk Management Professional competencies) and technical competencies. It is also imperative that practitioners continue to develop their competencies on an ongoing basis.

IMPORTANT CONCEPTS COVERED IN THIS CHAPTER

In this chapter, we reviewed the portfolio analysis tool within Project Server. The tool addresses cost constraints as well as resource constraints and allows an organization to leverage business drivers and the ranking and rating system to help select upcoming new work from a portfolio of choices.

We also reviewed the phases and stages that can be enabled to help leverage the native workflows in Project Server as well as the Department field, which can help a company to pivot and have multiple views, fields, and portfolio selection criteria in a single instance of Project Server. This happens without having to expose different objects, views, portfolio criteria, business drivers, or Project detail pages to any other group not associated with that department. This shelters all the customizations of one group from any other group using Project Server.

As organizations mature in a PM culture, simply managing project tasks and schedules is not enough. More vigorous management of project initiatives is needed.

This vigorous management takes the form of developing a stable program management practice where related projects are grouped together to reap benefits that can be realized only from this higher level of management. The next step in PM maturity is portfolio management, where strategic business goals and objectives are introduced and all project work is aligned in support of these measures.

Project Server 2010 provides a scalable tool to accommodate this organizational growth by providing organizations with a consolidated platform. This platform enables organizations to manage all aspects of PPM from tracking work at the project task level to performing project selection analysis through data-driven criteria application, while balancing internal and external constraints.

In this section, you completed a portfolio analysis. To take advantage of the different properties when defining a portfolio, you learned that you need to designate a primary cost constraint field and a resource role and that projects need to have a resource plan. The importance of force-in and force-out custom lookup fields was illustrated by how they allow you to designate a reason for selecting or unselecting projects in the portfolio.

Several topics illustrated how Project Server generates the priority levels using previously defined driver prioritizations and using the seven-point comparison scale against drivers. Once the portfolio was generated, you learned that you can modify the total available budget in a cost constraint analysis and modify the hired resources in a resource constraint analysis. Multiple scenarios in the portfolio can be saved and compared against before you choose the final scenario to commit and to advance the group of the projects to the next step in the workflow.

REFERENCES

O'Cull, Heather. 2009. "Project 2010: Introducing Demand Management." Accessed September 6, 2012, http://blogs.msdn.com/b/project/archive/2009/11/13/project-2010-introducing-demand-management.aspx.

U.S. General Accounting Office (GAO).1998. *Executive Guide: Measuring Performance and Demonstrating Results of Information Technology Investments*. AIMD-98-89. Accessed September 6, 2012, www.gao.gov/assets/80/76378.pdf.

CHAPTER 10

INTELLIGENT BUSINESS PLANNING AND REPORTING USING MICROSOFT PROJECT 2010

IN THIS CHAPTER

We explore business intelligence options such as Excel Services, PowerPivot, and SQL Reporting Services to help you take some basic steps in moving toward the business intelligence reporting capabilities of Project Server. We also demonstrate some functional steps in creating Pivot Reports using Project Server 2010 data.

What You Will Learn

- A better understanding of the business intelligence options for Dynamic Reporting
- Some helpful steps for Pivot Reporting and Pivot Formatting
- The flexibility of Office Web applications and ways to leverage these for data accessibility
- How to leverage the collaborative Business Portal (SharePoint) to assist in empowering your social reporting and communications network

WHAT IS DYNAMIC REPORTING . . .

. . . and, for that matter, what is business intelligence (BI)? Depending on whom you ask, you might get different answers, and based on their personal context, they may all be right.

For our purposes, we'll use the next definition of BI:

> **Business Intelligence**
>
> A category of methodologies and technologies for gathering, storing, analyzing, and providing access to data to help enterprise users make business decisions.

Businesses have been doing some version of BI for years, so what's so special about the BI version? Dynamic Reporting is one of the answers to that question.

In traditional reporting methodologies, you'd make a request to the information technology (IT) department to write a report that tells you something specific, and you'd get that specific report on a regular basis. If that report didn't slice things quite the way you needed them, you were pretty much out of luck until IT found the time to write you a new report. Another oft-used workaround was to get the details about some subset of data in a report written to Excel, then slice and dice via your own filtering and pivot tables. This alternative is still quite valid, and Excel, PowerPivot, and Excel Services help you leverage that approach. Using this workaround does, however, require a certain level of sophisticated knowledge about data structures and use of Excel.

A solution to this quandary can be found in something called an OLAP (online analytical processing) cube. The idea behind this is that rather than just using a table in Excel, you create a structure that adds more dimensions to the existing data. The base information, typically numeric in form, is called the measures. These measures are categorized and summarized in dimensions. In a traditional table, you're limited to rows and columns for dimensions. In the OLAP world, you can have many dimensions, and the dimensions themselves can have hierarchies. For example, you could have a time dimension, and within that dimension, you could have a hierarchy of year, quarter, month, and day. At the time the OLAP database is created, a summary process that aggregates the measures for these various dimensions is performed. So, when the summary data are requested from the cube, the processing overhead is minimized, giving the requestor a quick response time.

Once an analytical cube is available, users can view the data in different ways. They can filter for just the subset of data they would like within a dimension, then break it down within the hierarchies, or combine dimensions to group the results in a way that is of most interest. Once users have seen the data in a particular way, they may take an interest in a more specific area of the data or in reorienting or regrouping the way the data are arranged. This selection of data and subsequent reorientation is sometimes called slicing and pivoting. But ultimately what it amounts to is Dynamic Reporting, as the users are in control of asking the questions.

This all sounds pretty good, so why isn't everyone doing it? Well, there's a good news/bad news aspect to it. The bad news is that designing and building analytical cubes in OLAP databases calls for a specialized set of skills and a database server that can handle OLAP processing. Decisions have to be made about what dimensions and measures will need to be supported and what hierarchies and summary levels should be supported. The processing overhead to build the cubes can be considerable; in order to give end users a fast response, we have to take the hit for summary-level computations up front. Often, because of all the overhead and processing time cube building can take, it is done only periodically, perhaps weekly. The good news is that Project Server includes support for cubes. The complex design questions and the processes for building the custom Project Server cubes have already been addressed.

Slicing, Dicing, Drill Up/Down, and Pivot

Sounds like something out of a Vegematic commercial, doesn't it? But in the world of cubes and Dynamic Reporting, these things all have meaning. Let's take a brief look at each term.

- **Slicing.** A slice is a section of a cube that corresponds to a single value for one or more dimensions. For example, a cube containing resource utilization by work group for years 2005 to 2010 might have the year 2008 sliced out.
- **Dicing.** A dice is slices performed on more than one dimension. For example, after the slice for the year 2008 was performed, a second slice could be taken for a specific work group, resulting in that work group in 2008.
- **Drill up/down.** This is a technique whereby users navigate up or down from data that are more summarized to data that are less summarized, or vice versa. For example, you could drill down from 2008 resource utilization to the quarter-level utilization within 2008, then to the month level, and so on.
- **Pivot** (sometimes also called a rotate operation). The idea is that the data presented are pivoted to show a different dimensional orientation. For example, in our resource utilization cube, we might start by showing resources on the *x*-axis and years on the *y*-axis and then pivot showing years on the *x*-axis and work groups on the *y*-axis.

Project Server 2010 Cubes

The next analysis cubes are provided with Project Server:

- Project Non-Timephased data including project and master projects
- Assignment Non Timephased
- Assignment Timephased
- Task Non Timephased
- Resource Non Timephased
- Resource Timephased

- EPM Timesheet: timesheet and Enterprise Project Management (EPM) dimensions (project, task, and resource)
- Timesheet, including specific timesheet dimensions (Timesheet Project, Timesheet Task, and Timesheet Resource)
- MSP_Project_SharePoint: virtual cube that combines Project Non Timephased, Issues, Risks, and Deliverables cubes
- MSP_Project_Timesheet: virtual cube that combines Assignment Timephased, Resource TimePhased, and EPM Timephased cubes
- MSP_Portfolio Analyser: virtual cube that combines Assignment Timephased and Resource Timephased cubes (Comparable to the older cubes supporting Portfolio Server, which is now packaged with Project Server 2010)
- Risks: Windows SharePoint Services (WSS) Risk data
- Issues: WSS Issues data
- Deliverables: WSS Deliverables data

The term "timephased" simply means that the cube contains one or more time dimensions (such as time and fiscal time).

These descriptions are fairly generic. In general, the cubes contain most of the appropriate numeric fields as measures but are limited as to dimensions. The Task cube is surprisingly sparse; Assignments has the nitty-gritty details. A detailed description of each cube is beyond the scope of this chapter, but we encourage you to use the reporting tools to investigate further.

These cubes come out of the box with structure oriented around the fields that are native to Project Server. However, it is possible to extend them to include enterprise custom fields. There are some limitations in this area, however, particularly in the area of adding dimensions for text-based fields. Contact your Project Server administrator for more details.

CREATING EASY-TO-ACCESS REPORTING IN PROJECT SERVER/ SHAREPOINT BI

SharePoint has always had strength in the area of collaboration, and one of the key factors in collaboration is easy accessibility. Accessibility can be looked at in two ways: how easy it is to get to something and how easy it is for the average user to work with something. SharePoint 2010 and Project Server 2010 provide several ways to create and access BI reporting.

BI Center and PerformancePoint Services

When you work with BI in SharePoint, the obvious place to start is by creating a BI center site via the BI Center Site template. In the case of Project Server 2010,

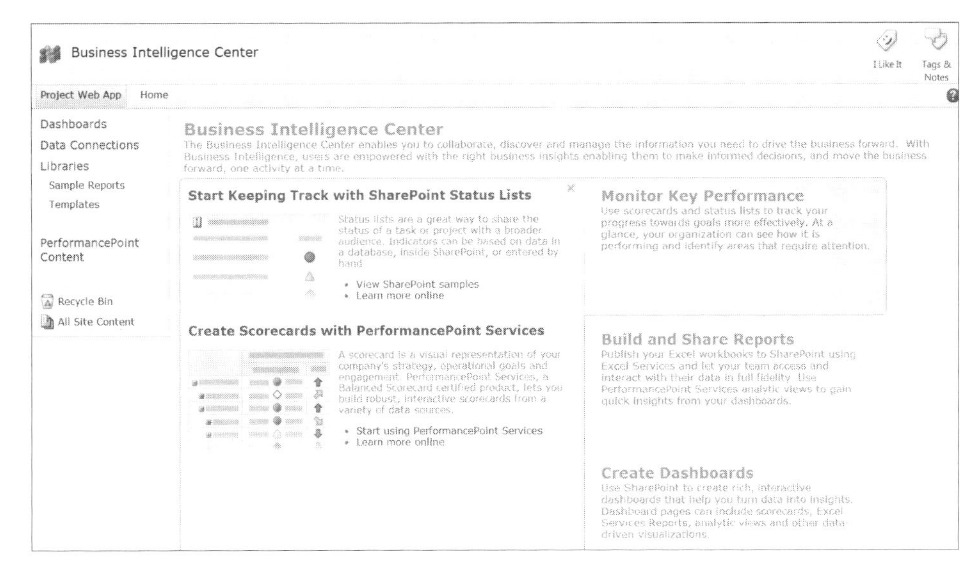

Figure 10.1 BI Center Source: Advisicon

the Project Web App (PWA) already includes a BI Center Site (see Figure 10.1), which is accessible via a link in the Quick Launch Bar. It doesn't get much easier than that!

Once you've arrived at the BI Center, you'll find that it has a front page that introduces you to different aspects of BI, organized into three sections: Monitor Key Performance, Build and Share Reports, and Create Dashboards. Within each of these sections there are links to samples and to learn more about a topic online. In addition, each of the sections contains a link to the PerformancePoint Server (PPS) sample page (see Figure 10.2).

The PPS sample page will help guide you into the world of PerformancePoint, with more links to help you learn about it. But possibly the most significant item on this page is the button that allows you to download and run the Dashboard Designer tool. Should you choose PerformancePoint and Dashboard Designer as a BI tool, this is the front door.

The BI Center that comes with Project Server has a more extended list of sample reports than that available when creating a BI Center via the out-of-the-box SharePoint BI Center template. These sample reports can demonstrate how you might access and display the appropriate information for various Project Report types and provide you a starting point for creating your own enterprise-specific versions of these reports. The samples are Excel-based reports. Click on the Sample Reports link on the Navigation bar on the left to see the list of samples (see Figure 10.3). Reports include:

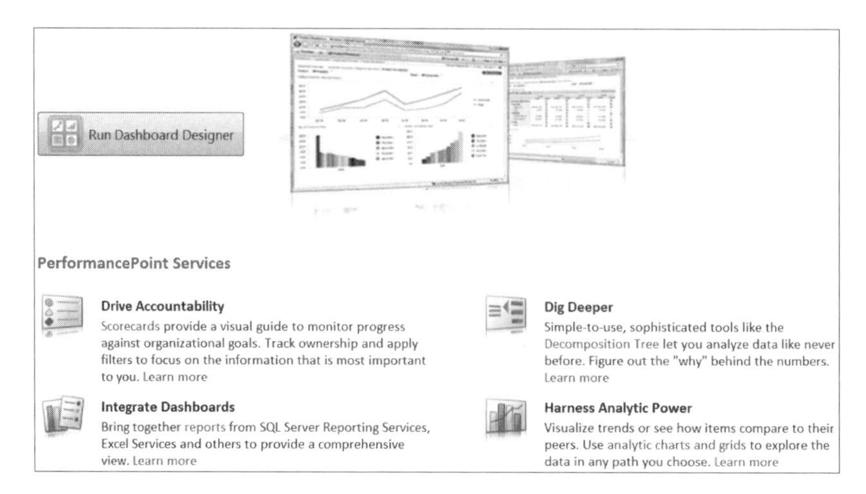

Figure 10.2 PPS Page Source: Advisicon

EPM

- Simple Projects List
- Milestones Due This Month
- Resource Capacity

Share Lists

- Deliverables
- Issues and Risks

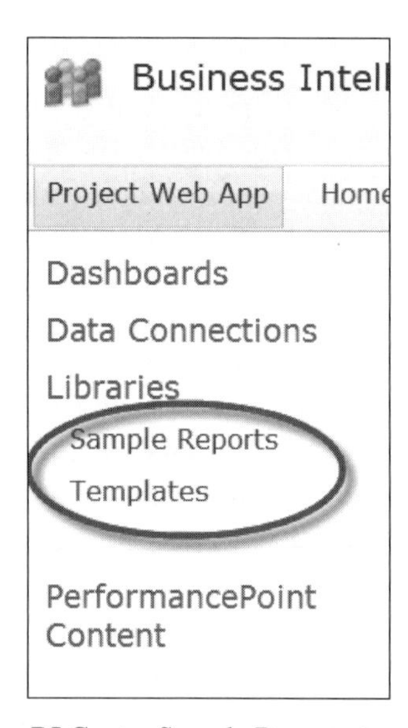

Figure 10.3 BI Center Sample Reports Source: Advisicon

Timesheet

- Timesheet Actuals

Portfolio

- Rejected Projects List
- Top Projects
- Workflow Chart
- Workflow Drilldown

In addition, there is a collection of Excel template files for both reporting DB access and OLAP cube access. The template files are essentially empty spreadsheets with predefined data connections/queries. Based on the data returned by those connections, you can customize your own reports. See the Templates link for more details.

PPS provides the key enterprise-level integration services between the SharePoint environment and BI-oriented functionality. Full advantage of this functionality can be taken via the Dashboard Designer. Dashboard Designer allows you to create indicators, key performance indicators (KPIs) at both a specific and overview level, scorecards to hold the KPIs, charts, graphs, strategy maps (Visio diagrams linked to KPIs), and links to reports created in Excel or SQL Server Reporting Services via integration with the Report Builder tool. Once you've designed and built the components required for your specific reporting needs, you can mash them all together into a dashboard. Dashboards can contain one or more pages of strategically grouped information, either grouped together on a page or on dedicated pages. Ultimately the goal is to give managers one-stop shopping for all their key reporting needs. All of this functionality is supported behind the scenes with SharePoint Web Part Pages and Web Parts along with connectivity to databases, analytic cubes, and various other data sources. Powerful stuff!

If the BI Center, PPS, and Dashboard Designer seem a bit overwhelming, don't worry. These tools are really aimed at integrated enterprise-level BI reporting, but you can keep it simpler. Let us consider some alternatives.

SQL Server Reporting Services

As you might guess, SQL Server Reporting Services (SSRS) reporting functionality comes packaged with MS SQL Server. It is used to create "traditional" reporting (i.e., with a specific report layout, going against specific data sources). Those data sources can include both analytical cubes and relational databases, among other items. In the Project Server context, you can access data both from the Analysis Service (the element of SQL Server that allows for data analytics against databases), cubes and the Project Server Reporting database. While it's possible to access both in the same report, that's probably not a good idea, as the Reporting database data has near-real-time data and the data in the analysis services cubes is rebuilt only periodically, often only weekly. Mixing the two could lead to inconsistent results.

The Project Server Reporting database is a MS SQL Server relational database that gets updated each time a project is published. While not all data about each project are available, some data is available here that is not available in the cubes, especially for custom fields. The information in the Reporting database is most easily accessed via Views, which resolve custom fields and lookup values.

Core Project Server functionality includes four relational databases: Draft, Published, Archived, and Reporting. Draft is used to store pending changes. Published is updated at the time a project is published for public consumption via the PWA, and includes all project and enterprise details. Archived supports project-level backup and restore as well other recovery components within Project Server. The Reporting database is also updated at publishing time but does not include all project details and is somewhat optimized for reporting queries. In addition to reporting data, it also contains staging data for the cube building process. Microsoft publishes schema documentation only for the Reporting database, as the other four are designed for internal use and could be redesigned without notice.

SSRS reports are created in one of two ways: by a developer using a tool called SQL Server BI Development Studio (BIDS) or by an end user using a tool called Report Builder. Report Builder has a MS Office look and feel. Report definitions can be saved locally or published to the Report Server. Reports published to Report Server can be displayed within SharePoint pages.

Reports are viewed either through the Report Builder tool or via a browser for reports published to the Report Server. Often the browser runs in a SharePoint context, but that is not strictly required. The report viewer allows you to preview and print one or more pages. Reports can include parameters to allow filtering at run time. Report presentation can be tabular, include graphs and charts, can be expanded and collapsed, and can include drill through to other reports. Reports also can be exported to PDF, Excel, and other formats, including formats that can be read as data sources to Excel 2010.

Similar to using the Dashboard Designer with the PerformancePoint Server, writing SSRS reports via BI Development Studio or Report Builder calls for some training and ideally some knowledge of database constructs. There are things that IT or an experienced user can create to make more complex queries accessible to newer users and wizards to help simplify or initiate basic report creation. SSRS Report design and creation is a significant topic in its own right and is outside the scope of this book.

With the advent of SQL Server 2012, SSRS includes a new tool called "Power View" (code-named Crescent). Power View is a Silverlight-based tool that runs from your browser, which enables easy creation of integrated dashboard-level pages containing pivot tables, charts, and filters. It uses a new SQL Server Analysis Services (SSAS) tabular data model as a data source. Another nice feature is that the SSAS can also use PowerPivot files as a data source, which is based on the same model, allowing exported reports to be part of additional report building. While the creation of the data models still may require specialized knowledge, once the data model connections are published to SharePoint, Power View will make creating complex graphical reports fairly straightforward for the average data analyst or information worker and doing ad hoc queries easy for the typical information consumer.

Excel and Excel PowerPivot

You've heard of a movie with a cast of thousands? Excel is a tool with a cast of millions. Even back in 2009, Microsoft estimated a worldwide number of users at more than 400 million. Some estimates suggest that 50 to 80 percent of enterprises still rely on stand-alone spreadsheets for critical applications like financial reporting or forecasting. Even in shops where BI is available, that last bit of analysis often is done by a chief financial officer with a spreadsheet. Needless to say, ignoring Excel as a reporting tool would be foolish.

You can get instant benefits from publishing some of your existing Excel spreadsheets to SharePoint, gaining the use of Excel Services. Benefits include:

- Reducing the number of copies in circulation as well as the potential variations on those copies introduced by each copy owner.
- The ability to use SharePoint's document management services.
- The ability to link the workbook to other SharePoint pages, including pages used to summarize BI information.
- Limiting what parts of your workbook are visible.
- The fact that users viewing published workbook components cannot change content. However, cells can be specified as parameters that viewers can enter values into. Excel formulas can key off those parameters to drive modification of other cells. This modification is nondestructive.
- Using published spreadsheets as data sources for other spreadsheets or reports.
- Server-level functionality that allows for enterprise-level efficiencies (i.e., rather than having a server open 20 different copies of the same big workbook for 20 different users, it knows to open one copy and share it).

PowerPivot

So what's this PowerPivot thing? As if people were not already making Excel jump through enough hoops with the data that currently can be stored in worksheets, PowerPivot is an add-on for Excel 2010 that allows a user to import data from various sources into "tables" within Excel. These sources include relational databases, OLAP cube databases, other Excel files, text files, cloud services, and feeds from reporting service reports.

Once imported, the disparate tables can be linked together via a relationship editor. One warning about how PowerPivot handles relationships: In the early release (and still true at the time this is written) many-to-many relationships are not supported. For example, if you have a task table and a resource table (a task can have many resources and a resource can have many tasks), even if you have an intersection table (Assignments), the aggregation results of something like task work by resource may not be what you expect. There is a rather technical workaround involving the use of the DAX query language, but for average mortals this is still an issue. Presumably Microsoft will address this in the next release.

The data are compressed and stored separately from the standard Excel data, although they still reside in the same Excel file. A special VertiPaq Analysis Services engine is used to process these data. Once in place, the data is displayed via Excel tables, pivot tables, pivot charts, filters, slicers, and others. In short, PowerPivot provides a do-it-yourself data warehouse with limited Analysis Services functionality, accessible via all the user interface functionality you've come to expect from Excel.

This begs the question: Why, if I'm a Project Server user and I already have all of these whiz-bang Project Server cubes already built for me, should I care about PowerPivot? The answer comes down to integration. Unless you already have a business process in place to define and update enterprise custom fields from other related business systems within your enterprise, you are likely to need an alternative for combining that data. PowerPivot provides an option there, and does so in a way that does not depend on the time and expertise of a team of IT specialists. As extending custom fields in MSP cubes does have some limitations and a data conversion may be needed to facilitate meeting these limitations, PowerPivot may provide a temporary workaround. You may not get the power of a full IT OLAP cube solution, but in these days of limited resources, it can get you started down the road to integration.

It's worth noting that the Vertipaq engine is the same engine used by SSAS in its "tabular" mode. In fact, when PowerPivot files are published to SharePoint, Excel Services coordinates with SSAS and a server-level PowerPivot service to transfer the Excel "database" into SSAS for processing. This enables enterprise-level efficiencies, as the same data can be shared by multiple users.

The Vertipaq engine takes advantage of two key concepts: (1) a more advanced data compression technology that allows a larger number of records to be stored in the same space and (2) including the entire data model in memory to speed up queries. This has pluses and minuses. On the plus side, because we're working at memory-based speeds rather than doing a lot of disk input/output, we can sacrifice some of the complexities that come with optimizing analytical cubes, which puts the data modeling more in reach of nonspecialists. On the minus side, you will need a lot of memory for bigger data models, and for really large models, you need to revert back to traditional OLAP. As such, PowerPivot is considered self-service BI, and if the data or data usage grows to a certain level, you may have to upgrade to a more traditional OLAP solution.

How to Create and Modify an Excel Pivot Report from a Project Server Cube

In this section, we demonstrate how to create a report that includes a pivot table and associated pivot chart, using out-of-the-box Excel 2010 and an out-of-the-box Project Server cube. Most of the same concepts can be applied to any SSAS cube or relational database.

1. Open Excel 2010.
2. Click on a cell in the new spreadsheet.
3. On the Insert tab, Tables group, click on Pivot Table (see Figure 10.4). If you'd like a chart as well, click the down arrow under the Pivot Table icon and select Pivot Chart.
4. A Create PivotTable with PivotChart dialog will come up (see Figure 10.5).

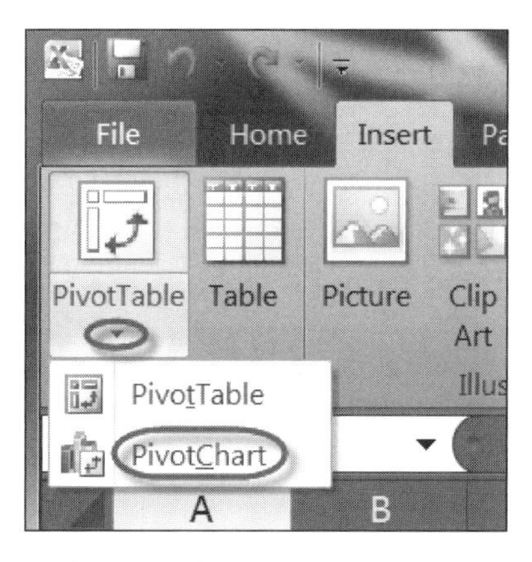

Figure 10.4 Insert Pivot Table

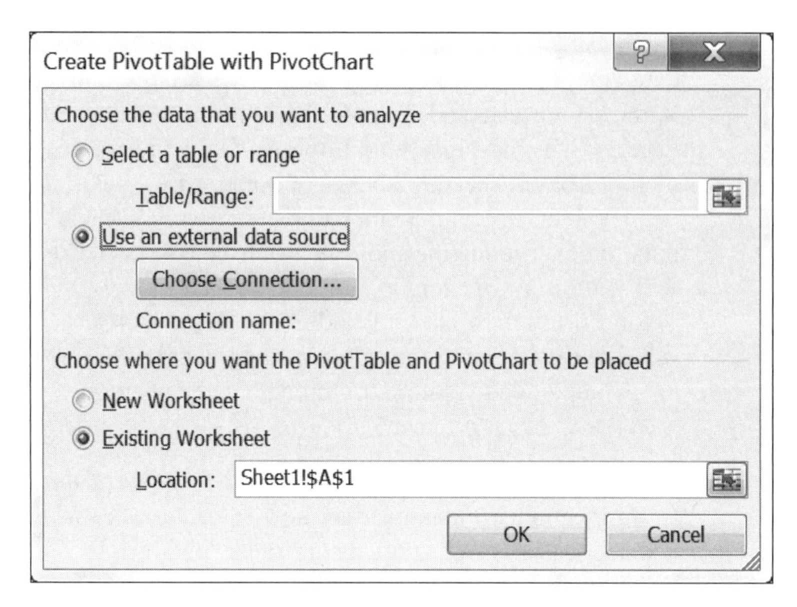

Figure 10.5 Create Pivot Dialog

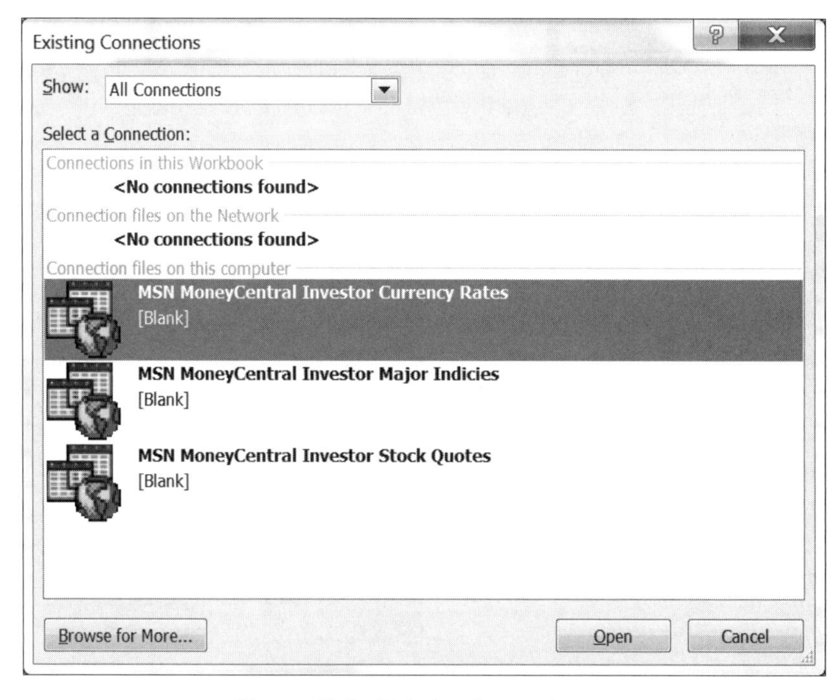

Figure 10.6 Existing Connections

5. Select the Use an external data source option and click the Choose Connection button. An Existing Connections dialog will open (see Figure 10.6). You can see existing connections in the workbook, on your computer, and on the network.

   Ideally you should use an.odc connection off the network so you can share your spreadsheet more easily later. If the report is going to be used in a BI Center context, data connections should be stored in the associated data connections library for that site. You may need to "Browse for More . . ." to find it.

   a. If you cannot find a connection for the data source you need, you'll need to create an.odc file first or have someone create one for you. See the section titled "How to Create an.odc" later in the chapter.
   b. For our example, we are using a connection to the Assignment Timephased cube in our Project Server Analysis Services Database server (see Figure 10.7).
   c. Note that the BI Center site that comes with a Project Server PWA installation includes a number of predefined connections that

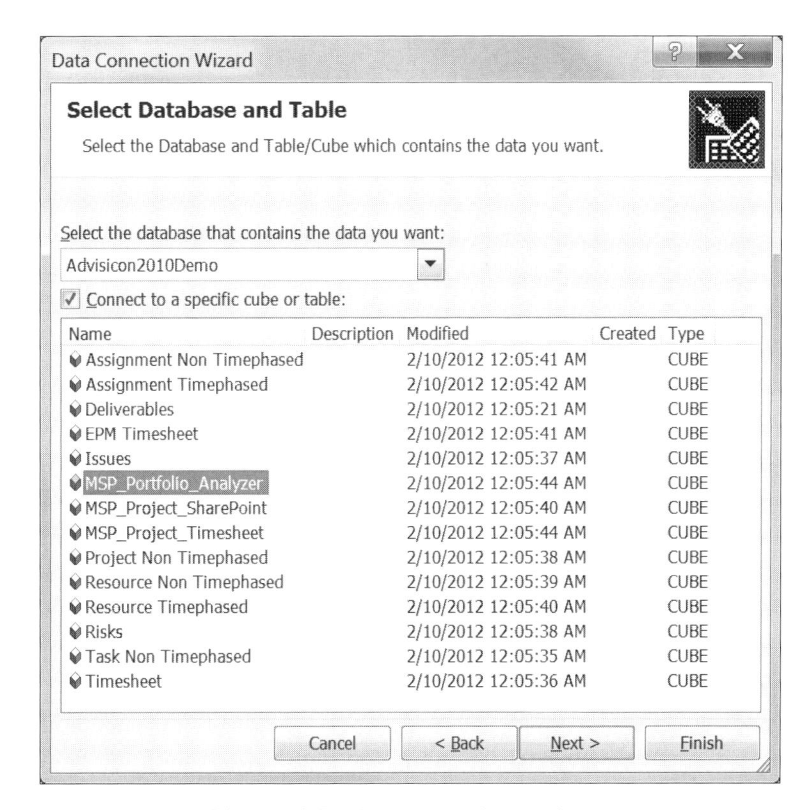

Figure 10.7 OLAP Data Connections

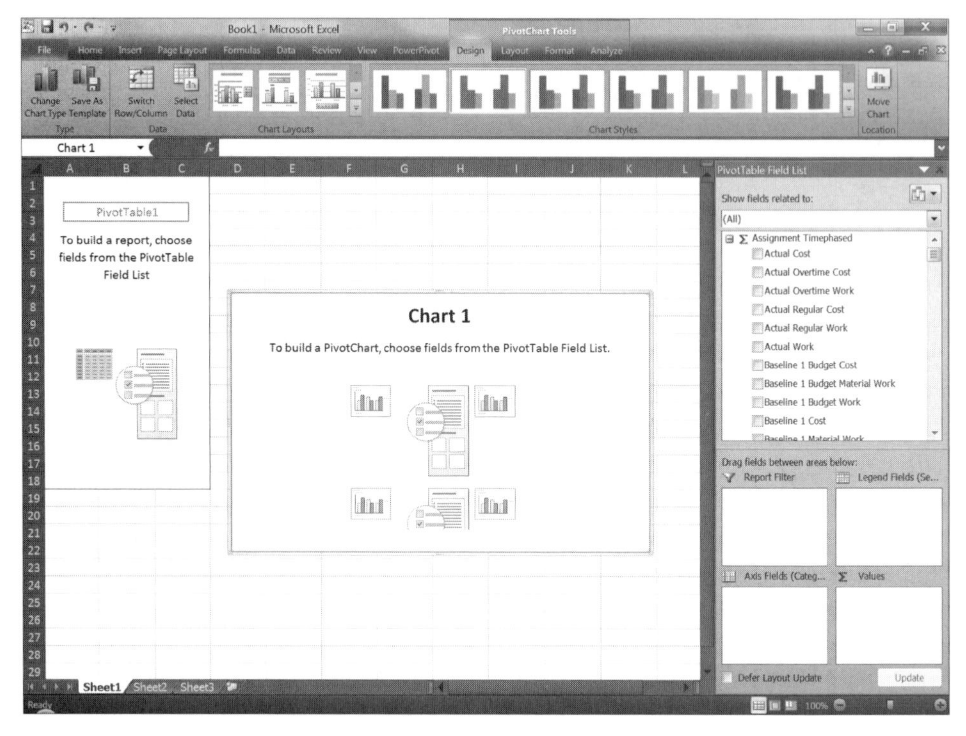

Figure 10.8 Pivot Table Setup

correspond to the sample reports, which access the Project Server Reporting database. If you opt to customize a sample report, your best bet is to create copies of both the sample report and the connection, as the samples potentially could be affected by future upgrades or service packs. Likewise a number of connections to the Analysis cubes have been predefined, with references from Template reports. These will be in a folder in the connections library that is likely named based on the server and OLAP database that contains the cubes.

6. Once you've chosen a connection and pressed Ok on the Create PivotTable with PivotChart dialog, you should get a pivot table and a chart. A PivotTable field list should be available on the right (see Figure 10.8).
7. Scroll down to and click on the check boxes next to the Capacity and Work field measures. These are the measures that we will be summarizing. By default, these will be placed in the Values box in the lower right. The pivot table (and chart) will already begin to fill in, but as we haven't selected any dimensions, all the capacity and work data are summarized into one total each.

> **Note**
>
> Including dollars and hours in the same table may not give the best results in a related chart, as the scale for one may not be appropriate for the other. Pivot tables work well for showing multiple columns of summarized data, but if you're going to do a chart, you should limit the number of values presented, both to avoid mixing apples and oranges and to keep the chart simple and to the point.

8. Scroll to the bottom of the field list and check next to the Time Dimension. By default, Excel placed this dimension in columns across the top, with subcolumns for capacity and work within each year and a grand total on the far right. Note that there are now two items in the Legend Fields box (also called series, or columns if the table is selected, or you might consider it the *x*-axis). The legends values will also provide color-coded legend references on some types of charts. The fields are Time (the dimension placeholder) and Values (the summarized values) (see Figures 10.9. and 10.10).
9. Let's try some dynamic reporting!
   a. A minor tweak: In the Legend Fields box, drag the Time field underneath the Values field. See what happens to the column headers (see Figure 10.11).
   b. Drag the Values field from the Legend Fields box into the Axis Fields box. This takes all those columns that used to contain Capacity and Work and turns them into rows. However, if you actually look at the summary values in the cell for say, 2010 Capacity, via both views, you'll find they're the same (see Figure 10.12).
   c. Undo your changes so Time and Values are in Legend Fields, then drag Time into the Axis fields. Now we see rows for each year and columns for Capacity and Work. If we have a chart, we'd notice the chart has also pivoted (see Figure 10.13).

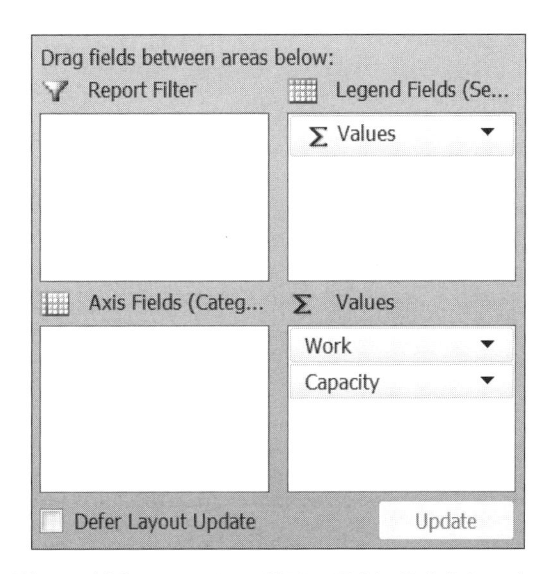

Figure 10.9 Drag Area if PivotTable Cell Selected

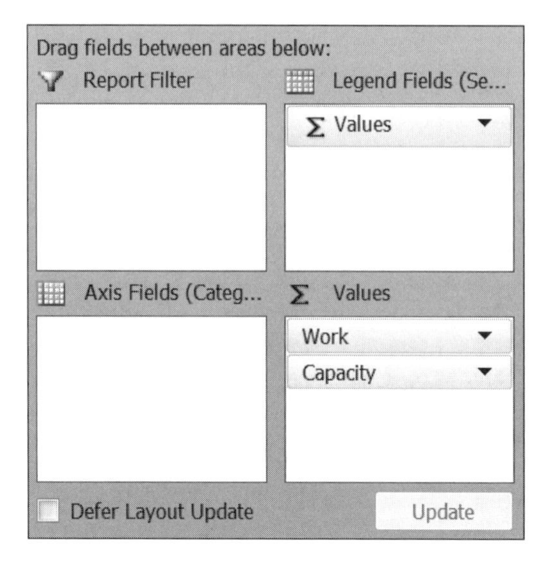

Figure 10.10 Drag Area if PivotChart Selected

d. You'll notice a + sign next to each year. This means that this dimension has a hierarchy that you can drill into (see Figure 10.14). Drill down as far as you can go. You should find breakouts for Year, Quarter, Month, Week, and Day. Drilling down also affects what is displayed on the chart. In Excel, the table (grid) and the chart are

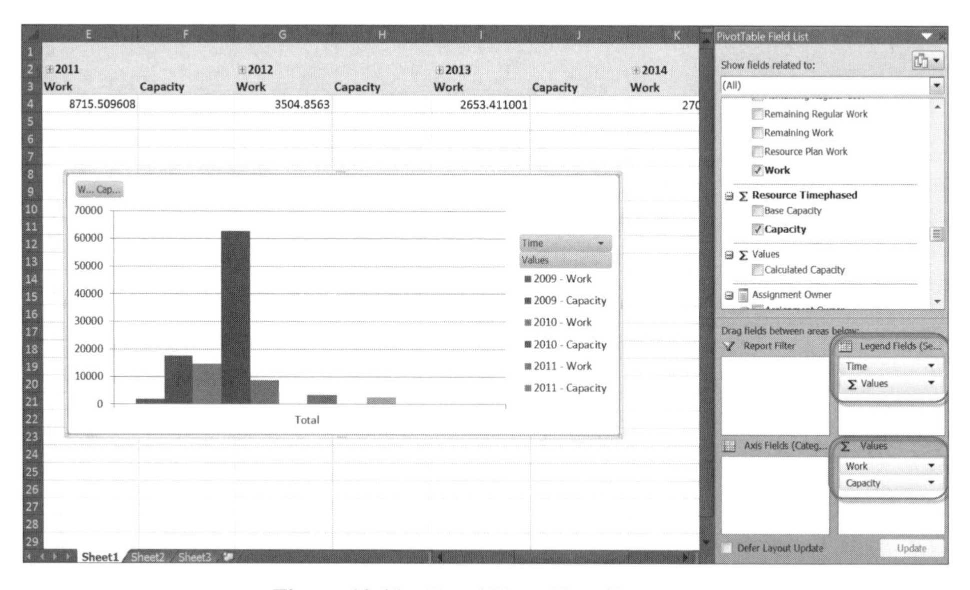

Figure 10.11 Excel Pivot Chart/Report

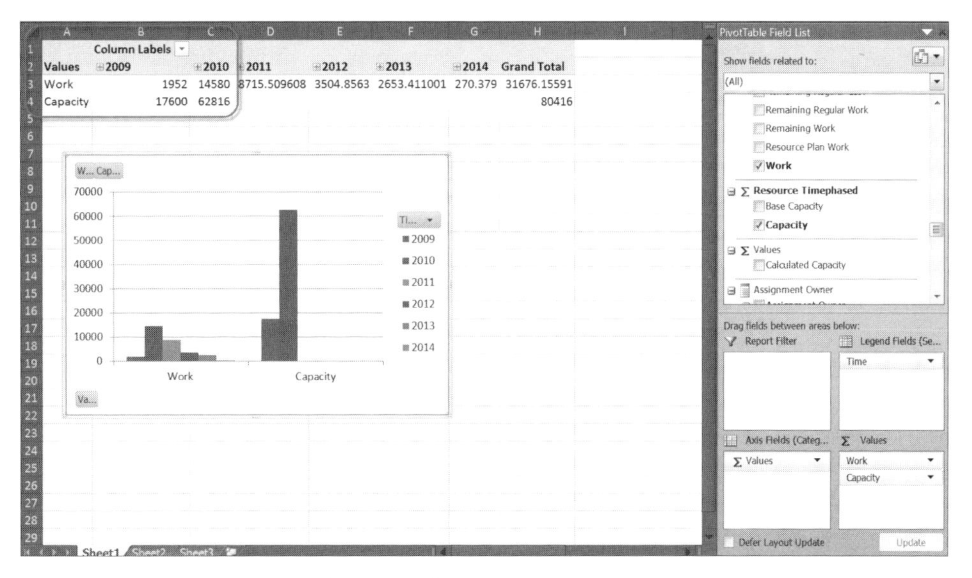

Figure 10.12 Excel Pivot Fields

tied together, and you control the chart drill down via the table. In other BI tools, such as a Performance Point Chart, the chart can stand alone and includes drilldown functionality. Click on the – sign at the top of the hierarchy you've expanded to collapse it again.

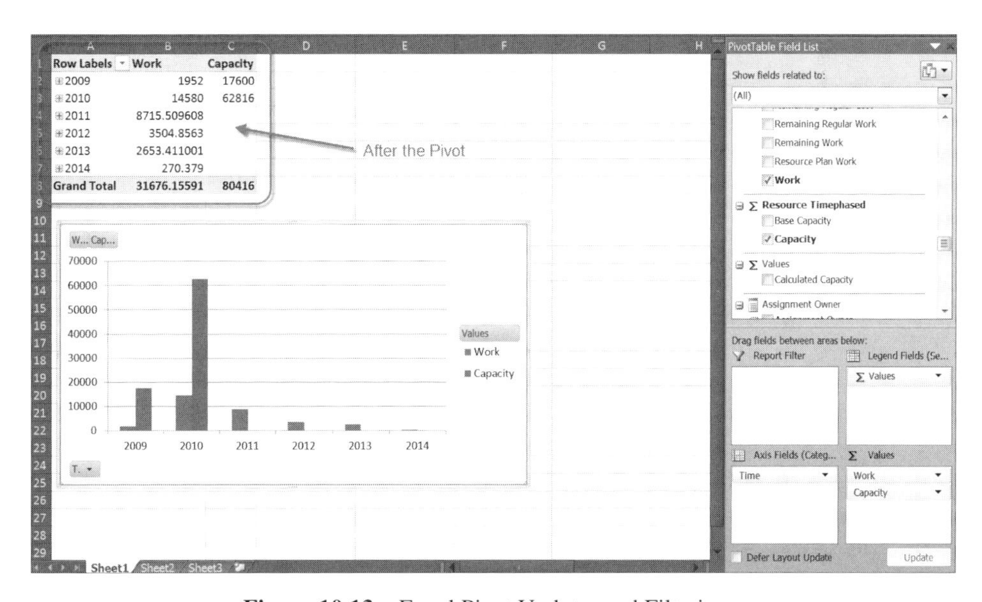

Figure 10.13 Excel Pivot Updates and Filtering

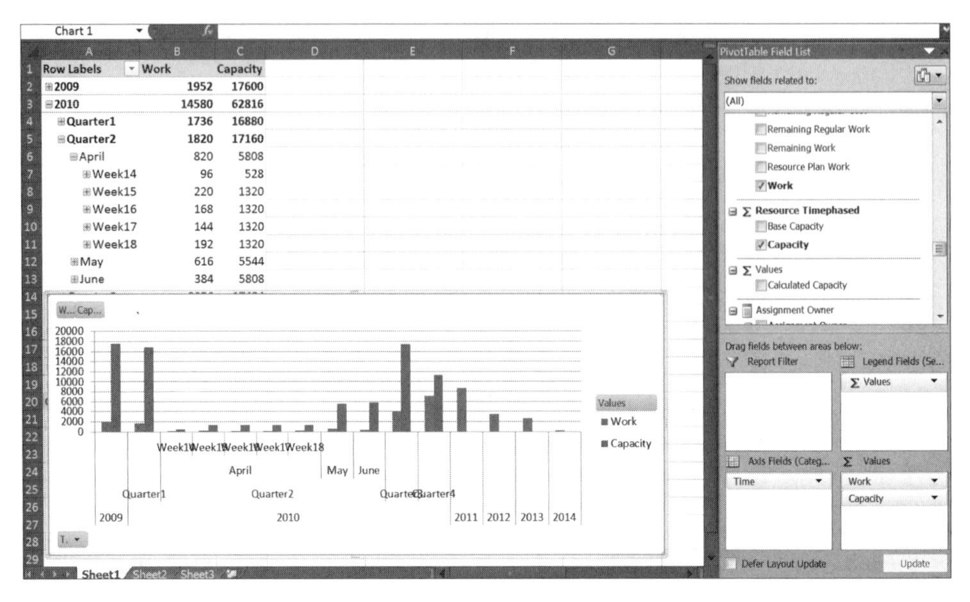

Figure 10.14 Pivot Drill-Down Row Fields

e. Add two more dimensions: Resource Is Generic and Resource List. These should land in the Axis Fields (rows or categories or the *y*-axis). You should now have groupings by Year, Resource is Generic flag, and Resource Name. Under the Resource is Generic if grouping = true, Resource Names should be generic assignment names; if false, Resource Names should be specific employees. Capacity and Work summarized values should go all the way down to per resource per year, and Resource is Generic flag per year. There should not be any resources in both the True and False groups.

f. Remove the Generic Flag group. You can do this by either unchecking it in the field picker or by right clicking on the item in the drag area and selecting Delete Field. Note that generic and nongenerics are now mixed together, and we no longer have Generic/Non Generic totals (see Figure 10.15).

g. With all the resources we have now, our chart is getting too busy. Let's do some filtering (slice and dice). There are a number of ways we can do this:

   i. In the field pick list, if you hover over a Dimension field, the field will highlight and a down arrow will appear on the right. Click the down arrow and a value pick list dialog will appear. Pick your desired values. If you use this option, you'll also notice filter icons are added to the Row Labels cell and the corresponding "button" in the chart.

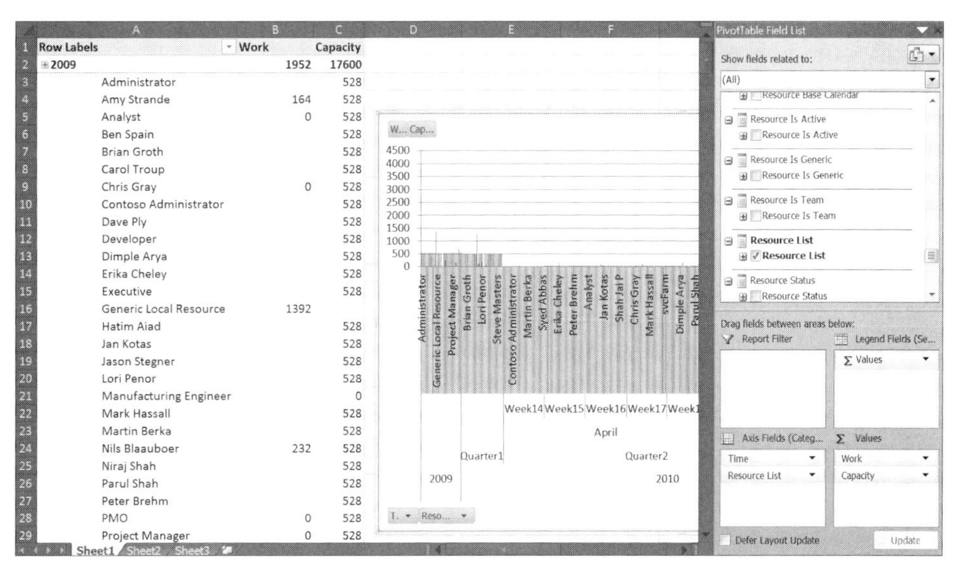

Figure 10.15 Dynamic Pivot Chart Update

ii. Use the flip side. Click on the down arrow in the Row Labels cell, get the value pick list, and pick values as above. In this case you'll also need to pick the field you're filtering from a drop-down list, as well as the values.

iii. In either of those cases, the dimension selected has to be selected for display; otherwise, it'll be ignored.

iv. Using one of the methods just described, filter to include only years 2009 and 2010.

v. Drag a Dimension from the field list into the Report Filter box in the drag area. This method is more useful to filter by a field that you are not displaying. If it is a field you're displaying, it gets converted from a grouping field to a pure filter field.) In this case, the field is added to a filter area at the top of the Pivot Table. For our example, drag Resource Is Generic into the Report Filter box, click on the down arrow in the new filter area field, and select True.

vi. Back when we were introducing Analysis Services terms, we mentioned "slicing." You might notice Excel has an "Insert Slicer" function. Is this the same thing? (See Figure 10.16)

Well, sort of. Slicing really comes down to filtering on a dimension, and yes, the "Slicer" helps you do just that. In our earlier filtering options, we mentioned a few methods that resulted in popping up

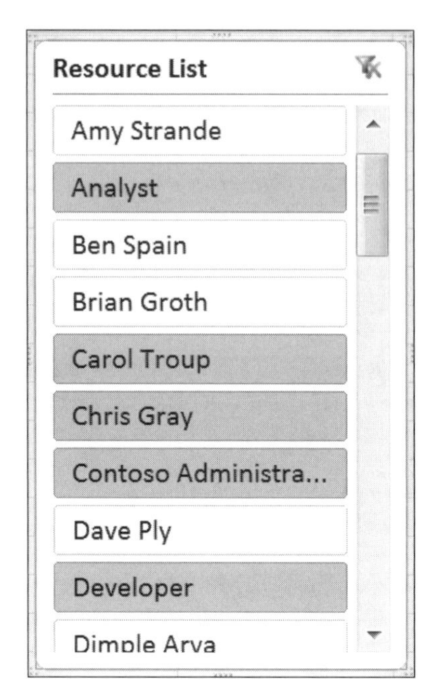

Figure 10.16 Field Slicer Dimension Fields

a value pick list for a particular dimension. The problem with that is, once the filter is applied, while an icon will show that a filter has been applied, it's not obvious what the filter is. The slicer provides a more user-friendly interface for a dimension filter, providing a button for each value, with a different button background color showing what the selected filter values are. More significantly, the slicer dialog remains up and available during report usage. It can also indicate which dimension values have no related aggregate values and shift those values to the end of the slicer.

The Insert Slicer button can be found in one of two places, depending on whether you've selected a cell in the PivotTable or have selected the PivotChart (see Figure 10.17) For the PivotTable, look in the PivotTable Tools group, Options tab, Sort & Filter subgroup. For the PivotChart, look in the PivotChart Tools group, Analyze tab, Data subgroup.

Go ahead and insert a slicer and, when prompted for a dimension, select Resource List. The slicer will show the available resource items, with the nongeneric items grayed out as we're already filtering on generic resources. The other resources should be selected. Deselect all but three by using Ctrl mouse click (see Figure 10.18).

10. Excel provides many different types of charts. To switch chart types, right click in a chart area and choose Change Chart Type, or go to the

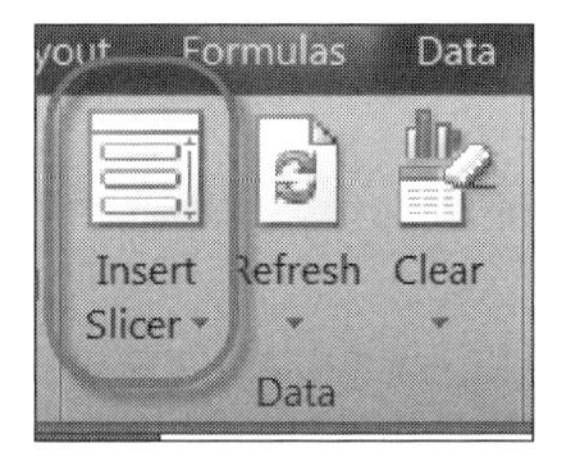

Figure 10.17 Insert Slicer Button

Design tab in the Pivot Chart Tools group in the ribbon (you'll need to have selected in the Pivot Chart area for that tab group to be visible). The PivotChart Ribbon group has Design, Layout, Format, and Analyze tabs, and each tab has various subgroups and options for choosing and configuring charts. Look around and experiment. You may wish to save your work before you start, to remember where you started from. See Excel Help, Available Chart Types for a rundown on the different chart types, when you may want to use a chart type, and what the limitations might be.

11. After you're done looking around, try a pie chart. One limitation of a pie chart is it only shows the first item in a series, so you'll only see the numbers for work or capacity, but not both, in the chart. You can control

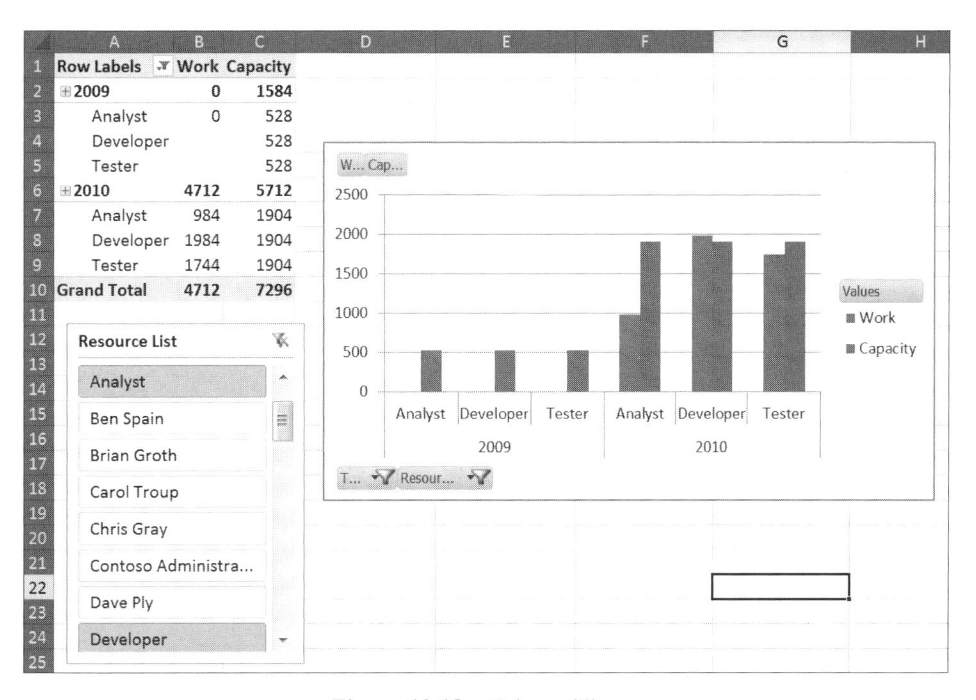

Figure 10.18 Using a Slicer

which of the values you see by moving their order in the Values section of the Drag area; the topmost value will be the one displayed in the pie chart. Once the pie chart is shared in SharePoint, your selection is locked in.

In our example case, we opted for a layout that shows the summed amount per grouping, no legend, and used the Layout tab and Data Labels option to show the labels on the outside end of the data points.

If you publish your Excel file to SharePoint for viewing via a browser, the position of charts and slicers in the browser becomes fixed. As your pivot table might expand due to removing filter criteria or expanding groups, you should make allowances when you place your slicers and charts. The illustrations show a good example of where not to place a slicer; it's there now only to fit into the illustrations.

12. Next, lets format the numbers in our table and chart.
    a. In the Drag area, click on the drop-down arrow for Capacity, and select Value Field Settings (see Figure 10.19).
    b. Click on the Number Format button (see Figure 10.20) to bring up the Format Cells dialog, and select currency, zero decimal positions. Press OK on both dialogs to set the format (see Figure 10.21).
    c. Likewise, select Work, Number Format, Number, and zero decimal positions. Add a comma to the work field.
    d. You can also use the standard Excel formatting controls available on the Home tab. Just remember to select all of the appropriate cells in the table before doing your formatting (see Figure 10.22).
13. When you're happy with your results, save them. Ultimately you will need to save them to SharePoint if you want to share your report, but you can save locally and upload if you prefer. Saving directly to SharePoint via Excel enables you to control what parts of your workbook are visible to those with whom you are sharing. See the section

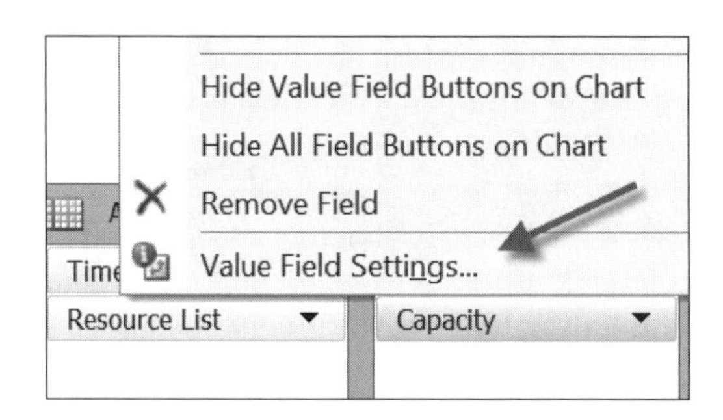

Figure 10.19 Value Field Settings Button

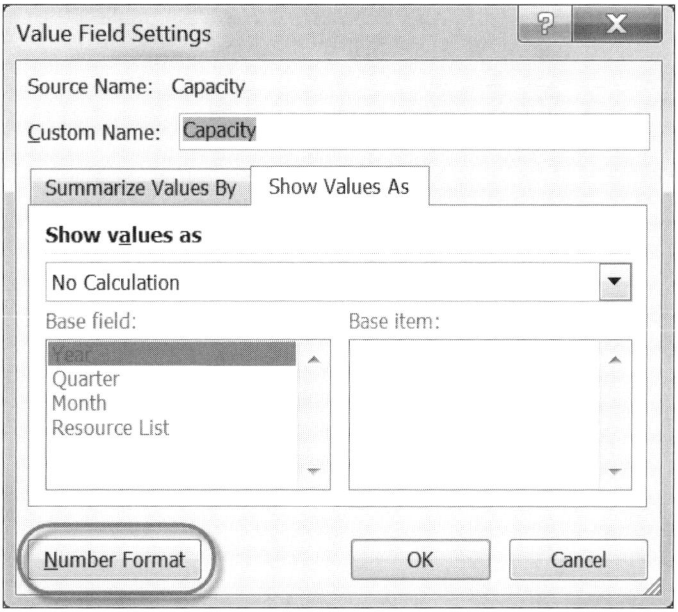

Value Field Settings

Source Name: Capacity

Custom Name: Capacity

Summarize Values By | Show Values As

Show values as

No Calculation

Base field: | Base item:

Year
Quarter
Month
Resource List

Number Format | OK | Cancel

Figure 10.20 Value Field Settings

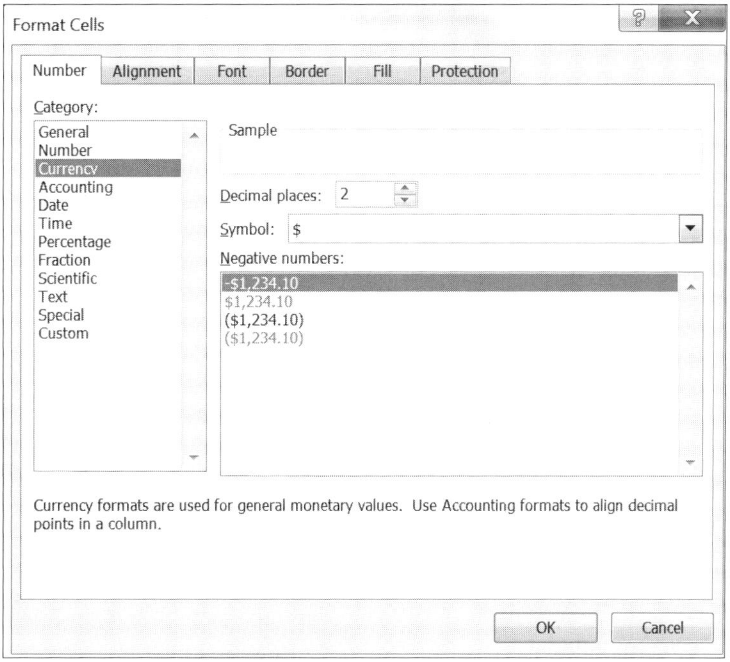

Format Cells

Number | Alignment | Font | Border | Fill | Protection

Category:

General
Number
Currency
Accounting
Date
Time
Percentage
Fraction
Scientific
Text
Special
Custom

Sample

Decimal places: 2

Symbol: $

Negative numbers:

-$1,234.10
$1,234.10
($1,234.10)
($1,234.10)

Currency formats are used for general monetary values. Use Accounting formats to align decimal points in a column.

OK | Cancel

Figure 10.21 Format Cells

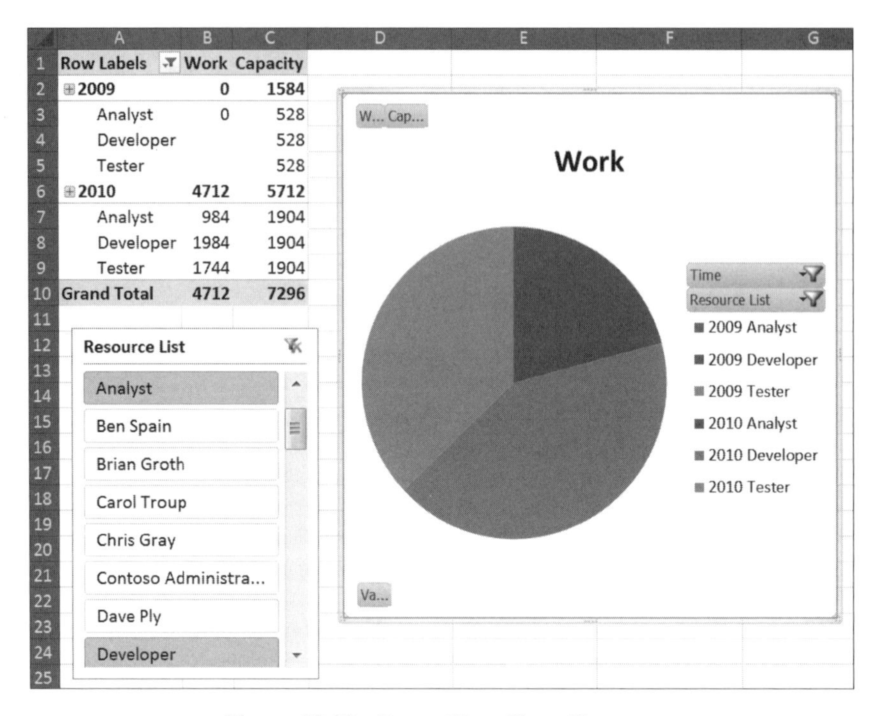

Figure 10.22 Power Pivot Chart Change

titled "How to Publish Only Parts of an Excel Report and Specify Parameters" for more details.

Office Web Apps

Depending on your environment, if a user navigates to a SharePoint library and opens an Excel file stored there, SharePoint may automatically open up a full page using the Excel Web App Web part. This may cause you a point of confusion, as Excel Web App is a Web part that is a separate but more fully functional version of Excel Web Access.

Earlier we worked with the Excel Web Access Web part, which is essentially read only. While spreadsheets displayed via that Web part can be manipulated to interact with pivot tables and charts and even to accept parameters into certain cells that can in turn trigger formula processing, the underlying spreadsheet remains unchanged.

Excel Web App is a part of a separately licensed suite of tools called Office Web Apps. Office Web Apps are designed to provide near-Excel client-level functionality for MS Office applications run from a Web browser. The Web part includes a view mode (much like Excel Web Access) and an Edit mode. Edit mode presents the workbooks and spreadsheets in a way that looks much like the client. There are some differences, however.

Included Functionality

- Data entry
- Formulas
- Basic formatting
- Tables, including sorting and filtering
- Interaction with existing pivot tables and pivot charts
- Perhaps the most interesting functionality: Multi-user coauthoring (This means if two users are working on the same document at the same time from different browsers, the changes that one user makes will be reflected on the other user's copy, in near real time. For those who think that this is not a good thing, that behavior can be prevented by using document control [i.e., checking the document out before editing].)

Not Supported

- Creating charts and pivot tables
- Query tables

Both Web Part versions allow you to open the document in the Client version. That option is configurable for Excel Web Access when you're defining the Web Part properties, as are a number of other options.

The overall suite of Office Web App tools includes online companions for Word, Excel, PowerPoint, and OneNote. As with Excel, each of these apps may have some differences with the client version. Usage requires an Internet connection and Internet Explorer, Firefox, or Safari browsers.

If You Like Dashboards, Wait Until You See PerformancePoint

Excel is a very useful tool for dynamic reporting and self-service BI, especially if you include the PowerPivot add-on. In combination with Excel Services and SharePoint, you can even extend its reach to the Enterprise. But at its heart, it's still a spreadsheet engine. SharePoint Enterprise also has some native Web Parts for key performance indicators (KPIs), KPI lists, and charts, but their functionality is rather limited. If you really want to do more sophisticated KPIs or build integrated dashboard pages using a range of reporting techniques, you should look into PerformancePoint Services.

With PerformancePoint, you create your dashboards using a custom Dashboard Designer tool. This is used to create a dashboard with one or more Web pages, each potentially containing one or more dashboard components. Components include KPIs, scorecards to arrange and roll up KPIs, KPI details, wrappers to contain Excel reports, as demonstrated earlier, as well as wrappers for SSRS reports or any external Web page. PerformancePoint also contains special components for charts and pivot tables customized for use with SSAS analytical cubes, which provides functionality beyond the reach of the Excel variety. These analytical charts and pivot tables also provide access to a decomposition tree that allows you to dynamically drill into the details of your cubes while displaying the paths you use to see the various facts and dimensions.

Business Users Can Run What-Ifs

Excel 2010 is chock full of functionality, much of which you may not be aware of. One of the lessor-known functions is what-if analysis available from the Data tab.

Doing what-if analysis does require a little work up front. You'll need to set the stage by creating a worksheet that contains a base scenario, where you can plug a value into one or more of your cells. Based on formulas or values in other cells, a target cell contains the results.

The What-If Analysis button provides three levels of functionality:

1. **Scenario Manager.** Given a worksheet with a parameter cell, a results cell, and a set of values for the parameter cell, the Scenario Manager shows a summary of comparative results.
2. **Goal Seek.** Given a worksheet with a results cell and a targeted results value, modify the parameter value to correspond to the target value.
3. **Data Table.** This is similar to the Scenario Manager, but instead of entering each input as a scenario name and generating a results report, it specifies a data range that contains a column of input values and a companion column to hold the results. A variation on this allows two inputs, where the data range has both columns and a top row to specify the two inputs, with the result values being placed in the remaining matrix.

This analysis is great for those who are doing spreadsheets with formulas and would like a way to investigate scenarios based on those formulas. But what does that have to do with Project Server? With Excel 2010, there's also an extension for what-if analysis that extends to pivot tables. Using this functionality, you can enter changes into the pivot table cells and have the changes reflected throughout the table. If write back has been enabled on the server, you can even publish your changes.

You'll need to enable this functionality before you can use it. Start by clicking in the pivot table area, then the Options tab under PivotTables Tools. Next click the What-If Analysis button in the Tools group, and then click the Enable What-If Analysis button.

Collaborative Business Intelligence

With the growth of social media and the continuing acceptance and use of collaboration portals, the natural outcome is to make BI data, reports, views, and information available in these environments.

SharePoint is the principal tool for collaboration and quick access to information, reports, and views, while social networks allow end users, stakeholders, and other interested parties to collaborate, review, and post information regarding projects and project information.

We have heard over and over again that the audiences and participants (both recipients and those actively involved in the projects) want to "tell the story" in a manner that is quick, easy, and Web accessible.

SharePoint Collaboration

The title "SharePoint Collaboration" is almost redundant; sharing and collaboration on business documents and processes is what SharePoint does best. From a BI perspective, we're using SharePoint and its partner services in these ways:

- To create Excel reports and share them via Excel Services
- For browser-based access to dashboards
- To integrate with PerformancePoint services for the creation and execution of dashboard components
- For storage and security for data connections and for reports generated via BI processes
- To create native SharePoint KPIs, KPI lists, and SharePoint charts
- For document control and alerts (i.e., native SharePoint functionality that can be applied to BI documents)

Social Networks

At first glance, you might think BI and social networks are odd bedfellows; after all, why would you want to share the very information that might give you a competitive advantage? But if you revisit the idea in the context of social networking techniques within a company, or even between business partners, the advantages begin to surface. In fairness, the term "social network" is probably too broad; "social sites" or "social applications" might be a more accurate description, as the scope is limited. In any case, the key advantage comes down to collaboration.

Once the BI reporting has been created, sharing it within your business environment and encouraging feedback and dynamic interaction can help you better respond to and tune that information. BI reporting is sometimes thought to be aimed at the executive level, but if you expand its reach to those better aligned to drill into and understand the details behind potential problems, they may be able to deal proactively with problems before they even get to the executives.

While formal social networking tools such as wikis, in-house blogs, and Facebook accounts can be used as avenues to communicate ideas or the existence of useful information, passive "following" can also be useful. This can be as simple as acquiring access to alerts for a SharePoint library of BI reports, even those that seem only peripheral to one's area of expertise. These alerts can provide related information and ideas on how to approach certain problems, or even become new data sources to augment BI reporting that you're already working up. Consider the PowerPivot model: One use is acquiring information from potentially different sources and mashing them together for a broader analysis.

Facilitating the social communication that empowers collaboration may be the key contribution of social networking to BI. The most important aspect simply may be encouraging those who have useful content to publish it to a wider interested audience and to encourage the wider audience to look beyond its little silo for information. This communication should be done with some care. If someone thinks they're pulling in the latest data but they're really getting two-year-old data because the poster didn't specify its currency, the results could be detrimental. As such, some standard for "tags" or other metadata should be included with a posting so the context can be interpreted appropriately.

Extending Reporting

A core component to extending reporting is the ability to open and connect to information from other data sources. This section helps you understand the potential and shows steps for connecting to other databases. It explains how reporting can be an easily attainable activity that will allow you to extend reporting to information from almost any source.

While connecting to external data sources is easy to do, it is absolutely critical to understand the thought processes of those who are storing or creating the data repository in order to creating good reports. This is where the business needs to work with the database administrator to ensure that information is compiled, collected, and stored in a manner that supports the continued and extended reporting capabilities of Excel, Excel Services, Performance Point, and PowerPivot.

How to Create an.odc Connection File

1. Open Excel 2010.
2. Within the Data tab, click the Get External Data button. An initial pop-up menu will appear, showing From Access, From Web, From Other Sources, and Existing Connections. Click the From Other Sources button (see Figure 10.23).

   As you can see, you have many options for external data sources. Each of them has different characteristics and settings and will have

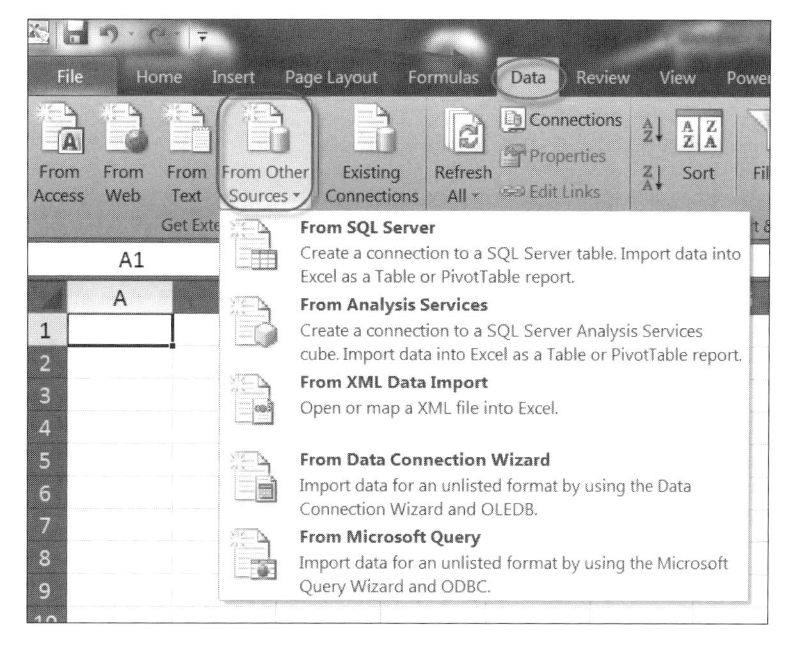

Figure 10.23 Getting External Data

corresponding dialogs to control those settings. Some may involve building queries appropriate to the data source. Exploring them all is beyond the scope of this book. We'll use From Analysis Services for our example, as that has been the source of data for the rest of our Excel reporting presentation.

3. Enter the name of the server that contains the cubes. Normally Use Windows Authentication is the preferred option, but select and enter a user name and password if that is your company's preference. Click Next (see Figure 10.24).
4. There may be several Analysis Services databases on the server. Choose the appropriate one. A list of cubes associated with the selected database should display. Choose the appropriate cube and click Next (see Figure 10.25).
5. Optionally override the file name and friendly name, and enter a description to help others understand what the connection points to and how it might be used.

   If you're always going run spreadsheets that use this connection from your computer, you can leave that. However, if you want to share those spreadsheets, you should use a data connection file that will need to be stored in a trusted Data Connection Library on SharePoint. Use the Browse . . . button to navigate to the appropriate

> **Note**
> By default, the new. odc file is stored on your computer under My Data Sources.

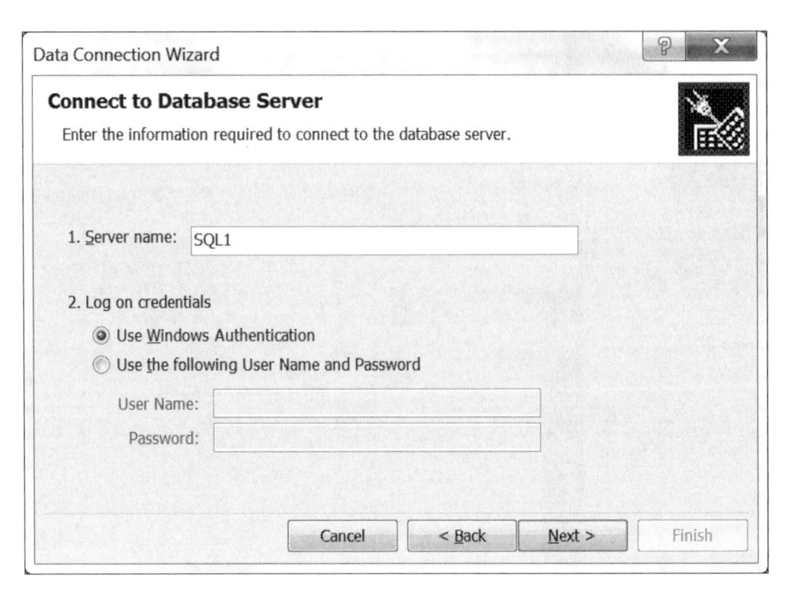

Figure 10.24 SQL Connection Wizard

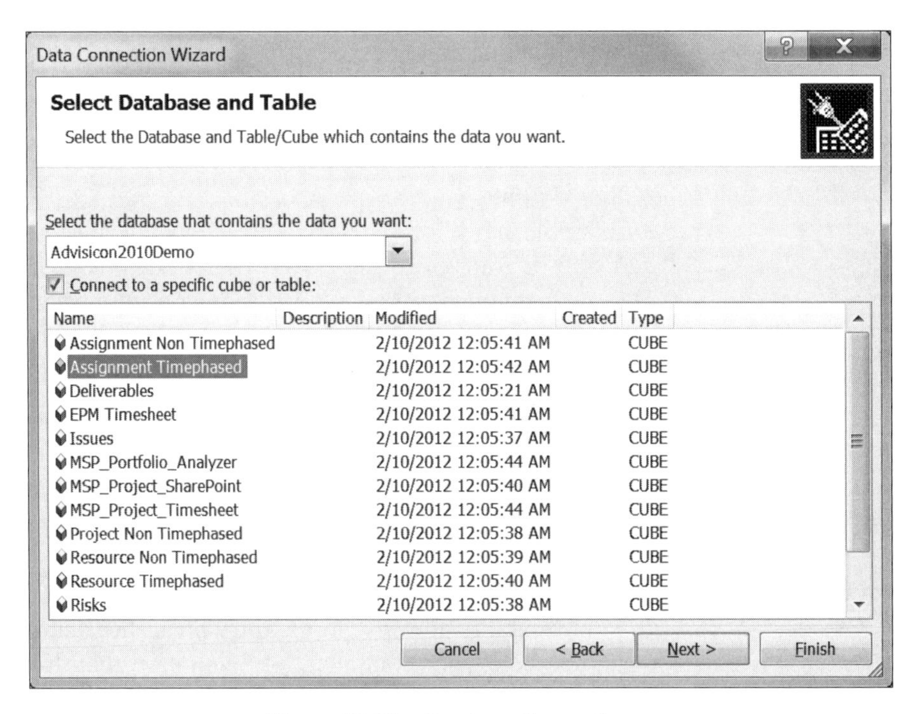

Figure 10.25 Database Connection

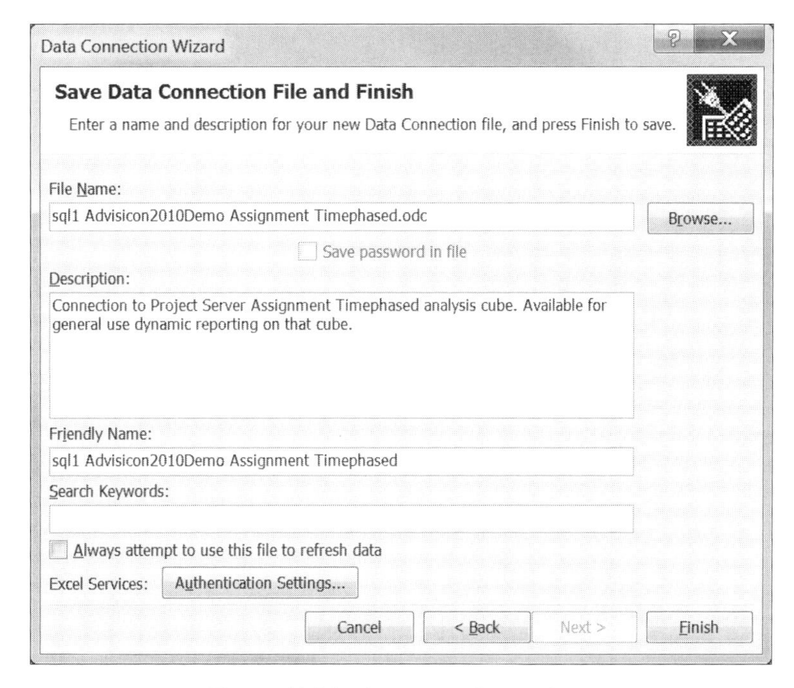

Figure 10.26 Save Data Connection

Data Connection Library (see Figure 10.26). These connection files can be reused.

6. Click Finish.

Note that in this case we've named an entire cube. If you had selected SQL Server as a data source, it would have led you down a path to select a table. Frequently, queries to SQL involve multiple tables and should include some filtering on rows and columns to cut down on overhead. Should this be the kind of functionality you need, you might want to look into the Microsoft Query Wizard (available via From Other Sources). It's not the friendliest wizard, so if that doesn't work out, you might consider consulting with your local SQL super user or IT applications developer to help you tune up the query or create a view that you can select directly.

If you want a native SQL query rather than a view, you can still have the SQL embedded in the connection file. You'll need to have figured out the quirks of the Microsoft Query Wizard or used some other method to come up with a workable SQL query. Once you have it, getting it into the.odc file takes a bit of a kludge. You actually have to start with an existing.odc file (preferably one that's already pointing at the database server using the preferred credentials). You can start down that path by creating the.odc using the method we just described, only

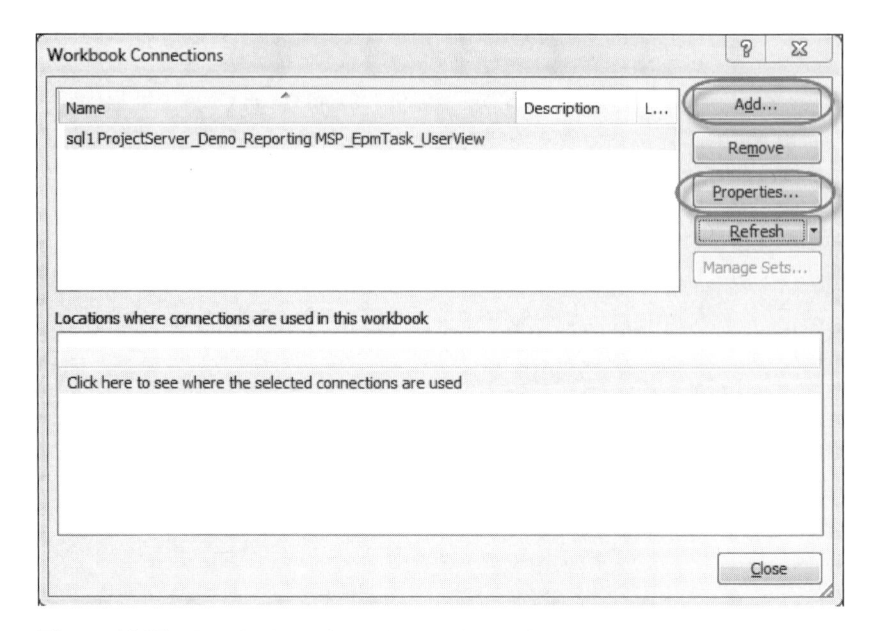

Figure 10.27 Use the Workbook Connections Dialog to Add, Delete, Modify, or Refresh Connections

using a relational database rather than a cube and picking any table or view to get things started. An.odc created and stored locally is fine at this point.

Once that connection exists, you can access it via the Connections button on the Data tab. This will open the Workbook Connections dialog, where you should click on the Add . . . button. This will bring up an Existing Connections dialog, where you should select and open the appropriate connection. After it's opened, you'll return to the Connections dialog with your connection file listed (see Figure 10.27).

From here, press the Properties . . . button, which will bring up a Connection Properties dialog. This dialog has two tabs: Usage and Definition.

The Usage tab includes some check boxes of interest: Enable background refresh and Refresh data when opening the file. These are of particular interest in the context of Excel Services, as they help ensure that workbook data are refreshed or can be refreshed when being accessed from SharePoint.

The Definition tab contains the path to the connection file, an Always use connection file check box, Connection string, Command type, Command Text, Authentication Settings, and Export button (see Figure 10.28).

In this context, we're interested in the command type and text. You'll need to click the drop-down for command type and change "Table" to SQL. In the command text, insert (paste) the SQL that you've worked up. Note that if you already have a SQL query, the command type will be locked in and the Edit Query

Figure 10.28 Use the Connection Properties Dialog to View or Modify Various Aspects of a Connection

button will become enabled. That button will take you back to our old friend, the Microsoft Query Wizard.

The last item of interest is the Export Connection File button. This is what we'll use to export the connection file from our local machine to the SharePoint document library that is being used to house the data connections. This needs to be a library designated as "trusted" by Excel Services, and naturally you will need to have SharePoint permissions to save to that library. The button opens a Save File dialog, where you can enter the URL for the site that contains your library. From there the dialog will assist you in navigating to the Data Connection Library, and you can save the connection file for general use.

How to Display an Excel Report from a SharePoint Page

Once you've created your Pivoting masterpiece, it's only fair that you share it with the rest of the world (or at least the rest of your work group). You have four options:

1. E-mail it to all your friends.
2. Drop it into a shared network folder.
3. Save it to a trusted document library in a SharePoint collaboration site.
4. Include it in a Web Part on a page where it adds value and context to other items on the page.

All of these options may be valid, depending on the circumstances. But since we're talking about BI, it's really the last option that provides the most value added. Options 2 and 3 are somewhat similar, with the exception that, with option 3, you also can take advantage of document control (check out, check in, version history), Excel Services, and more. Option 3 also sets up your report so it can be referenced in other SharePoint pages, which brings us back to option 4.

Displaying an Excel report on a SharePoint page is really quite simple. Since we are using SharePoint, the obvious solution is to use a Web Part designed for showing Excel, and those clever folks at Microsoft have built just such a thing.

There are three variations for viewing your Excel report from SharePoint:

1. Navigate to the document library that contains the Excel report and open it directly.
2. Include the report in a page or section of a page within a PerformancePoint dashboard.
3. Create a Web Part page in a site or workspace, and set up the Excel Web Access part to display your Excel report. We will walk through this option.

Steps for Displaying Excel in SharePoint

1. If your spreadsheet contains a Data Connection (as our How To example does), you need to ensure that the Data Connection exists in a Data Connection Library on SharePoint and that your file is pointing at that connection. That Connection Library also has to be designated as "trusted" by Excel Services, which calls for intervention by a SharePoint administrator. In general, your organization should have a set of data connections predefined in a Data Connection Library for targets of interest, with the appropriate security context being used for the connections. Since the Web Part routes its requests through Excel Services, security has to be considered in that context.

2. Your Excel file also needs to have been saved to a SharePoint document library that has been configured as "trusted" by Excel Services. Contact your IT department if you do not know which libraries have been configured in this way. Ideally, this library should be in the same site as your Web page, although it is possible to display a file from a trusted library in a different site if you know the URL.
3. If you do not have a page, you can create one by going to the site of interest, clicking the Site Actions button in the upper left corner, and choosing the New Page button.
4. Go to the SharePoint page you like to include the report in, select Edit, and open the Insert tab in the Editing Tools group on the ribbon (see Figure 10.29).
5. Click on Web Part (see Figure 10.30). In the Categories list, select Business Data. In the Web Part list, find and select Excel Web Access (see Figure 10.31), and click the Add button.
6. Click on the link: Click here to open the tool pane (see Figure 10.32). The Excel Web Access properties should pop up.
7. In the Workbook Display section, enter the location of your Excel file. Click on the ... button (see Figure 10.33). This will bring up a Select

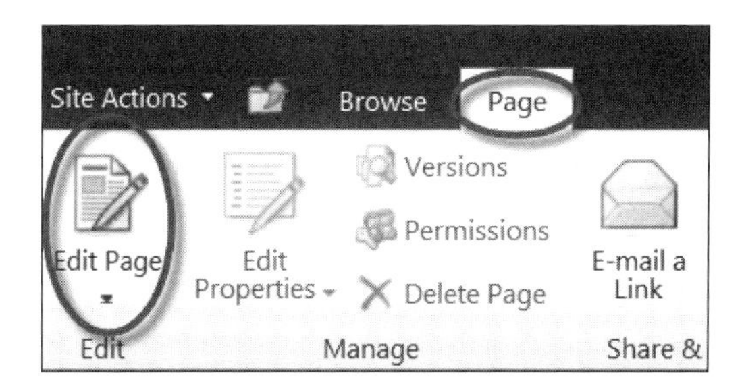

Figure 10.29 Editing SharePoint Page

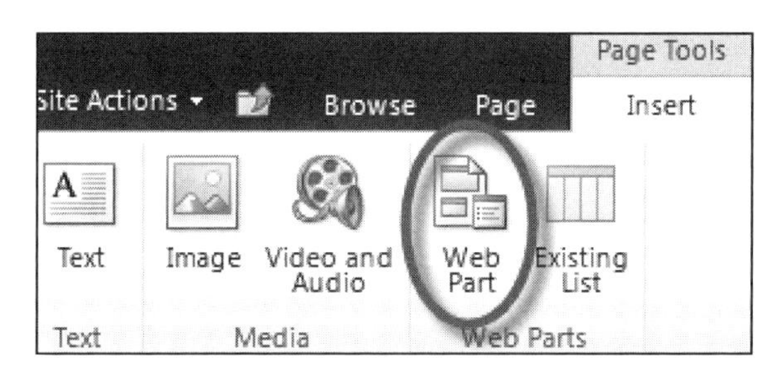

Figure 10.30 Inserting a Web Part

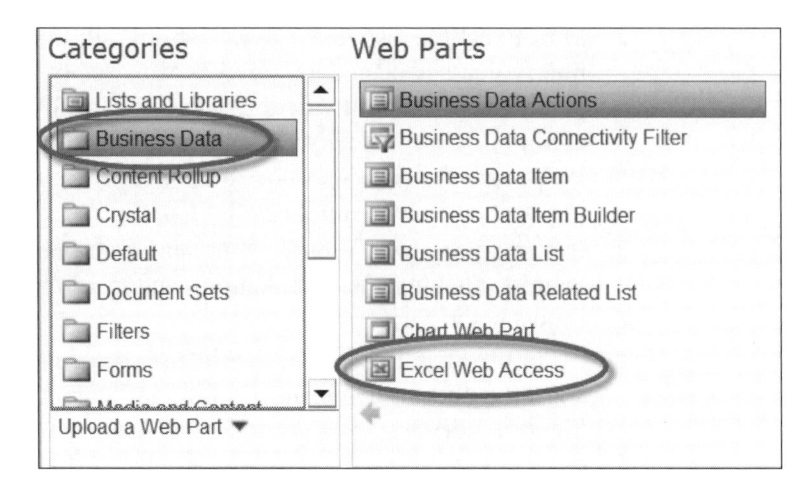

Figure 10.31 SharePoint Web Part Selection

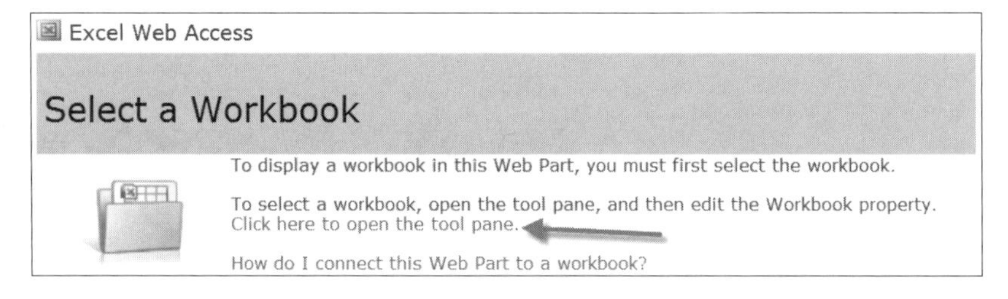

Figure 10.32 Linking to Excel Workbook

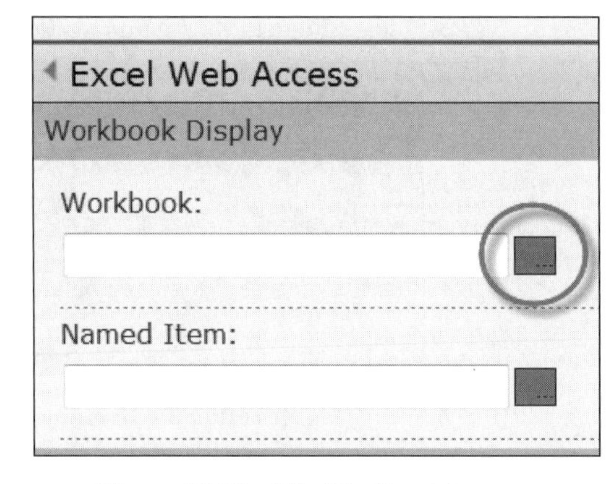

Figure 10.33 File Workbook Locator

an Asset dialog to help you navigate to the library where you've saved your spreadsheet. Navigate, select your file, and hit OK.

8. Leave the "Named Item" blank. Note that you have the option of restricting what portion of a workbook gets shown in the Web Part via this mechanism. This can be useful for security purposes. You could use this to restrict the displayed item to a pivot table, a chart, or a named range. You can also control what parts of a workbook can be displayed at the time you publish the workbook to SharePoint from Excel.
9. Review the options for toolbar and title bar. Note that if you want to modify the title, you will need to uncheck the Autogenerate Web Part Title check box here, but the actual title is set in the Appearance section.
10. Click on the + for Appearance, and change the height to 600 pixels.
11. Keep all the remaining default options for the remainder of the Web Part properties, but take a moment to look them over. Click the OK button when you're finished.Barring data connectivity issues, you should now have a working Web Part that will dynamically refresh when you bring it up and, in our sample case, allow you to filter, drill down, and the like.

How to Publish Only Parts of an Excel Report and Specify Parameters

One of the advantages of publishing an Excel workbook to SharePoint is you gain the ability to limit what parts of the workbook are visible. This mechanism can help protect sensitive information. For example, you might have a worksheet that contains raw data that are used to feed a pivot table on another worksheet. Using the publish options, you can limit what worksheets are seen or even which pivot table on a worksheet can be seen.

A second consideration is that when the Excel report is viewed via the browser, it cannot be modified directly. However, you can configure certain cells to act as parameters that viewers can enter data into. Excel formulas or filters that point to those parameter cells can trigger changes to the content of the remaining worksheet cells. These changes are nondestructive (i.e., the base workbook file is unaffected).

To take advantage of this functionality:

1. From the File tab, use the Save & Send option.
2. Choose Save to SharePoint (see Figure 10.34)
3. In the upper right corner, press the Publish Options button. This will bring up a Publish Options dialog.
4. In the Show tab, pick the level to share at (Entire Workbook, Sheets, Items in the Workbook). (See Figure 10.35)

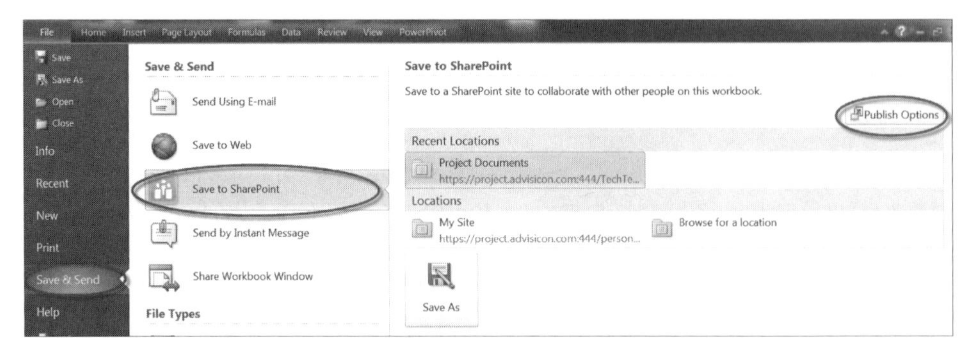

Figure 10.34 Save to SharePoint

> **Note**
> The View drop-down is a configurable item for the Web Part.

Items would include pivot tables, charts, and named ranges (see Figure 10.36). If you opt for Items in Workbook, they end up being displayed individually in the browser, where a View drop-down box allows you to look at the specific item.

5. To specify a cell as a parameter value, you must first have replaced its default cell name (such as A1) with a named range name. To add the parameter, select the Parameters tab, click the Add… button, and select the named range name (or names) of interest (see Figure 10.37). These cells can be accessed via a Parameters pane in the Excel Web Part (if that is enabled) or used in a dashboard to connect Web Parts.

Some additional considerations:

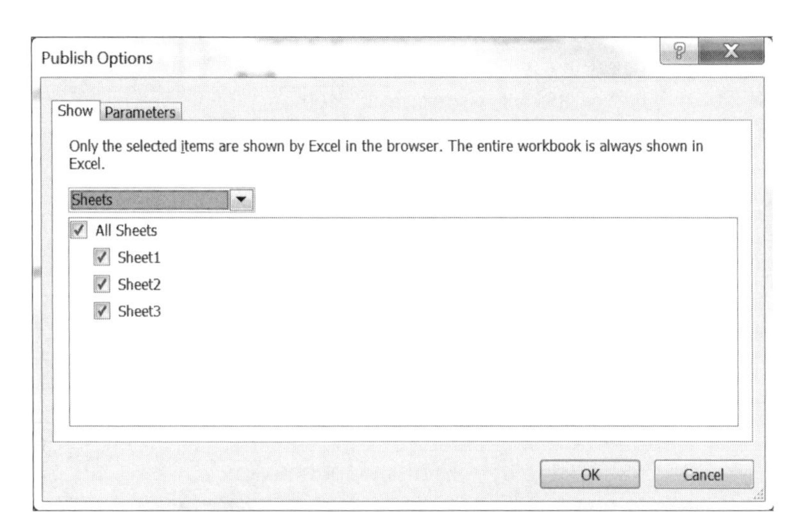

Figure 10.35 Selecting Specific Worksheets

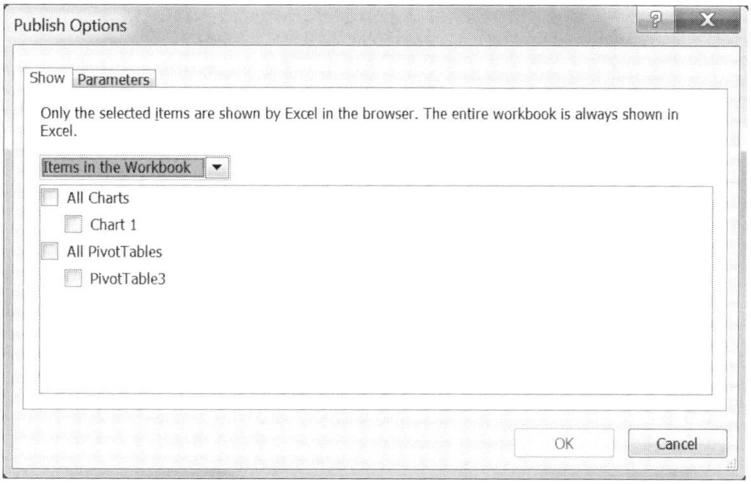

Publish Options
Show
Parameters
Only the selected items are shown by Excel in the browser. The entire workbook is always shown in Excel.
Items in the Workbook
All Charts
Chart 1
All PivotTables
PivotTable3
OK
Cancel

Figure 10.36 Publish Specific Objects

Publish Options
Show
Parameters
Parameters allow you to specify cells that are editable when viewing the workbook using Excel in the browser. Determine which parameters you want to make available by adding them and deleting them from the list below.
Add...
Delete
Delete All
Name
Add Parameters
Name
Comments
Value
Refers To
Slicer_Resource_List
{...}
Resource List (sheet: Sheet1)
OK
Cancel

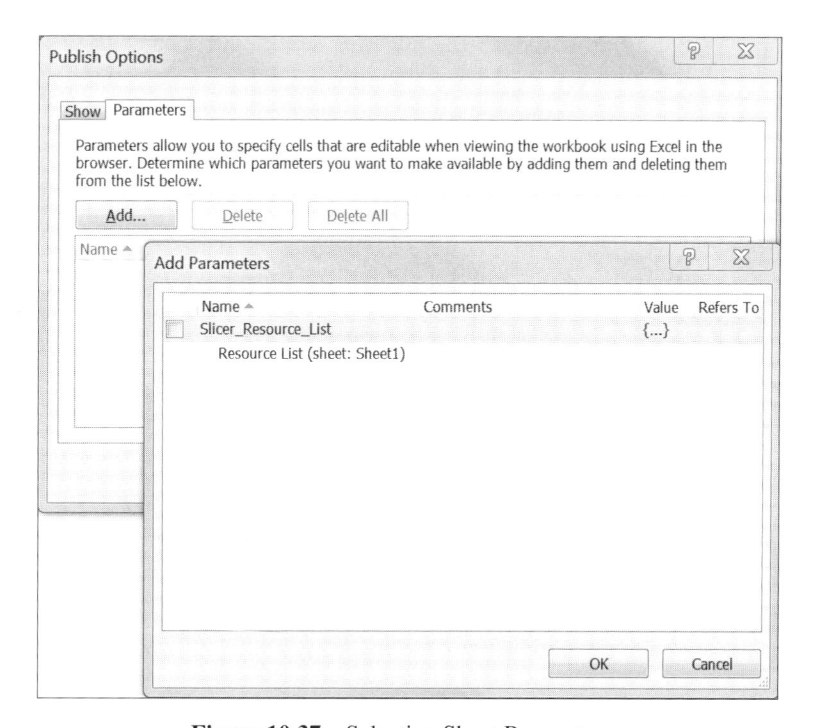

Figure 10.37 Selecting Sheet Parameters

- A named cell can be a single cell or a merged cell. There is one cell for each parameter; a parameter cannot represent a range that contains two or more cells.
- The named cell must be a cell in the workbook. The cell cannot be located in a table or in a pivot table, and you cannot use an external cell reference.
- A named cell can contain a report filter field from a pivot table report. However, the report filter field will not appear in the Parameters pane. To change the range filter field, you can use a filter Web Part.
- The named cell must not reference another named or absolute cell.
- The cell cannot have data validation defined, cannot be locked, and cannot be on a protected worksheet.

6. Press the Save As button, navigate to the SharePoint document library that is to contain your published report, and press Save.

Printing

Printing from the Excel Web Access Web Part is a bit problematic. Your best bet may be to do any filtering, drill up/down, and so on, until you get the view you like, then use the File button to download a snapshot to your Excel client. From that point, you'll have the full power of Excel client printing rather than being limited to browser-based printing.

Mobile Applications

Office Web Apps can extend MS Office tool functionality into the mobile world. Some of the mobile functionality requires Office Mobile 2010, which is sold separately. For SharePoint, SharePoint Workspace Mobile (part of Office Mobile) can be accessed by a Windows Phone 7.

SharePoint Workspace Mobile helps ensure that the latest versions of your documents stored on a SharePoint site are also available on your phone. When you're connected to the Internet, you can:

- View content hosted on a SharePoint 2010 site.
- Open and edit Word, Excel, PowerPoint, and OneNote files hosted on a SharePoint 2010 site.
- Browse SharePoint 2010 sites, lists, and document libraries.
- Obtain secure remote access to corporate resources through a Forefront Unified Access Gateway (UAG), if your company uses one.

You can also use SharePoint Workspace Mobile to take your SharePoint 2010 files offline and put them on your phone. You can then open and edit the documents and save them back to the SharePoint site when you're back online.

IMPORTANT CONCEPTS COVERED IN THIS CHAPTER

In this chapter, we reviewed the reporting capabilities that come with a PPM 2010 solution and got a taste of the power of BI, including a walk-through and exploration of these features:

- PowerPivot and using the slicer for reporting
- Extending reporting to other data sources, and mapping and modeling in Excel
- Displaying reports in different formats in SharePoint to extend user accessibility
- Localizing what you present (limiting views and parts of Excel reports in SharePoint)
- Overall understanding of the BI capabilities that come with Project Server and SharePoint together

Index

A

Active updating, usage, 219–220
Activities, network (tasks connection), 145
Activity Network, Microsoft Project Gantt Chart (usage), 139
Actuals
  estimates, contrast, 203
  fluency, 93–95
Adhocracies, 41
Ad hoc task captures, 191
Advanced workflow, 209–213
Agile planning, 156–160
  EPM 2010, usage, 191
Agile project planning, Project Desktop (usage), 157–159
Analysis functionality, 241
Analysis primary cost constraint, 298–299
Analytical competency, construction, 4
Analyze cost/resources views, 311
Application lifecycle management (ALM), 55, 56, 170
Application service provider (ASP), usage, 6
Appreciative inquiry, knowledge, 53
Approvals, 107f, 122–124
Automated project processes, 187
Automation, importance, 18–20
Auto-scheduled tasks, 145
AX 2012, 128–130

B

Backstage, 171. *See also* Project Backstage; Project Desktop
Balanced scorecard, 314–315
Baseline cost scenarios, 294–295
Baseline resource scenario, 298
Blogs, usage, 65, 217
Boolean search, 209
Bottlenecks, control, 90–92
Bottom-up planning, 95
Budget resource, 125–126
Burndown charts, generation, 157
Business
  accountability, 92–93
  activities source, 23f
  focus, 37–38
  lifecycle, 8–13, 201
  metrics, monitoring, 241
  plan, flowchart, 253f
  planning/control, Project 2010 (usage), 251
  platform, design, 101–104
  PPM component, 186–187
  practices, leverage, 201
  priorities, alignment, 258
  process/forms, 232
  requirements, project relevance, 188
  results, translation, 206f
  strategy, EPM element, 100
  success (delivery), project completion (impact), 188
  system initiation, launch, 71
  technology, leverage, 201
  value (creation), project management processes (usage), 188
Business case
  components, 5
  creation, 196
  development, 107
  importance, 4–6
  questions, 83
Business drivers
  calculated prioritization, creation, 266
  creation, 261–262

Business drivers (*continued*)
defining, 258, 260–268
deletion, 262–263
departmental association, 308
priorities, review, 266–267
prioritization, 264–265, 288
projects. comparison, 295
technology, 261
understanding/building, 259–268
usage, value, 259–260
Business influencers, profiles, 26
Business intelligence (BI). *See* Collaborative business intelligence
analysis, 260
capabilities, 208
Center, 325f
sample reports, 326f
service, 324–326
components, 113
decision support, 240
definition, 321
distribution/control, 242–243
effectiveness, 242
examples, 3
functionalities, 241
graphical techniques, 242
improvement, 2
information technology (IT), impact, 2–3
people, impact, 3–4
processes, impact, 2–3
project dashboard, 115f
project summary, 115f
reporting, 112–115
Business Intelligence Development Studio (BIDS), 328
Business users
benefits, 104
business objectives, connection, 32
customized pages, 218
empowerment, tabs (usage), 177–178
logical architecture, usage, 100
needs, 40–41
search capabilities, usage, 208–209
network, usage, 29–32
what-ifs, usage, 3

C

Calculated prioritization, creation, 266
Calendar
exceptions, impact, 151f
No Calendar Exceptions, 151f
project calendars, 150
Capacity
forecasting/management, 188
optimization, portfolio resource optimization, 5
Capacity planning, PPM lifecycle component, 120
Capital, expenses, approval, 41
Cash flow investment curve, 314f
Change management, impact, 49f
Chief financial officer (CFO)
actuals, fluency, 93–95
bottlenecks, control, 90–92
bottom-up planning, 95
business accountability, 92–93
constraints, management, 90–92
control options, 95
enterprise resource planning (ERP), integration, 96–97
financial data, integration, 96–97
integration, 83–84
long-term solution, 95–96
management effectiveness, 80
needs, 79
PMO attention, 79–87
proposals, validation, 92–93
responsibility, 80
structuring strategies, delivery, 93
top-down planning, 95
Christensen, Clayton, 29
Code mask, 275
Collaboration
infrastructure, 317–318
integration, 203–208
leverage, 200
SharePoint 2010, usage, 214
Collaborative business intelligence, 347–348
Collaborative project management, 187
Collaborative workspace, 205–206
Collins, Jim, 93
Commit portfolio selection dialog box, example, 311f
Committed planned end/start dates, 310

Committed portfolio selection decisions, 310–311
  dates, 311
Communication skills, 227
Component-based development, usage, 246–247
Connected teams (creation), alerts/notifications (usage), 220–221
Connection Properties Dialog, usage, 353f
Consistency ratio
  example, 268f
  significance, 267–268
Constraint
  analysis, performing, 258
  deletion process, 146f
  Project Finish-No-Earlier-Than Constraint, 146f
Consumer (stakeholder), 50
  business solutions, usage, 51–52
Content, tagging/rating, 219–220
Continuous integration (CI), steps, 247
Cooper, Robert, 194
Corporate candidates, profiles, 27
Corporate economic environments, pull/driving force, 42
Corporate-wide PPM system, presence, 252
Cost/benefit analysis, 314
Cost constraints, 297
  analysis, cost constraint view, 297f
Cost limits, modification, 299
Cost of ownership, 313
Cost scenarios. *See* Baseline cost scenarios
  analysis, 293
Critical success factors (CSFs), 133, 142–143
  columns, change, 180f
  Project Desktop client, usage, 159–160
  usage, 170–171
Critical Tools PERT Chart Expert Add-in, 147f
Critical Tools WBS Chart Pro Add-in, 147f
Cultural adoption, enabling, 168–171
Cultural changes, 60–67
Customer
  satisfaction, 241
  status reviews, 229
Customer relationship management (CRM), 56
Custom field. *See* Enterprise custom field
  usage, 307–309
Customized Ribbon menu, 176f

D

Dashboard Designer, usage, 328–329
Dashboards
  access, components, 19
  creation, 114
  example, 19f
  usage, 18–20
Data
  assumptions, 43–44
  elements, 2
  table, functionality level, 346
Database connection, 350f
Decision support, 240
Decision threats, 39, 60–67
Deficit reports, viewing, 303
Deliverable-based structure, utilization, 146
Deliverables
  addition, 239f
  management, 239–240
Demand
  capture, 258
  derivation, 13
  hierarchy, 14f
Demand management (DM), 230
  importance, 15–16
  mastery, Project Server (usage), 268–286
  planning, 269–273
  PPM lifecycle component, 119
  process, 24
  Project Server 2010 concept, 232
  value, 269
  variations, 13, 15
Demand-side resource management, 5–6
Departmental project management office, 224
Development, roll-out (decoupling), 6
Dicing, 323
Discounted cash flow, 313
Discussion boards (setup), SharePoint 2010 (usage), 207
Divisional PMO, information demands, 317–318
Document management, 118
Drilldown, example, 20f
Drill up/down, 323
Drivers. *See* Business drivers
Dynamic Pivot chart update, 339f

Dynamic reporting
  attempt, 335–340
  defining, 321–324
  terminology, 323
Dynamic schedules, usage, 144–146
Dynamic worksites, 205–207

E

Earned value (EV)
  capability, 86
  impact, 64–65
Efficient frontier
  curve, 292f
  defining/usage, 292–296
  strategic alignment, relationship, 299
  understanding/using, 293
Effort, strategic objectives (synchronization), 93–97
Emerging markets, volatility, 42
Emerging stakeholders, 41–43
End-to-end PPM capabilities, 256
End users
  critical success factors (CSFs), 133
  cultural adoption, enabling, 168–171
  management, 45
Enterprise, project management (usage), 134–143
Enterprise architecture, EPM element, 100
Enterprise custom field
  behavior, determination, 277
  creation, 275
  entities, 273–277
  lookup tables, 275
  name field, example, 276f
  options, 274f
  types, 273–275
Enterprise governance, 119
Enterprise platform, usage, 116–118
Enterprise project management (EPM), 99, 326
  collaborative strategies, factors, 168–169
  elements, 100
  framework, 101f
  implementation, 171
  interoperability, 129
  platform, PPM capabilities, 102
  Project Center, usage, 166–168
  project costs summary, 128f
  solution components, 103f
  solution hosting, 168
  technologies, leverage, 187
  variance scorecard, 113
Enterprise project management office (ePMO/EPMO), 224, 316
  information demands, 317–318
Enterprise project management 2010
  drilldown ability, 114–115
  usage, 126–128
Enterprise project manager, Project Server 2010 (usage), 161–171
Enterprise project portfolio management, Project Server (usage), 140–143
Enterprise Project/Program Management (EPM) 2010, change, 186
Enterprise project scheduling, 161–162
Enterprise project types (EPTs), 103, 203, 270, 307
  addition, 272f
  components, 270
  creation, 271–272
  department, example, 91f
  modification/deletion, 272–273
  selection, 105f
  usage, 25
Enterprise reporting, 241
Enterprise resource planning (ERP)
  deployment, 48
  integration, 96–97
  interoperability, 129
  Project Server 2010, usage, 164–166
  solution, 44
  system, 69
    PPM system, impact, 244
Enterprise Resource Pool, impact, 163f
Excel. *See* Pivot; PowerPivot
  charts, provision, 340–341
  data, exportation, 291
  display steps, SharePoint (usage), 354–360
  report, 329, 354, 357–360
  spreadsheets, publishing, 329
  workbook, linking, 356f
  worksheets, selection, 358f
Excel Services, 205
Execution bottlenecks, control, 90–92

Extending reporting, 348–360
Extensible development platform, 124
External customers, EPM element, 100
External data, obtaining, 349f
External information stream, usage, 70
External stakeholders, impact, 45, 46–47

F

Facebook accounts, usage, 348
Faceted search, 209
Field slicer dimension fields, 340f
File Workbook Locator, 356f
Financial data, integration, 96–97
Financial management, examination, 81
Financials, 124–128
  budget resource, 125–126
  management, project business case, 82–83
  project cost types, 125
  rate tables, 125
Fluent project management, critical success factors, 180–181
Fluent user interface, usage, 180–181
Fluent User Interface (UI), usage, 171–181
Force-in/force-out options, configuration, 289–290
Format cells, 343f
Forms, control, 122–124
Fowler, Martin, 247
Frontstage, 171. *See also* Project Frontstage
Functional silos, 52

G

Gantt chart, 300
  view, requirements details (contrast), 301
Gartner Group, 316
Global business climate, change, 190
Global economic situations, impact, 45
Global team, communication tools, 228
Goal seek, functionality level, 346
Governance workflow, example, 119f
Governed environment, planning, 13–15
Governmental organizations, Project Server 2010 (usage), 304
Graphical techniques, 242
Growth path, identification, 40–41

H

Hired resource reports, viewing, 303
Hiring resources, 303
Human resource capacity, management, 88
Human resource information system (HRIS), 56

I

Ideation, usage, 191
Independent solution partners, 38
InfoPath, 210f, 211
Information
  acquisition, 17–18
  collection, 215–216
  content, 227
  finding/presenting, 179
  gathering, example, 215f
  impact, 16–20
  management, 227
  requirement, 277
  sources, 229
  updating/sharing/connecting, options, 179–181
Information technology (IT), win-win solution, 240
IN function, 172. *See also* Project Frontstage
In-house blogs, usage, 348
Innovation, differentiator, 259
Insert Slicer button, 341f
Integrated communities, leverage, 213–215
Intelligent business planning/reporting, Project 2010 (usage), 321
Internal customers, EPM element, 100
Internal rate of return (IRR), 313
Internal stakeholders, impact, 45, 46–47
Investments, project cost value derivation, 85–87
Islands of excellence, 41
Issue management, benefits, 236f

J

James, William, 201

K

Kaplan, Robert, 314
Key performance factors, 254
Key performance indicators (KPIs), 241, 327
  impact, 113
  usage, 188

Knowledge assets (building/management), integrated communities (leverage), 213–215
Knowledge management (KM), 18
KPI web part, example, 94f

L

Legacy systems, data assumptions, 43–44
Line of business (LOB), 44, 59
Logical architecture, 100
Lookup tables, 275
  assignation, 276
  creation, 275–277
  usage, 290
  values, entry, 275

M

Management, 240–242
  procedures, sources, 229
Mandatory projects, options, 5
Manual prioritization, creation, 264–265
Metadata, management, 45
Microsoft-certified PPM partner, utilization, 59
Microsoft Developments Network (MSDN) licenses, usage, 74
Microsoft Dynamics AX 2012
  EPM/ERP interoperability, 129
  integration, 128–130
  Project Server 2012, integration, 130f
  third-party data exchange solutions, 129–130
Microsoft Human Resources, support, 28
Microsoft Project Gantt Chart, usage, 139f
Microsoft Project Server 2020, whitepaper, 234
Milestones Chart, usage, 139f
Mobile applications, 360
MySites, 214

N

Name field, example, 276f
Net present value (NPV), 196
New calculated prioritization, 266
New manual prioritization, 264–265
Nintex Workflow, 211
No Calendar Exceptions, 151f
Norton, David, 314

O

Objects, publication, 359f
ODC connection file, creation process, 348–353
Office Fluent UI, 171
Office Mobile 2010, 360
Office web applications, 344
Online Analytical Processing (OLAP)
  cube, 322, 327
  database data, 307
  data connections, 333f
  processing, 322–323
  products, 242
Open source platform, 124
Optimization, organizational evolution, 54f
Order of magnitude estimates, 196–197
Organizational maturity, project management (relationship), 134–138
Organizations
  audiences, 169–170
  department initiatives, management, 89
  long-term intent, defining, 259
  performance, strategic objectives (mapping), 254f
  products/services, development/delivery, 259
  success, work management (importance), 87–93
Original equipment manufacturer (OEM), example, 8
OUT function, 172. *See also* Project Backstage
Ownership, cost, 313

P

Partner (stakeholder), 50
  solutions/consulting services, 51
  sourcing, advantages, 70–71
Payback Period, 313
People Connections, 214–215
Performance, strategy (relationship), 253–256
PerformancePoint
  analytics, 115
  services, 324–326
  usage, 345–346
PerformancePoint Server (PPS), 325
  page, 326f
Performance point server drilldown, 94f
Personal networks, impact, 64–65
PERT Chart EXPERT for Project, 147

Phases. *See* Workflow phases
representation, 11–12
Pivot, 323
chart/report, 336f
drill-down row fields, 338f
dynamic chart report, 339f
fields, 337f
report (creation/modification), Project Server Cube (usage), 331–344
updates/filtering, 337f
PivotChart, 154
selection, drag area, 336f
PivotDiagram, 154
Pivot Dialog, creation, 332f
PivotTable
cell, drag fields, 335f
insertion, 331f
Pivot Table setup, 334f
Planning bottlenecks, control, 90–92
Planning disciplines, understanding, 6
Points of entry (POEs), 40–44
identification, 40–41
Portfolio, 327
baseline, 304
business strategy, interface, 260f
drivers, department (example), 91f
importance, 191–192
leadership, project management (impact), 189–191
planning scenarios, 303–306
planning, constraints (usage), 296–303
project lifecycle, overview, 258f
project value, addition, 189
selection scenarios, comparison, 305f
selection views, creation, 308–309
strength, communications (usage), 189
Portfolio analysis
commitment, 310–312
cost constraints, 297
force-in-/force-out options, configuration, 289–290
knowledge topic map, 252f
prioritization types, 289
project selection view, creation, 309
properties, defining, 288–289
resource constraints, 297–303
scatter chart, usage, 295
views, creation, 287–288
Portfolio lifecycle, 256
creation/management, 256–259
example, 257f
governance/workflow, 256–258
planning, 258–259
Portfolio management, 140–141
EPM element, 100
importance, 15–16
project management, unification, 55f
Portfolio of Projects, concept, 255–256
Portfolio resource
categories, 6
optimization, capacity optimization, 5
Portfolio scenarios, 305
comparison, 306
creation, 304–306
management, 310
saving, 305–306
PowerPivot (Excel), 329–344
chart, change, 344f
connections, existence, 332f
usage, 330–331
Powershell, usage, 212
Practitioners, 38
Predictive analysis, usage, 241
Present value (PV), 313
analysis, 314
Primary stakeholders, 41–44
types, 50
Prioritization types, 289
Private sector enterprises, Project Server 2010 (usage), 305
Process mapping, relevance, 196
Process systems, 40–44
Product backlog, management, 157
Production reporting, 241
Productivity, increase, 200
Professional associations, impact, 65
Program
analysis (performing), key performance indicators (usage), 188
collaboration, business results, 230
description, 224

Program/project/task interdependencies/ relationships, 204
Project
  analysis, 187
    key performance indicators, usage, 188
  budget actual comparison, 127f
  budget resource, 127f
  business case, 82–83
    creation, 82
  business driver, relationship, 120f
  business life cycles, alignment, 8–13
  closure, PPM lifecycle component, 120
  coach model, 316
  completion, 188
  components, governing, 118–122
  control parameters, 197
  data
    collection, 187
    SharePoint, synchronization, 233–235
  defining, 104–107
  description, 224
  effectiveness, increase, 243
  environment, social qualities, 217–221
  execution, PPM lifecycle component, 120
  failure, risk mitigation, 190
  forcing, 296
  governance environment, internal/external stakeholders (impact), 46–47
  history, 229
  impact statements, understanding, 263–264
  initiatives, example, 25–26
  intelligence, transition, 61
  iteration, goal, 197
  manage stage, 112f
  network diagram, 145f
  optimal mix, selection, 258
  path, prediction, 188
  plan, 271
  portfolio management lifecycle, 312f
  priorities, review, 291, 296
  prioritization, 192, 291
    PPM lifecycle component, 120
    selection, 195f
  process, design, 83–85
  repository, 316
  requirements/changes, control, 198–199
  risks, attributes, 236–237
  saving, 235f
  scalability flowchart, 36f
  schedule, planning, 110f
  scope, changes, 197
  selection
    criteria, building, 286–291
    PPM lifecycle component, 120
  separation, 188
  site, deliverables (addition), 239f
  source, department (example), 89f
  status reviews, 229
  strategies, 24–26
  team, building, 109f
Project Backstage
  OUT function, 172–174
  Project Server, connection, 175f
Project-based work items, acceptance/ prioritization, 194
Project calendar
  example, 150
  usage, 150
Project Center, usage, 166–168
Project costs
  determination/forecasting, 138
  rate table, 126f
  summary (EPM), 128f
  tracking, 86
  types, 125
  value, derivation, 85–87
Project dependencies, 286–291
  configuration, 290–291
  creation, 291
  dropdown, 287f
  example, 287f
Project Desktop, 139–140
  Backstage, 174f
  Critical Tools PERT Chart Expert Add-in, usage, 147f
  Critical Tools WBS Chart Pro Add-in, 147f
  dynamic schedules, 144–146
  Format tab, 173f
  iterations, scope (assignation), 159f
  No Calendar Exceptions, 151f

Project Tab, 173f
Resource Graph, 148f
Ribbon, 176
Schedule Impact, calendar exceptions (impact), 151f
Schedule Updates, 168f
Task Inspector, 149f
Task Tab, 173f
Team Planner View, 149f
usage, 21, 143–150
View Tab, 173f
Project Desktop Client, usage, 138–139, 143–160
Project detail pages (PDPs), 277–280
availability, 278–279
components, 279
creation, 279–280
defining, 278
department, example, 90f
determination, 103
modification/deletion, 280
project attribute data, 257
proposal data, 122
Project/Excel Visual Report, 155f
Project Finish-No-Earlier-Than Constraint, 146f
Project Frontstage, IN functions, 172
Project initiation
Project Desktop Client, usage, 143–160
Project Server 2010, usage, 161–168
Project lifecycle
flow, 12f
management, 230–231
support, 190
Project lifecycle management (PLM), 13
synchronization, 232–233
Project management (PM), 111–112
closing phase, 138
culture, construction, 56
emerging economic systems, modeling, 7
environment, stakeholders (impact), 21–29
EPM, usage, 100
execution phase, 137
expansion, 192
fluency, 171–181
future, 185–187
impact, 189–191
initiation, Project Desktop Client (usage), 143–160
initiation phase, 135–136
integration, 191
Lifecycle, 161f
lifecycle, explanation, 7–16
methodology, technical installation (contrast), 57–58
monitoring/controlling phase, 137–138
organization, Project Server 2010 approach, 88
organizational maturity, relationship, 134–138
planning/analysis tools, 136
planning phase, 136–137
portfolio management, unification, 55f
processes, 135f, 188
Project Server 2010, usage, 161–168
relationships, challenges, 31
roles, traits, 30
shortcoming, 194
software, impact, 194
standardization, 243–245
training, coordination/development, 224
usage, 134–143, 205
workforce, support ability, 243
Project Management Book of Knowledge (PMBOK), 30
Process Areas, 134, 135f
Project management office (PMO)
application development teams, collaboration gap, 171
business component, 186
challenges, 225–227
competencies, 318
creation, 11
dashboard creation, 114
delivery expectations, 81
EPM element, 100
governance, 317Project Server optimization, 315–318
impact, 79–87
internal/external stakeholders, impact, 45, 46–47
knowledge, 39, 44–49
pages, usage, 175
potential, achievement, 223

Project management office (PMO) (*continued*)
primary/secondary stakeholders, 41–44
purpose, 224–225
socialization/connectivity, leverage, 190
strategic role, usage, 316
success, information (impact), 16–20
usage, 245
Project managers
differences, 30
empowerment, 230–249
interaction, challenges, 30–31
partnership, success, 31–32
responsibilities, 226
roles, 30
Project momentum, negativity (avoidance), 187
Project-oriented business activities, 194
Project planning, 110
PPM lifecycle component, 120
Project Portfolio, strategic linkage, 255f
Project portfolio lifecycle, 121f
Project portfolio management (PPM)
business case, 4–6
business component, 186–187
business imperative, 120–122
competencies, 39
components, maximization, 2–6
cultural practices, 187
culture, change, 60–67
decisions
interrelated data, absence, 193–201
threats, 60–67
development, stages, 38f
efficiency quadrant, 198f
environment, 60, 64
facts/opinions, 65
implementation, 6, 192
problems, 10
workflows, usage, 231–232
initiative, 193
lifecycle, 188–201
adoption, 201
components, 119–120
expansion, 191–193
movements, human input (integration), 7–8
prevalence, 38
processes, adoption, 188
process mapping, relevance, 196
solution, selection (challenges), 73
stages, deliverables, 37f
success, 11
success, steps, 44
supply chain, 50–52
system, delivery speed (increase), 244
technology, leverage, 85–86
traction, problems, 10
views, management, 218
Virtual Showcase, 70
Project portfolio management (PPM) system, 1
optimization/automation ability, 61–62
problems, 63
Project Professional
client, usage, 137
deliverables, addition, 239f
usage, 122
Project/program
approach, consolidation, 21–23
governance, 23–24
phases/stages, 11–13
Project proposals, 24–26, –105
defining, 105–107
form, example, 106f
review/approval, 107–115
Project Proposed Resources, 165f
Project-related information, integration, 203–208
Project Resource 2010
Booking Type, 165f
Project Schedule, 145
Project Scrum Ribbon Tab, 158f
Project Server Demand Management, white paper, 72
Project Server Enterprise Project Type, 270–271
Project Server Interface (PSI), 122
Project Server Portfolio, example, 140f
Project Server Reporting, 328
Project Server 2007, Portfolio Server 2007 (combination), 22
Project Server 2010
acquisition, 50
add-ins, 146–147
agile planning, 156–160

analysis cubes, 323–324
Build Team Menu, 163f
components, 11
cubes, 323–324
elements, relationship, 122–123, 123f
enterprise project manager usage, 161–171
functionality, 345
lifecycle usage, 23
Microsoft Dynamics AX 2012, integration, 130f
Nintex Workflow, 211
optimization, 315–318
perspectives, 39
portfolio lifecycle, creation/management, 256–259
project detail pages, availability, 278–279
Proposals features, combination, 25
reporting, creation, 324
resource management, 162
results, 21–26
results forecast, 37
Ribbon(s), 175f, 176–177
scheduling engine, 54–55
SharePoint
functions, 142
leverage, 205
Solution Starter Kit, workflows, 210
stages, availability, 281–282
tagging, SharePoint (usage), 219f
usage, 185, 251–256
SharePoint, impact, 140–143
workflow phases, availability, 284
Project Server 2010 workflow
components, 257f
design, 102
list, 213f
Project teams
discussion, 207–208
success, Project Server 2010 (usage), 185
Project Web Application (PWA), 174–175, 230
Detail Pages, customization, 177
field options, 274f
last publish, status, 174
usage, 105, 122, 137
user interface, extension, 124
Web Parts, location, 123
Project Workspace, 211
Project 2010
acquisition, roles, 39, 49–53
business justifications, 69
challenges/assumptions, 39, 67–75
consumers, solutions, 50
costs, 61
entry points, 39, 40–44
implementation/leveragability, ease, 39, 54–59
platform, extensibility, 122–130
results, 21–26
selection/deployment/support, options, 69
succession planning, 28–29
worksites, SharePoint Server 2010 (usage), 203
Project 2010 scalability, 28–29, 39, 40–44
flowchart, 37
Proofs of concept (POCs), 47–48
pilot, conducting, 47
projects, activities (phases), 47–48
test/evaluation, 47–48
utilization, 48–49
Proposals. *See* Project proposals
validation, 92–93
Public utility organization, PPM example, 43

R

Rate tables, 125
Really Simple Syndication (RSS), 217
feeds, subscription, 204
Reporting. *See* Dynamic reporting; Extending reporting
Requirements details, Gantt chart view (contrast), 301
Research, impact, 65
Resource (resources). *See* Hiring resources
account status, 302
assignment owner, usage, 302
assignments, 300–301
constraints, 297–303
costs, 125
custom fields, 307
department, example, 90f
estimation, 107–109
increase, 200
management, 138, 162

Resource (resources). *See* Hiring resources (*continued*)
optimization, achievement, 5
planning, 148–150
Project Desktop, usage, 143–150
plans, 109f, 166
reduction, 199
role custom field, 300
scenario. *See* Baseline resource scenario.
settings, 301–302
types, tracking, 171
Return on investment (ROI), 192
analysis, 313
balanced scorecard, 314–315
delivery, 195f
identification, 53
improvement, 71
leverage, 193
measurement, 309, 312–315
provision, 292
reduction, 59
Ribbon, 171–181
customization, 178f
Risk
benefits, 236f
uncertainty, connection, 237
Role-based PPM technology, 187

S

Sample Proposal Workflow, 210
Sarbanes-Oxley Act, passage, 224
Save Data Connection, example, 351f
Scatter chart, usage, 295
Scenario manager, functionality level, 346
Scenarios. *See* Portfolio scenarios
Schedule
impact, calendar exceptions (impact), 151f
planning, 110f
tracking/forecasting, 137
Scrum backlog, management, 157
Scrum Solution Starter, 157
Scrum Solution Starter for Project, 158f
Search functionality, usage, 241
Secondary stakeholders, 41–44
Services/training providers, 38
Share lists, 327
SharePoint
business intelligence (BI), 324, 345
collaboration, 21, 347
functions, 142
impact, 71
integration, 150–153
Lists, 204
lists, usage, 25
page, usage, 354
project data, synchronization, 233–235
Project Task list, 152f
risks/issues, management, 235–238
saving, example, 358f
schedules, usage, 28
Solutions, 101
Task Lists, inclusion, 153f
Team Site, 152f
usage, 140–143, 204
Web Part selection, 356f
workflow, creation (example), 212f
SharePoint Enterprise, leverage, 69
SharePoint Server 2010
business capabilities, 208–216
contextual search, 208f
discussion, example, 207f
Project Server 2010, relationship, 206f
steps, Project 2010 list, 234f
task list, example, 233
usage, 54, 203
SharePoint Workspace Mobile, usage, 360
Slicer
field slicer dimension fields, 340
usage, 341f
Slicing, 323
Small business, project management (usage), 134–143
Project Desktop Client, usage, 138–139
Social computing/communication, 217–218
Social feedback/organization, 214
Social media, integration, 203–208
Social networking tools, usage, 348
Social networks, 347–348
impact, 64–65
Software development, approach, 246–247

Software Development Kit (SDK), components, 124
Software solutions, requirements, 247–249
Special-purpose project management office, 224
Sprint Burndown Chart, 160f
  button, usage, 159
Sprint Timeline View, 159f
SQL analytical tools, 242
SQL Connection Wizard, 350f
SQL Server Analysis Services (SSAS), tabular data model, 329
SQL Server Reporting Services (SSRS), 327–329
Stage-Gate technique, 194–195
Stages. *See* Workflow stages
  representation, 12
Stakeholder classes, 27–28
  example, 9f
Stakeholder-class scalability (enablement), long-term solution (usage), 95–96
Stakeholders
  change management, impact, 49f
  classification, 38
  emergence, 41–43
Strategic alignment, efficient frontier (relationship), 299
Strategic analysis
  application, 306–309
  custom fields, usages, 307–309
Strategic initiatives, delivery, 226
Strategic objectives
  effort, synchronization, 93–97
  mapping, 254f
Strategic planning
  defining, 252
  importance, 252–253
  understanding, Project Server (usage), 251–256
Strategic plans, alignment, 192
Strategy, defining, 252
Structuring strategies, delivery, 93
Surplus reports, viewing, 303
System development life cycle (SDLC), 134

T
Tabs, customization, 178f
Tasks, linking, 146f
Team Foundation Server (TFS), 170
Team Foundation Server 2010, integration, 55
TechNet, examples/samples, 212
Technical project lifecycle, example, 57f
Technology
  delivery, 245–246
  differentiator, 259
  effectiveness/usage, 62–63
  options, leverage, 68
  perception, 58
  planned/organic growth, 40–44
Test-driven development (TDD), concept, 247
Third-party data exchange solutions, 129–130
Time-adjusted rate of return (TARR), 196
Timephased data, 300
Time scales, change, 6
Timesheet, 327
Time-sheeting capabilities, 137
Timesheet manager, usage, 302
Top-down planning, 95, 192
Total cost of ownership (TCO), 61, 313

U
Unified natural models, 128
User empowerment, project management (impact), 17–18
User Interface (UI), 171
  extensibility platform, span (extension), 171–175
  information, 179–181
User profiles, 214

V
Value field settings, 343f
  button, 342f
Value proposition, 35
  maximum, standards (promotion), 245
Vendor (stakeholder), 50
  software products, 50–51
Vertipaq engine, 330–331
Visual reporting, usage, 153–155
Visual Reports
  Project/Excel Visual Report, 155f
  Project/Visio Critical Tasks Visual Report, 155f
  Project Visual Reports, 154f
  usage, 153–155

Visual Studio, 211
  usage, 212

W

Waterfall
  approaches, 248
  development, 247
WBS Chart Pro
  Critical Tools WBS Chart Pro Add-in, 147f
  usage, 146–147
Webpage dialog box, message (example), 263f
Web Parts
  insertion, 335f
  location, 123
  selection, 356f
  usage, 177
What-if analysis, 192
What-if scenarios, usage, 346
Wikis, usage, 217, 348
Wildcard search, 209
Workbook Connections Dialog, usage, 352f
Work breakdown structure (WBS) chart, 146–147
Workflow, 102–103, 203
  control, 122–124
  deep programming abilities, absence, 209–213
  define stage, 108f
  impact, 277
  integration, 123–124
  Sample Proposal Workflow, 210
  status, 106f, 278
Workflow phases, 283–286
  availability, 284
  creation, 284–285
  inclusion, 11
  modification/deletion, 285
  naming convention decisions, 285–286
Workflow stages, 280–283
  approval, 108f
  availability, 281–282
  creation, 282
  defining, 281
  modification/deletion, 283
Work management
  collaborative work management, SharePoint integration, 150–153
  defining, project components (governing), 118–122
Work management, importance, 87–93
Work planning, 143–144
Work portfolios, commitment, 309–315
Work scheduling, 143–144
  calendar exceptions, impact, 151f
  dynamic schedules, 144–146
  No Calendar Exceptions, 151f
  Project Desktop, usage, 143–150
Workspace, collaboration, 205–206